Pentachlorophenol

Chemistry, Pharmacology, and
Environmental Toxicology

Environmental Science Research

Recent Volumes in this Series

A Continuation Order Plan is available for this series. A continuation order will bring
delivery of each new volume immediately upon publication. Volumes are billed only upon
actual shipment. For further information please contact the publisher.

Pentachlorophenol

Chemistry, Pharmacology, and Environmental Toxicology

Edited by
K. Ranga Rao

University of West Florida
Pensacola, Florida

Plenum Press • New York and London

Library of Congress Cataloging in Publication Data

Main entry under title:

Pentachlorophenol: chemistry, pharmacology, and environmental toxicology.

Proceedings of a symposium held in Pensacola, Fla., June 27-29, 1977; sponsored by the U.S. Environmental Protection Agency and the University of West Florida.
Includes index.
1. Pentachlorophenol – Environmental aspects – Congresses. 2. Pentachlorophenol – Toxicology – Congresses. 3. Aquatic animals, Effect of water pollution on – Congresses. I. Ranga Rao, K., 1941-   II. United States. Environmental Protection Agency. III. University of West Florida. [DNLM: 1. Chlorophenols – Congresses. 2. Chlorophenols – Poisoning – Congresses. 3. Environmental pollutants – Congresses. QV223 P419 1977]
QH545.P37P46   574.2′4   78-300
ISBN 978-1-4615-8950-1   ISBN 978-1-4615-8948-8 (eBook)
DOI 10.1007/978-1-4615-8948-8

Proceedings of a Symposium held in Pensacola, Florida, June 27–29, 1977

© 1978 Plenum Press, New York
Softcover reprint of the hardcover 1st edition 1978
A Division of Plenum Publishing Corporation
227 West 17th Street, New York, N.Y. 10011

Preface

Pentachlorophenol and its salts are used as biocides. Although they are used mainly for preservation and treatment of wood, their antimicrobial, antifungal, herbicidal, insecticidal and molluscicidal properties led to a widespread application of PCP formulations. Bevenue and Beckman reviewed the literature up to 1967 on the chemistry, toxicology and environmental residues of PCP *(Residue Rev.,* **19:** 83-134, 1967). Significant advances in analytical methodology, recurrent incidents of mortalities of non-target organisms exposed to PCP, regulatory actions pertaining to PCP usage in countries such as Japan and Sweden, and detection of the ubiquitous distribution of PCP in the environment, added to the wealth of recent literature on pentachlorophenol. In spite of the usage of PCP as an antimicrobial agent in drilling and packer fluids during oil-drilling operations in the marine environment, little is known of the toxicity of PCP to marine and estuarine organisms. The purpose of this volume is to present up-to-date information (including a number of new studies on marine and estuarine organisms) on the chemistry, pharmacology and environmental toxicology of pentachlorophenol.

This volume is a collection of papers presented at an international symposium sponsored by the U.S. Environmental Protection Agency and The University of West Florida, held at Pensacola Beach, Florida, June 27-29, 1977. I am grateful to Norman L. Richards, Associate Director for Extramural Activities, Environmental Research Laboratory, Gulf Breeze, Florida, who suggested the timeliness of the symposium and helped greatly in coordinating the participation of representatives of the U.S. Environmental Protection Agency. I am thankful to Thomas W. Duke, Director of the Environmental Research Laboratory, for his interest and encouragement. I am grateful to D.P. Cirelli (U.S. Environmental Protection Agency), R.L. Johnson (Dow Chemical Company), E.E. Kenaga (Dow Chemical Company), E.J. Kirsch (Purdue University) and J.I. Lowe (U.S. Environmental Protection Agency) for serving as session chairmen at the symposium. I am indebted to my research associates Angela and Frank Cantelmo, F.R. Fox, P.J. Conklin, A.C. Brannon and W. Plaia for their generous help in coordinating the symposium activities. The assistance of W. Moran (University of West Florida) in recording the entire proceedings is gratefully acknowledged.

The papers in this volume are assembled in three major sections. The first section deals with the usage and environmental fate of PCP. The patterns of PCP usage in Canada and the United States of America are described by R.A.W. Hoos and D.P. Cirelli, respectively. The entrance of PCP into the aquatic environment, especially in runoff water and wood-treatment plant effluents, is of concern. The photolysis of PCP in water is discussed by A.S. Wong and D.G. Crosby. The herbicidal and molluscicidal uses of PCP elicit much interest in determining the degradation of PCP in soils. The degradation of PCP in soil and by soil microorganisms is reviewed by D.D. Kaufman. The fate of PCP in three laboratory model ecosystems is reported by Po-Yung Lu, L. Cole and R.L. Metcalf. Detailed lists of PCP degradation products and their identification by thin layer chromatography are to be found in these papers. The fate of PCP in a lake after two spills from a wood-treatment plant near Hattiesburg, Mississippi, is the subject of a paper by R.H. Pierce, Jr., and D.M. Victor.

The second section deals with comparative toxicology and pharmacology of PCP. While the previous section contained papers on metabolism and degradation of PCP at the ecosystem level, studies on the metabolism of PCP in selected organisms are presented in subsection IIA. The topics covered include the metabolism of PCP by bacteria (E.A. Reiner et al.), blue crabs (A.K. Bose and H. Fujiwara), rainbow trout, (J.J. Lech et al.), a review of PCP metabolism in fish and shellfish (K. Kobayashi) and a study of dechlorination of PCP in rats (U.G. Ahlborg). The capability of fish, birds and mammals to metabolize hexachlorobenzene and pentachlorobenzene to PCP is demonstrated by G. Koss and W. Koransky.

The section IIB includes papers on toxicology and pharmacology of PCP. The toxicity of PCP to selected estuarine organisms, both larval stages and adults, was studied using static and flow-through tests (P.J. Borthwick and S.C. Schimmel; Schimmel et al.). The latter group also studied the uptake and depuration of PCP by selected invertebrates and fish. The effects of PCP and related compounds on the establishment of macrobenthic and meiobenthic communities in an experimental flow-through system are reported (M.E. Tagatz et al., F.R. Cantelmo and K.R. Rao). Further work done at our laboratory deals with an analysis of physiological and biochemical basis for the toxicity of PCP to crustaceans. We studied the toxicity of Na-PCP to grass shrimp at different stages of the molt cycle (P.J. Conklin and K.R. Rao), inhibition of limb regeneration as a sensitive, sublethal bioassay (Rao et al.), variations in exoskeletal calcium (A.C. Brannon and P.J. Conklin) and oxygen consumption (A.C. Cantelmo et al.). Although the grass shrimp, Palaemonetes pugio, is used extensively in aquatic toxicology studies, the ultrastructure of tissues such as gills, hepatopancreas, hindgut and midgut have not been previously reported. A detailed description of normal gills and pathobiology of gills from shrimp exposed to Na-PCP is presented by D.G. Doughtie and K.R. Rao. That uncoupling of oxidative phosphorylation may not be the sole basis for the toxicity of Na-PCP is demonstrated by a comparative analysis of the effects of Na-PCP and DNP on hepatopancreatic enzymes of blue crabs (F.R. Fox and K.R. Rao). The studies on survival, yolk utilization and energy metabolism of steelhead trout alevins confirm the disruptive effects of Na-PCP on metabolism (G.A. Chapman and D.L. Shumway). The behavioral toxicology of Na-PCP was studied utilizing a lugworm feeding activity bioassay (N.I. Rubinstein). The effects of PCP on Na^+, K^+-ATPase, oligomycin-sensitive and oligomycin-insensitive ATPases from rat tissues are reported by D. Desaiah. The effects of chronic ingestion of PCP by rats for two years are described by B.A. Schwetz *et al.*

When considering the toxicity and environmental impact of PCP, it is important to evaluate the nature and effects of impurities in commercial preparations. The nature and extent of contamination seem to vary with the preparations tested. C.-A. Nilsson *et al.* discuss the chemistry of chlorophenol contaminants, while R. Fahrig *et al.* report the genetic activity of chlorophenols and chlorophenol impurities. Chlorinated dibenzo-*p*-dioxins, predioxins and chlorinated dibenzofurans are some of the non-phenolic impurities in commercial PCP. The inclusion of two papers on 2,3,7,8-tetrachlorodibenzo-*p*-dioxin (health effects — J.R. Allen and J.P. van Miller; renal effects — J.B. Hook *et al.*) does not indicate or suggest the presence of this dioxin in PCP samples. These papers are included merely to show the need to study the effects of other dioxins which indeed are present in PCP. Hexa-, hepta-, and octachlorodioxins are present in PCP samples in considerable quantities (see review by Dougherty) while tetra- and trichlorodioxins are present in ppb (parts per billion, μg/Kg) quantities.

What is the current level of environmental pollution by PCP? After presenting Negative Chemical Ionization Mass Spectrometry as a tool for detecting environmental pollutants, R.C. Dougherty reviews the literature on human exposure to pentachlorophenol. This review and the results of a preliminary analysis of urine samples from the general population (F.R. Kutz *et al.*) show the ubiquitous distribution of PCP. The sources of PCP contamination and its occurrence as a residue relative to other pesticides are also discussed.

In a general discussion of the environmental impact of PCP, Don Isleib of Michigan Department of Agriculture presented the problems related to PCP-exposure of farm animals. This aspect and incidents related to PCP exposure in other countries are summarized by P.J. Conklin and F.R. Fox who edited the transcripts of the round table discussion.

It is hoped that this volume serves as a source of up-to-date information pertinent to PCP for researchers and students in environmental toxicology, pesticide manufacturers and consumers, and regulatory agencies.

This volume is a result of the dedicated efforts, promptness and cooperation of the contributors. It has been my privilege to work with the contributors and I am grateful to them. I am also indebted to the following individuals for reviewing several of the manuscripts and for their constructive criticism: Drs. J.R. Baylis, W.O. Berndt, C.W.J. Chang, J.A. Couch, M.A. Hood and J.P. Riehm. My research associates, F.R. Fox, P.J. Conklin and A.C. Brannon have generously assisted in reviewing and editing the manuscripts. The assistance of Mrs. Mary Bonifay in typing several manuscripts is gratefully acknowledged.

The camera-ready copy of this volume was prepared at The University of West Florida and I am thankful to the following personnel. Ms. Anita G. Barbo served as an editorial assistant and supervised the preparation of camera-ready pages. Typesetting of the manuscripts was done by Ms. Laurel A. Simpson while graphic art work and preparation of camera-ready pages were done by Ms. Diana R. Manderson. The cooperative efforts of Mr. R.L. Chuites (Duplicating Services) and Mr. Dan F. Dull (Instructional Media Center) are gratefully acknowledged.

K. Ranga Rao

Contributors

U.G. Ahlborg, Laboratory of Toxicology, National Food Administration, Uppsala, Sweden

J.R. Allen, Department of Pathology, University of Wisconsin Medical School, Madison, Wisconsin

K. Andersson, Department of Organic Chemistry, University of Umeå, Umeå, Sweden

P.W. Borthwick, Environmental Research Laboratory, U.S. Environmental Protection Agency, Gulf Breeze, Florida

A.K. Bose, Department of Chemistry and Chemical Engineering, Stevens Institute of Technology, Hoboken, New Jersey

A.C. Brannon, Faculty of Biology, University of West Florida, Pensacola, Florida

A.C. Cantelmo, Faculty of Biology, University of West Florida, Pensacola, Florida

F.R. Cantelmo, Faculty of Biology, University of West Florida, Pensacola, Florida

G.A. Chapman, Western Fisheries Toxicology Laboratory, U.S. Environmental Protection Agency, Corvallis, Oregon

J.P. Chu, General Motors Corporation, Warren, Michigan

D.P. Cirelli, U.S. Environmental Protection Agency, Washington, D.C.

L.K. Cole, Department of Entomology, Institute for Environmental Studies, University of Illinois, Urbana, Illinois

P.J. Conklin, Faculty of Biology, University of West Florida, Pensacola, Florida

D.G. Crosby, Department of Environmental Toxicology, University of California, Davis, California

D. Desaiah, Department of Pharmacology and Toxicology, University of Mississippi Medical Center, Jackson, Mississippi

R.C. Dougherty, Department of Chemistry, Florida State University, Tallahassee, Florida

D.G. Doughtie, Faculty of Biology, University of West Florida, Pensacola, Florida

L.F. Faas, Environmental Research Laboratory, U.S. Environmental Protection Agency, Gulf Breeze, Florida

R. Fahrig, Central Laboratory for Mutagen Testing, Freiburg, West Germany

F.R. Fox, Faculty of Biology, University of West Florida, Pensacola, Florida

H. Fujiwara, Department of Chemistry and Chemical Engineering, Stevens Institute of Technology, Hoboken, New Jersey

A.H. Glickman, Department of Pharmacology, Medical College of Wisconsin, Milwaukee, Wisconsin

E.A. Hett, Department of Chemistry, Florida State University, Tallahassee, Florida

J.B. Hook, Department of Pharmacology, Michigan State University, East Lansing, Michigan

R.A.W. Hoos, Department of Fisheries and Environment, Kapilano, British Columbia, Canada

C.G. Humiston, Toxicology Research Laboratory, Dow Chemical Company, Midland, Michigan

J.M. Ivey, Environmental Research Laboratory, U.S. Environmental Protection Agency, Gulf Breeze, Florida

D.D. Kaufman, Pesticide Degradation Laboratory, U.S. Department of Agriculture, Beltsville, Maryland

P.A. Keeler, Toxicology Research Laboratory, Dow Chemical Company, Midland, Michigan

E.J. Kirsch, School of Environmental and Civil Engineering, Purdue University, West Lafayette, Indiana

W.M. Kluwe, Department of Pharmacology, Michigan State University, East Lansing, Michigan

K. Kobayashi, Laboratory of Fishery Chemistry, Kyushu University, Fukuoka, Japan

J. Kociba, Toxicology Research Laboratory, Dow Chemical Company, Midland, Michigan

W. Koransky, Institute of Toxicology and Pharmacology, Marburg, Federal Republic of Germany

G. Koss, Institute of Toxicology and Pharmacology, Marburg, Federal Republic of Germany

F.W. Kutz, Ecological Monitoring Branch, U.S. Environmental Protection Agency, Washington, D.C.

J.J. Lech, Department of Pharmacology, Medical College of Wisconsin, Milwaukee, Wisconsin

P.-Y. Lu, Department of Entomology, Institute for Environmental Studies, University of Illinois, Urbana, Illinois

K.M. McCormack, Department of Pharmacology, Michigan State University, East Lansing, Michigan

R.L. Metcalf, Department of Entomology, Institute for Environmental Studies, University of Illinois, Urbana, Illinois

R.S. Murphy, National Center for Health Statistics, United States Public Health Service, Rockville, Maryland

C.-A. Nilsson, Chemical Division, Department of Occupational Health in Umeå, Umeå Hospital, Umeå, Sweden

Å. Norström, Department of Organic Chemistry, University of Umeå, Umeå, Sweden

J.M. Patrick, Jr., Environmental Research Laboratory, U.S. Environmental Protection Agency, Gulf Breeze, Florida

R.H. Pierce, Jr., Institute of Environmental Science, University of Southern Mississippi, Hattiesburg, Mississippi

J.F. Quast, Toxicology Research Laboratory, Dow Chemical Company, Midland, Michigan

K.R. Rao, Faculty of Biology, University of West Florida, Pensacola, Florida

C. Rappe, Department of Organic Chemistry, University of Umeå, Umeå, Sweden

E.A. Reiner, 3M Company, St. Paul, Minnesota

N.I. Rubinstein, Faculty of Biology, University of West Florida, Pensacola, Florida

S.C. Schimmel, Environmental Research Laboratory, U.S. Environmental Protection Agency, Gulf Breeze, Florida

B.A. Schwetz, Toxicology Research Laboratory, Dow Chemical Company, Midland, Michigan

D.L. Shumway, Bureau of Power, Federal Power Commission, Washington, D.C.

C.N. Statham, Department of Pharmacology, Medical College of Wisconsin, Milwaukee, Wisconsin

S.C. Strassman, Ecological Monitoring Branch, U.S. Environmental Protection Agency, Washington, D.C.

M.E. Tagatz, Environmental Research Laboratory, U.S. Environmental Protection Agency, Gulf Breeze, Florida

M. Tobia, Environmental Research Laboratory, U.S. Environmental Protection Agency, Gulf Breeze, Florida

J.P. van Miller, Department of Pathology, University of Wisconsin Medical Center, Madison, Wisconsin

D.M. Victor, Institute of Environmental Science, University of Southern Mississippi, Hattiesburg, Mississippi

A.S. Wong, California Analytical Laboratories, Inc., Sacramento, California

Contents

B. Comparative Toxicology and Pharmacology

III. Chlorophenols and Their Contaminants: Chemistry, Toxicology, Environmental Residues and Environmental Impact

I

Usage Patterns and Environmental Fate of Pentachlorophenol

Patterns of Pentachlorophenol Usage in Canada — an Overview

RICHARD A.W. HOOS

Abstract—At the present time, approximately five to six million pounds of pentachlorophenol are consumed per year in Canada. Although the chemical also serves as a herbicide and insecticide for agricultural purposes, its primary use is in the protection and preservation of wood products produced by the forest products industry. This paper reviews, in general terms, the use of pentachlorophenol in Canada, and measures that are being taken to control the release of this chemical to the Canadian environment.

Introduction

Pentachlorophenol (PCP), a chlorinated hydrocarbon, is generally used as a wood preservative, insecticide and herbicide. Canada's current consumption of this compound is estimated to range between five and six million pounds per year (Environmental Protection Service, estimate) and is expected to increase.

The purpose of this paper is to review, in general terms, the use of pentachlorophenol in Canada, and steps that are being taken to control the release of this chemical to the Canadian environment. As is the case in the United States of America, in excess of 80% of PCP is used for the purpose of wood preservation, and other activities associated with the forest products industry. The bulk of this presentation will therefore be directed to this area of PCP usage.

After the production of food crops, wood constitutes Canada's most valuable renewable resource (Smith, 1977). Wood destined to be utilized for long term

---

RICHARD A.W. HOOS • Environmental Protection Service, Department of Fisheries and the Environment, Kapilano 100, Park Royal, West Vancouver, B.C., V7T 1A2, Canada.

purposes such as construction lumber and timbers, telephone poles, railway ties, etc. generally requires some form of protection against biological attack in order to maintain structural strength and thereby prolong the effective period of service.

Depending on the ultimate use of the lumber, the processes involved in treating wood can vary from superficial applications such as brushing, spraying and dipping, through diffusion and soaking techniques, to more complicated procedures involving the use of pressure and vacuum cycles to obtain the deep penetrations and the optimum retentions of preservatives required by standards to provide satisfactory performance of wood for the different end uses (Shields and Stranks, 1977). For the purpose of this presentation it is useful to differentiate between the longer term treatments which will be referred to as *preservation* techniques and the short term treatments which serve to *protect* wood from fungal infections etc. for short periods of time.

Wood Preservation in Canada

(adapted from Shields, 1976; Shields and Stranks, 1977)

Canadian wood treating (preservation) operations first officially began in 1910 in British Columbia and in 1911 in Ontario employing creosote to treat commodities such as railway ties and road paving blocks. Preservative-treated ties, imported from the U.S.A., had already been in use since 1906. Today, railway ties, poles, posts and structural timbers are important treated commodities.

Examples of the magnitude of the wood preserving industry may be noted in the approximate 3 million treated ties required by the major Canadian railways as annual replacements during the last few years. There are in excess of 150 million ties in service in Canada. Treated poles are another essential commodity in the maintenance of Canada's communication systems. One hydroelectric company alone required an estimated 30,000 treated replacement poles or poles for new lines during 1974. Presently they have about 1 ½ million poles in service. Estimates on the total number of treated poles produced annually in Canada vary from 300,000 to 400,000 and of these 5 to 10% are required to replace decayed or otherwise damaged poles (Roff and Kvzyewski, 1972). Somewhere in excess of 4 million treated posts and more than 16 million cubic feet of treated timbers, based upon partial estimates, were produced in Canada during 1975. According to Statistics Canada (Anonymous, 1974) the value of shipments from the Canadian wood treating industry during 1974 was close to $80 million.

In addition, over the last few years there has been growth in the use of preserved wood foundations for new house construction in North America. Over 2,000 such houses have been erected in Canada and are expected to be a satisfactory economical alternative to conventional foundation systems in many parts of the country.

Monetary savings to users of treated wood products are substantial over the longer service life obtained when compared to much shorter service and greater maintenance requirements when untreated wood is used. Some examples are given in the report by Roche (1965). A simple example is commercial pressure-treated jackpine poles, which will give excellent service for at least 35 years under normal conditions as compared to an average useful life of approximately 7 years for untreated jackpine poles.

Another important aspect is the general reduction of wastage of our forest resources. By increasing the length of useful service for wood products the conservation of these resources is automatically being achieved.

Canada's Wood Preserving Industry — Current Status

(adapted from Shields, 1976)

At present there are 50 wood-preserving plants in operation in Canada, 37 of which use pressure retorts. The distribution of plants by type and location is shown in Table 1. Nineteen of the pressure-treating plants are located in Western Canada, evenly distributed in British Columbia, Alberta and Saskatchewan. In contrast, most of the non-pressure plants (9) are located in Ontario. Quebec and the Atlantic Provinces together account for only 7 of the treating plants, all of them employing pressure processes.

Table 2 shows an inventory of plants using different preservative treatments by pressure in Canada. Amongst the 37 pressure plants, water-borne preservatives and fire retardants are used exclusively in 10 plants. Twelve plants (except one which employs pentachlorophenol in methylene chloride) use only penta-petroleum, accounting for one-third of pressure plants, and all except the one using penta in methylene chloride are located in Western Canada.

TABLE 1. Wood-preserving Plants in Canada by
Locality and Type

Region	Number of plants	
	Pressure	Non pressure
British Columbia	7	2
Saskatchewan	6	0
Alberta	6	2
Manitoba	3	0
Ontario	8	9
Quebec	3	0
Atlantic Provinces	4	0
Total	37	13

From Shields, 1976

TABLE 2. Preservatives Used by Canadian Pressure Treating Plants
by Locality

Region	Penta-petroleum	Water borne preservatives and fire-retardant	Penta-petroleum/ creosote	Penta-petroleum/ creosote/ waterborne preservatives
British Columbia	3	2	0	2
Saskatchewan	4	1	1	0
Alberta	4	0	1	1
Manitoba	0	2	1	0
Ontario	0	4	2	2
Quebec	1	0	0	2
Atlantic Provinces	0	1	3	0
Total	12	10	8	7

From Shields, 1976

Of the remaining 15 plants, both creosote and penta-petroleum preservative are used in 8 plants and the other seven are based on creosote, penta and water-borne preservatives. In all of these last seven plants using both oil and water-borne preservatives, some retorts are strictly reserved for water-borne preservatives. All the pressure-treating plants, except those which use water-borne preservatives exclusively, use pentachlorophenol as the principal preservative.

Table 3 summarizes waste treatment and disposal methods as currently practised in 36 pressure-treating plants in Canada. One recently constructed plant in Ontario was not surveyed. This shows that the main approach to the pollution problem taken by wood-preserving companies is to store their waste water on company property (containment and evaporation). This is by far the most popular method of handling waste water, accounting for 17 plants of the 36 surveyed. Ten have little waste water, as they use water-borne preservatives exclusively. Three use incineration, while 1 plant disposes of its waste water by a special method, namely the use of clarified effluent from oily waste for makeup of water-borne preservative solution. The above 31 plants (of 36) claim to have no discharge and either do not generate any waste water or produce small volumes of waste water that are easily disposed of by incineration or containment and evaporation. The

TABLE 3. Disposal of Waste Water by Pressure Treating Plants in Canada

Disposal method	Number of Plants
1. INCINERATION - No discharge Separation of bulk oil in gravity separation tanks, evaporation/evaporation, incineration mixed with bunker fuel oil.	3
2. CONTAINMENT AND EVAPORATION - No discharge (a) Separation of bulk oil in gravity separation tanks, spray evaporation/evaporation by heating and store in tank/open pit, and sludge trucked by commercial disposal company (8 plants)	17
(b) Same as in (a), but wood shavings etc. dumped in open pit and sludge incinerated (3 plants)/sludge disposed of by burial, land fill or cattle bedding (3 plants)/sludge disposal by commercial disposal company (3 plants).	
3. LAGOONING - Discharge Separation of bulk oil in gravity separation tanks, flocculation with coagulant and setting (1 plant), /API separation and hay filtration (1 plant)/API separation and air flotation (1 plant), lagooning in ditch and discharge.	3
4. SECONDARY BIOLOGICAL TREATMENT - discharge Separation of bulk oil in gravity separation tanks, and activated sludge biological treatment.	1
5. ACTIVATED CARBON PHYSICO-CHEMICAL TREATMENT - discharge Separation of bulk oil in gravity separation tanks, filtration through sand/polyurethane, and absorption on activated carbon.	1
6. NO WASTE WATER From plants exclusively on water-borne preservative.	10
7. OTHER - No discharge Separation of bulk oil in gravity separation tanks, filtration through sand and effluent reuse for make-up water-borne preservative solution.	1
	36

From Shields, 1976

adequacy of such waste disposal systems is unknown since pertinent data on water pollution from seepage, leaching, run-off, and on air pollution associated with incineration are unavailable.

According to the surveys, the remaining five Canadian plants give their waste the equivalent of secondary treatment before discharge into natural waterways. These plants (four of which use steam conditioning and the other uses Boultonizing) generate waste water at 2000 to 7000 gallons (9 to 32 cubic meters) per day. Of the five plants, three are presently giving extended primary treatment, e.g. API separation, air flotation, flocculation with flocculants, settling, and filtration prior to natural lagooning in a ditch. Two of these three plants plan to install activated carbon systems in conjunction with extended primary treatment to meet Provincial water quality standards.

Of the remaining two plants, one, located in Nova Scotia, has been treating waste water by extended primary treatment and activated carbon absorption (physico-chemical treatment) since 1973, and is producing effluent which meets both Federal and Provincial water quality standards that include 96-hr toxicity tests using salmon. The other plant, situated in Ontario, has treated its waste water by an activated sludge biological system since 1968. The system produces effluent within limits acceptable to Provincial authorities.

Federal Regulations/Wood Preserving Plants

To date, there are no specific Federal regulations controlling the discharge of pollutants from the wood preserving industry. Any actions of a regulatory nature that have been undertaken in the past have utilized the general provisions of the Federal Fisheries Act. However, the government through the Department of Fisheries and Environment, spearheaded by the Environmental Protection Service, are now taking the preliminary steps leading up to the development of specific effluent regulations for this industrial sector.

The general rationale and process through which Regulations/Guidelines have been developed for other industries has been described by Allard (1977). "For the development of water pollution control effluent standards, there are two main approaches available; the first based on comprehensive effluent standards and the second based on comprehensive water resource management (receiving water quality standards). The Federal government has chosen the first option and establishes so called "end of the pipe" effluent standards, involving the removal of contaminants from effluent to the greatest degree practicable. This approach requires the containment and treatment of polluted waste waters at their source through the application of best practicable technology, which implies demonstrated technology that is both technically and economically feasible. This approach is the more simplistic of the two but the Federal government is convinced that it will result in a significant clean up at existing industrial sites and will establish a floor for pollution at all new sites."

National base line standards are being developed to ensure that all major industrial water pollution sources limit their liquid effluent "end of pipe" loadings to a level consistant with "Best Practicable Technology" which will result in:

1. A significant abatement of discharges of deleterious substances from existing sources.
2. A significant limitation of discharges of deleterious substances from new sources.

The Enrivonmental Protection Service has normally taken the lead in acquiring sufficient data to determine the magnitude of the pollution problem resulting from the activities of a particular industrial sector. Where appropriate the Environmental Protection Service has prepared preliminary discussion papers which address the following items:

1. Definition of the Plants included in the activity.
2. Effluent sources, contaminants and potential environmental problems.
3. A review of wastewater control and/or process technology that may constitute Best Practicable Technology.
4. Economic aspects of implementing water pollution control.

After this preliminary review of an industrial sector, the Federal government has generally called a joint industry/government Task Force. The value and benefit of working in a cooperative atmosphere with technical experts from both industry and the provincial regulatory agencies has therefore been recognized. The joint government/industry approach is felt to be an integral part of ensuring that the control package that is developed is reasonable and the recommendations of this task force are given very serious consideration in the formulation of the final control package. (Adapted from Allard, 1977). It is anticipated that a similar approach will be adopted leading to the development of appropriate Regulations/Guidelines for the Wood Preservation Industry.

Short-term Wood Protection

(largely adapted from Smith, 1977)

The preceeding has described wood preservation which generally reflects treatment of lumber for the purpose of producing a product which will have a long period of service such as railway ties, pilings, hydro poles, etc.

However, a substantial proportion of processed wood does not require long term preservation in order to satisfy its end users. For example, lumber which will be used for the construction of homes etc., where the wood will be isolated from external, natural elements by other barriers such as cement, brick, aluminum, etc. does not require the same degree of protection as that afforded by preservation. Similarly, in the production of groundwood pulp, the product is preferred to be bright and clean without the use of bleaching chemicals, and unsightly growths of darkly pigmented fungi in the pulp are unacceptable (Smith, 1977). However, both still require protection during the transportation and storage stages, in order to reduce or eliminate the incidence of fungal, bacterial, insect and/or other micro-organism attacks.

The results of infestations by one or more of the aforementioned agents can lead to afflictions such as dry rot, wet rot, sap stains, moulds, honeycombing, etc. Many of these problems are only cosmetic in nature, but the economic impacts can be significant.

Citing from Smith (1977), "Sap-stain fungi cause serious economic deterioration problems for green lumber in transit. Although no serious loss in strength of the lumber occurs following attack by these fungi, its overall appearance may be so unattractive and dirty that it is condemned and a financial loss is accepted by the shipper. Although the use of sap-stained lumber may be acceptable for certain constructional purposes, it must be remembered that it contains viable fungi. These fungi could resume growth, should the lumber become subsequently rewetted in service, and may cause the blistering of paint surfaces or unsightly growth in or below the paint surface, although this belief has been questioned (Butin, 1965).

In recent years, with the advent of bulk packaging of green lumber, the problems of sap-stain, as well as mould and decay, have intensified (Roff, 1962). This, together with the longer transit times necessitated by softer world markets and increased labour strife, has placed great demands on the anti sap-stain treatments normally applied by the producer mills.

Sap-stain preventives are widely used on fresh-cut green lumber to prevent its attack on sap-stain fungi and moulds (Roff et al., 1974). The most successful and widely used chemicals are the sodium salts of either tetrachloro or pentachlorophenol. However, recently Captafol has gained considerable acceptance in New Zealand, where it is considered less of an environmental risk than the chlorinated phenols (Butcher, 1974).

Sap-stain preventives are applied to the green lumber directly after cutting, either by passing the lumber through a spray tunnel, or by dipping. The former method is greatly favoured by industry, but it has been shown to be considerably less successful than dipping as a control measure for deterioration due to sap-stain and mould fungi (Cserjesi and Roff, 1966).

Chemicals also are used to control the biodeterioration of wood products in some more unusual situations. The protection of wood chips during outside chip storage remains a difficult problem. Here, because of the enormous surface area of wood to be protected and the relative low value of the raw material, a successful chemical treatment must be very inexpensive (Smith and Hatton, 1971). Also because of the open exposure of chip piles, leaching of the chemical treatment into the adjacent environment remains a constant threat. A variety of chemicals incuding PCP have been tried (Springer et al., 1975), but so far without any constant success.

Currently considerable effort is being made, particularly by lumber exporting countries, to find alternative chemicals to the chlorinated phenols (Cserjesi and Roff, 1975). But so far all trial chemicals have proven to be either inferior or too costly at effective concentrations.

Summary

To summarize, it is apparent that for the foreseeable future pentachlorophenol and its sodium salt will continue to play an important role in the forests products industry and therefore the economy of Canada. Federal regulations are being developed to control and reduce losses of contaminants including PCP from the wood preserving industry to the receiving environment. Through increased awareness on the part of users of PCP regarding the potential deleterious effects of the chemical, improved housekeeping procedures are continuing to be adopted, thereby further reducing the potential for losses to the environment; and researchers are continuing to examine superior alternatives for wood treatment consistent with enhanced environmental protection.

References

Allard, G.A., 1977. The Role of the Environmental Protection Service in Establishing Effluent Controls. Technology Transfer Seminar — Timber Processing Industry, March 1977, Toronto, Ontario. 11 pp.

Anonymous, 1974. Miscellaneous Wood Industries. Statistics Canada Catalogue 35-208 Annual.

Butcher, J.A., 1974. Captafol - an acceptable anti-sapstain chemical. *New Zealand Wood Ind., 20:* 9-11.

Butin, H., 1965. Untersuchungen zur Okologie einiger Blauepilze an verabeitetem kiefernbolz. *Flora,* **155:** 400-440.

Cserjesi, A.J., and J.W. Roff, 1966. Sapstain and mold prevention by spraying and dipping. *British Columbia Lumberman,* **50:** 64-66.

Cserjesi, A.J., and J.W. Roff, 1975. Toxicity tests of some chemicals against certain wood-staining fungi. *Int. Biodetn. Bull.,* **11:** 90-96.

Roche, J.N., 1965. Wood preservation - an important factor in conservation. *Proc. Wood Preserv. Assn.,* **61:** 11-15.

Roff, J.W., 1962. Reduction of decay in packaged lumber. *British Columbia Lumberman,* **46:** 61-62.

Roff, J.W., A.J. Cserjesi, and G.W. Swann, 1974. Prevention of sapstain and mold in packaged lumber. Department of Environment Canada. Forest Service, Pub. 1325.

Roff, J.W., and J. Krzyzewski, 1972. In-service assessment of load capacity, decay and re-treatment of utility poles. Eastern Forest Products Laboratory Information Report OPXU5, 11 pp.

Shields, J.K., 1976. Control of preservative wastes from wood treatment. Eastern Forest Products Laboratory Report OPX163E. 26 pp.

Shields, J.K., and D.W. Stranks, 1977. Wood preservatives and the environment. Technology Transfer Seminar - Timber Processing Industry, March, 1977, Toronto, Ontario. 30 pp.

Smith, R.S., 1977. Protection and preservation of wood against attack by fungi. Technology Transfer Seminar- Timber Processing Industry, March, 1977. Toronto, Ontario, 20 pp.

Smith, R.S., and J.V. Hatton, 1971. Economic feasibility of chemical protection for outside chip storage. *Tech. Assn. Pulp Paper Industry,* **54:** 1638-1640.

Springer, E.L., W.D. Feist, L.L. Zoch, Jr., and G.J. Hajny, 1975. Evaluation of the treatments to prevent the deterioration of wood chips during storage of pile simulations *Tech. Assn. Pulp Industry,* **58:** 128-131.

Patterns of Pentachlorophenol Usage in the United States of America — an Overview

DANIEL P. CIRELLI

Abstract—Pentachlorophenol (PCP) and its salts are widely used as biocides in the United States of America. Although the principal use of PCP is for wood preservation, the versatility of this compound can be seen by the great diversity of uses.

Introduction

Pentachlorophenol (PCP) and its sodium salt (Na-PCP) are probably the most versatile pesticides now in use in the United States. In fact, collectively they are the second heaviest used pesticides in the country. They are properly called biocides because they are lethal to a wide variety of living organisms, both plant and animal. PCP is registered by the U.S. Environmental Protection Agency (EPA) for use as an insecticide (termicide), fungicide, herbicide, algicide, disinfectant, and as a ingredient in antifouling paint. This versatility is due in large part to the solubility of the pentachlorophenols in both organic solvents (PCP) and water (Na-PCP). Thus, PCP can be applied to such diverse materials as agricultural seeds (for non-food uses), leather, masonry, wood, cooling tower water, rope, and paper mill systems. The various trade names used are as follows: Chem-Penta, Chemtrol, Chlorophen, Dowicide EC-7, Dowicide G, Durotox, Lauxtol A, Na-PCP, PCP, Penchlorol, Penta, Penta-kil, Pentanol, Pentasol, Permacide, Permaguard, Permasem, Permatox, Sinituho, Term-l-trol, and Weed-Beads.

DANIEL P. CIRELLI • United States Environmental Protection Agency, WH-566, 401 M. Street, S.W., Washington, D.C. 20460, U.S.A.

Production and Usage of PCP

PCP is produced in the United States by Dow Chemical Company at Midland, Michigan; Monsanto Company at Sauget, Illinois; Reichold Chemical Company at Tacoma, Washington; and Vulcan Materials Company at Wichita, Kansas. Monsanto sells all of its PCP production to the Koppers Company, a major wood preserving firm. Total PCP production in 1974 amounted to 52.4 million pounds (U.S. International Trade Commission, November 1975). There are plans to expand production to 80 million pounds in 1977 (Stedman, 1976).

Pentachlorophenol is a buff colored crystal which is produced in the United States by chlorination of molten phenol in the presence of a catalyst. Industrial production of PCP is a two stage process. In the first stage at a reaction temperature of about 105°C, isomers of tri and tetrachlorophenol are formed. In the second stage the temperature is gradually increased to keep the reaction mixture molten, and tri- and tetrachlorophenols are further chlorinated to form pentachlorophenol. This reaction is not quantitative; tetrachlorophenols persist during the reaction and are carried with PCP during subsequent processing. The result is that technical grade PCP contains 4 to 12% tetrachlorophenols. Tetrachlorophenols are registered pesticides (fungicides) with the U.S. Environmental Protection Agency.

1 Northeast region
2 North Central region
3 Southeast region
4 South Central region
5 Rocky Mountain region
6 Pacific region

Figure 1. Distribution of wood treatment plants in the United States. Modified from a distribution map, Division of Forest Economics and Marketing Research, Forest Service, U.S. Dept. Agriculture, 1974.

PCP is used in the following manner: 78% by the wood preserving industry, 12% in production of Na-PCP, 6% in plywood and fiberboard waterproofing, 3% in home and garden uses (mostly termite control) and 1% as a herbicide for use on rights-of-way, industrial sites, etc. (source: Midwest Research Institute).

The EPA has registered 578 products containing PCP. These are manufactured or formulated by 240 registrants. Registrations for 196 products containing Na-PCP are held by 88 registrants. In addition, there are 5 products containing the potassium salt of PCP with 4 registrants, and 6 products containing the lauric acid ester of PCP with 2 registrants (source: EPA computer files).

Since most of the PCP produced is used in wood preservation, there are a large number of treatment plants. The wood preservation industry was composed of 299 companies with 468 separate plants in 1974. These plants are distributed in 45 states. Most of the industry is concentrated in the south, southeast, and northwest (Figure 1), due to the availability of preferred timber species such as southern pine, Douglas pine, and western red cedar. Of the 468 plants treating wood, 194 treat with pentachlorophenol. More than half of all plants treat with more than one preservative, the others being creosote and the water based arsenical salts. More than 98% of all wood processed is treated by pressure using these three preservatives. Table 1 shows the quantities and types of wood products treated with PCP in 1974.

TABLE 1. Quantity of Wood Products Treated with
Pentachlorophenol (PCP), 1974

	Wood treated		Proportion treated with PCP
	Total	With PCP	
	Million cu ft		%
Railway ties	82.4	0.3	0.4
Poles	73.1	42.3	57.5
Piling	13.3	0.1	0.8
Lumber[a]	80.3	21.2	26.4
Fence posts	17.3	9.6	55.5
Other	8.5	2.5	29.4
Total	274.7	75.8	27.6

Source: American Wood-Preservers' Association.
Wood Preservation Statistics, 1974.
[a]Includes timbers and crossarms.

TABLE 2. Industrial/Commercial Use Distribution for Pentachlorophenol (PCP), 1972.

Region	PCP Used by Wood Preservation Plants	Distribution of PCP-Treated Wood
Northeast	0.4[a]	9
North central	3.6	11
Southeast	11.0	6
South central	14.0	6
West	9.0	6

Source: Midwest Research Institute estimates

[a]All figures in millions of pounds/year active ingredient

About 0.23 kg of PCP is required for each cubic foot of wood preserved. PCP is usually applied to wood products after dilution to a 5% solution with solvents such as mineral spirits, No. 2 fuel oil or kerosene. However, liquid petroleum gas or methylene chloride have also been used when the availability of the other solvents is restricted.

Estimates of where PCP-treated wood was used in 1972 are shown in Table 2. The relative use distribution has probably not changed in succeeding years, although the amount of PCP consumed has increased. The distribution of other PCP uses, i.e., herbicide, insecticide, is not known. However, Figure 2 illustrates the estimate of total PCP use in the United States for 1972.

The 12 percent of PCP converted to the sodium salt has many uses and is widely distributed. It is a wide spectrum fungicide and bactericide used for the treatment of sap stain in freshly sawn logs and unseasoned timber. Other large uses are addition to cooling towers at electric plants to control algae and fungi and in the production of pressed board and insulation board. Twenty to eighty parts per million (ppm) Na-PCP is added to the water cooling systems for control measures. It is added to adhesives based on starch, vegetable protein and animal protein to protect against their deterioration. Construction materials such as asbestos shingles, roof tiles, brick walls, concrete blocks, insulation, pipe sealing compound and wallboard are protected by Na-PCP. Leather is prevented from deterioration during tanning and protected from molding by Na-PCP. It is added to paint as a preservative and to photographic solutions as a slime and fungus control. Pulp and finished paper products are protected against mildew, rot and termites when Na-PCP is present. Na-PCP is used in the textile industry to preserve the processing materials and protect the finished products against mildew. The petroleum industry uses Na-PCP as a bactericide in drilling muds, gypsum muds and packer fluids. The Food and Drug Administration has allowed its usage under the food additive

Figure 2. Distribution of PCP usage (millions of lbs.) and location of production plants. ▲ Production Plants. Modified from a distribution map prepared by the Midwestern Research Institute, based on data for 1972.

regulations. The following are applications under this license: Slimicides, defoaming agents used in the manufacture of paper and paperboard, resinous and polymeric coatings, adhesives, component of paper and paperboard in contact with dry and aqueous food, closures with sealing gaskets for food containers, as a preservative for ammonium alginate employed in the manufacture of polyvinyl chloride emulsions, textile and textile fibers and rubber articles intended for repeated use. Finally, Na-PCP is used in secondary oil recovery to control microbial growth and obtain maximum recovery. It is injected at a concentration of 15-40 ppm.

Pentachlorophenol was formerly widely used as a herbicide, frequently in combination with other herbicides. It is used on non-crop areas or dormant crops as a pre-harvest desiccant. The home and garden applications are many. It can be bought in any hardware store and is used in paints for porch and lawn furniture, trailers and boats. It has also been used as a bird repellant. It discourages woodpeckers when mixed as a pellet and plugged into the holes drilled by the bird.

The textile industry uses PCP to preserve rope binder twine, burlap, cable coverings and canvas. It is applied in rubber cable covering and is used as a termite preventative in homes. The leather industry applies it to upper leather of shoes to provide mold resistance.

In summary, although the wood preserving industry uses the greatest amount of PCP, the versatility of this compound can be seen by the great diversity of uses.

References

Midwest Research Institute, 1975. Production, Distribution, Use and Environmental Impact Potential of Selected Pesticides. For Environmental Protection Agency, Office of Pesticide Programs.

Sittig, M., 1974. Pollution control in the organic chemistry industry. Pollution Technology Review No. 9, 182-184. Nayes Data Corporation, Park Ridge, New Jersey.

Stedman, H.A., 1976. Manager, Agricultural Chemicals, Vulcan Materials Company. Letter to Loren F. Dorman, President of the American Wood Preservers Institute. November, 1976.

U.S. International Trade Commission, 1975. Synthetic Organic Chemicals. U.S. Production and Sales. November, 1975.

Photolysis of Pentachlorophenol in Water

ANTHONY S. WONG and DONALD G. CROSBY

Abstract—When a dilute aqueous solution (100 ppm) of pentachlorophenol (PCP) was irradiated with sunlight or UV light, the photodegradation products were found to be chlorinated phenols, tetrachlorodihydroxyl benzenes and non-aromatic fragments such as dichloromaleic acid. Subsequent irradiation of the tetrachlorodiols resulted in the formations of hydroxylated trichlorobenzoquinones, trichlorodiols, dichloromaleic acid and non-aromatic fragments. Dichloromaleic acid, when irradiated, produced chloride ions and carbon dioxide. Prolonged irradiation of PCP or its photodegradation products yielded colorless solutions containing no ether-extractable volatile materials; and evaporation of the aqueous layer left no observable polymeric residue such as humic acid. Octachlorodibenzo-*p*-dioxin was formed when a high concentration of the sodium salt of PCP was irradiated.

Introduction

The environmental impact of pentachlorophenol (PCP), a pesticide widely used for wood preservation, has been the subject of many investigations in recent years. With an annual production of 50 million pounds in this country alone (von Rumker *et al.*, 1974), the substance often finds its way into the aquatic environment, especially in runoff waters and wood-treatment plant effluents (Bevenue and Beckman, 1967).

DONALD G. CROSBY • Department of Environmental Toxicology, University of California at Davis, Davis, California 95616, U.S.A.

ANTHONY S. WONG • present address, California Analytical Laboratories, Inc., 401 North 16th Street, Sacramento, California 95814, U.S.A.

PCP absorbs sunlight readily in the ultraviolet region of sunlight (λ_{max} = 320 nm), and its photochemical behavior has been reported previously. Kuwahara and co-workers (1966a, 1966b) found that a 2% solution of the sodium salt of PCP exposed to sunlight suffered a 50% loss of parent pesticide within ten days, and the major products were identified as chloranilic acid (2,5-dichloro-3,6-dihydroxybenzoquinone), tetrachlororesorcinol and several complex chlorinated benzoquinones. Hamadmad (1967) irradiated an aqueous suspension of PCP with sunlight and found chloranil (tetrachlorobenzoquinone) and "humic acid" as two of the degradation products. Our paper will present further findings from irradiation of a dilute aqueous solution of PCP with sunlight or artificial ultraviolet (UV) light which simulated the UV portion of sunlight (Crosby and Wong, 1973a).

Experimental Procedures

Standard solutions of PCP (100 mg/l in pH 7.3 borate-phosphate buffer) were irradiated in borosilicate glass containers in a UV photo-reactor (Crosby and Wong, 1973b) or under outdoor sunlight during summer months in Davis, California. After time intervals up to 100 hours, the yellow solutions were acidified with sulfuric acid, extracted with ether, the combined organic phases were dried and evaporated to a small volume, and the photodegradation products were separated and collected by gas chromatography (GLC). Most separations employed a 2 ft x ¼ inch stainless steel column packed with 1% DEGA on 60-80 mesh Chromsorb G, nitrogen carrier gas at 20 ml/min, and an oven temperature programmed from 100° to 200° at 10°/min. Aqueous solutions of tetrachlororesorcinol (IV), tetrachlorocatechol (VI), tetrachlorohydroquinone (V), and 2,3-dichloromaleic acid (XI), also were irradiated and worked up in the same manner.

Photodegradation rates for buffered solutions of the same compounds were measured at 26° C in 100 ml borosilicate glass flasks irradiated indoors in the UV reactor. Flasks were removed at regular intervals, the contents acidified, and unreacted chemicals extracted into hexane or ether, methylated with fresh and ethereal diazomethane, and quantitatively analyzed by GLC. Product identifications were based on comparisons of physical properties and spectral data with those obtained from authentic standards.

Results

The photodegradation rate of PCP in aqueous buffer at pH 7.3 was rather rapid (Fig. 1). Under simulated sunlight, the time needed for the total disappearance of ionized PCP was approximately 20 hours, while in sunlight, total degradation of PCP was achieved within five to seven days. The photolysis rate for unionized PCP (pH 3.3) was much slower, and irradiation of PCP in distilled water gave similar results.

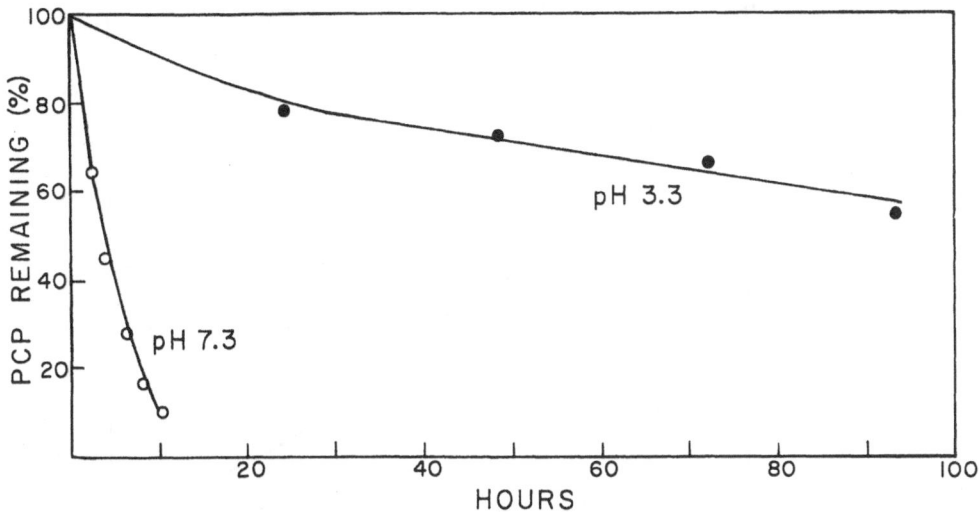

Figure 1. Photodecomposition rate of PCP at pH 3.3 and 7.3.

When the irradiation was interrupted at a point where 50 to 75% of the PCP had been degraded, three types of degradation products were isolated: (1) lower chlorinated phenols, (2) chlorinated dihydroxybenzenes (diols), and (3) nonaromatic fragments. The small proportion of lower chlorinated phenols was found to consist primarily of 2,3,4,6-tetrachlorophenol (II) and 2,3,5,6-tetrachlorophenol (III) together with trichlorophenols. Of the three possible tetrachlorodiols, tetrachlororesorcinol and tetrachlorocatechol were isolated from the photolysis mixture and identified, the former being a major product; tetrachlorohydroquinone presumably could be formed but was never detected due to its instability towards both UV light and air oxidation.

The formation of non-aromatic fragments from PCP was somewhat unexpected, although; tetrachloroquinones were known to form smaller molecules under harsh conditions such as treatment with peroxy-acids (Karrer and Testa, 1949). PCP provides the first reported instance of photochemical cleavage of a chlorinated aromatic ring in sunlight. The principal non-aromatic fragment proved to be dichloromaleic acid (XI), isolated by preparatory GLC as dichloromaleic anhydride. Several minor products also were isolated and tentatively identified as chlorolactones. When the irradiation was allowed to proceed until more than 95% of the PCP had been degraded (approximately seven days), dichloromaleic acid was the sole product detectable by GLC; after 20 days of outdoor exposure, the colorless PCP solution had become yellow and then colorless again, nothing volatile could be extracted and evaporation of the aqueous phase after extraction left no residue at all: PCP had been photodecomposed completely to chloride ions, carbon dioxide, and possibly small aliphatic fragments. Dichloromaleic acid underwent photolysis much more slowly than PCP or the other intermediates (Fig. 2), accounting for its

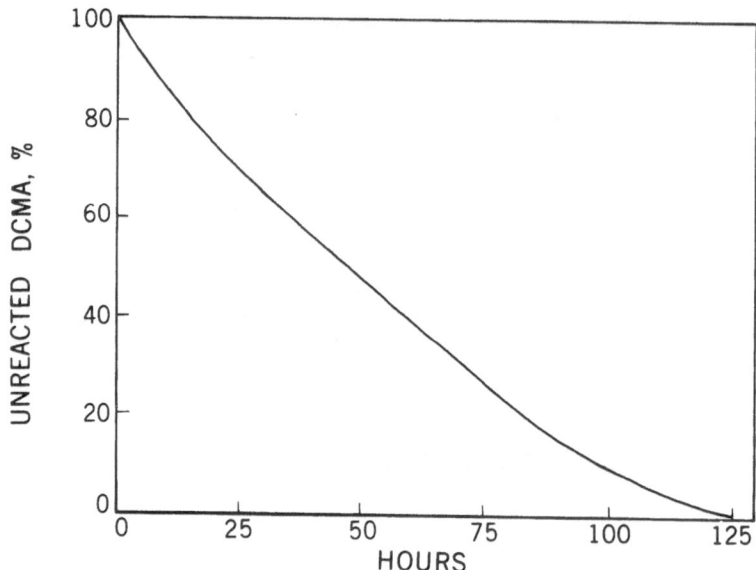

Figure 2. Photodecomposition rate of 2,3-dichloromaleic acid (DCMA) at pH 7.3.

accumulation as the sole product after all the PCP was gone and its prolonged irradiation also produced only hydrogen and carbon dioxide.

Irradiation of aqueous solutions of each of the tetrachlorodiols in turn resulted in their complete loss in less than 8 hours to give trichlorodiols, trichloroquinones, and dichloromaleic acid among the complex mixture of products. While the dark control solutions of tetrachlororesorcinol and tetrachlorocatechol showed no change on standing, the tetrachlorohydroquinone decomposed even in the dark in the presence of air to form tetrachlorobenzoquinone (chloranil), 2-hydroxy-3,5,6-trichlorobenzoquinone (VII), 2,5-dichloro-3,6-dihydroxybenzoquinone (chloranilic acid, X), and dichloromaleic acid.

Discussion

A route from PCP to dichloromaleic acid is proposed in Fig. 3 which involves the missing but logical intermediate, 2,3-dichloro-5,6-dihydroxy benzoquinone (IX). Judging from the yields and the subsequent photochemical reaction of the products, the initial and rate-limiting reaction is the photonucleophilic replacement of PCP chlorine atoms by hydroxyl groups (Crosby *et al.,* 1972). The chlorine substituent is displaced as chloride ion, and the rate of the reaction increases with increasing pH (Fig. 1). The formation of brominated compounds when PCP was irradiated in a solution of potassium bromide (Crosby and Wong, 1976) provide strong evidence of an ionic mechanism, and further support for a nucleophilic displacement reaction coming from the formation of cyanide and the closely related *p*-chlorophenol under similar conditions (Omura and Matsuura, 1969).

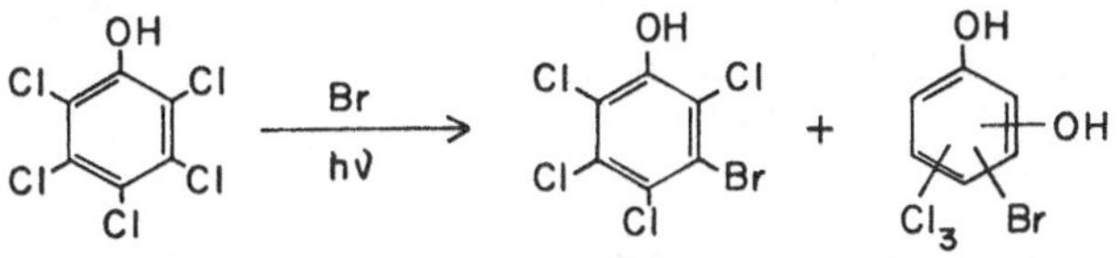

The possible photonucleophilic displacement of chloride by chloro-phenoxide ion in *ortho*-chlorophenols raises the intriguing question of the photochemical generation of chlorinated dibenzo-*p*-dioxins. Indeed, exposure of an aqueous solution of the sodium salt of PCP to sunlight wavelengths provided identifiable amounts of octachlorodibenzo-*p*-dioxin (OCDD) (Crosby and Wong, 1976); none of the extremely toxic 2,3,7,8-tetrachlorodibenzo-*p*-dioxin (TCDD) could be detected, perhaps because of its rapid photoreduction (Crosby *et al.*, 1971).

The air-oxidation of the resulting tetrachlorohydroquinone and tetrachlorocatechol to the corresponding quinones is rapid. Tetrachlororesorcinol cannot form a quinone and so is more stable in concentrated solutions; it reacts with the quinones to produce some of the complex products observed by Kuwahara

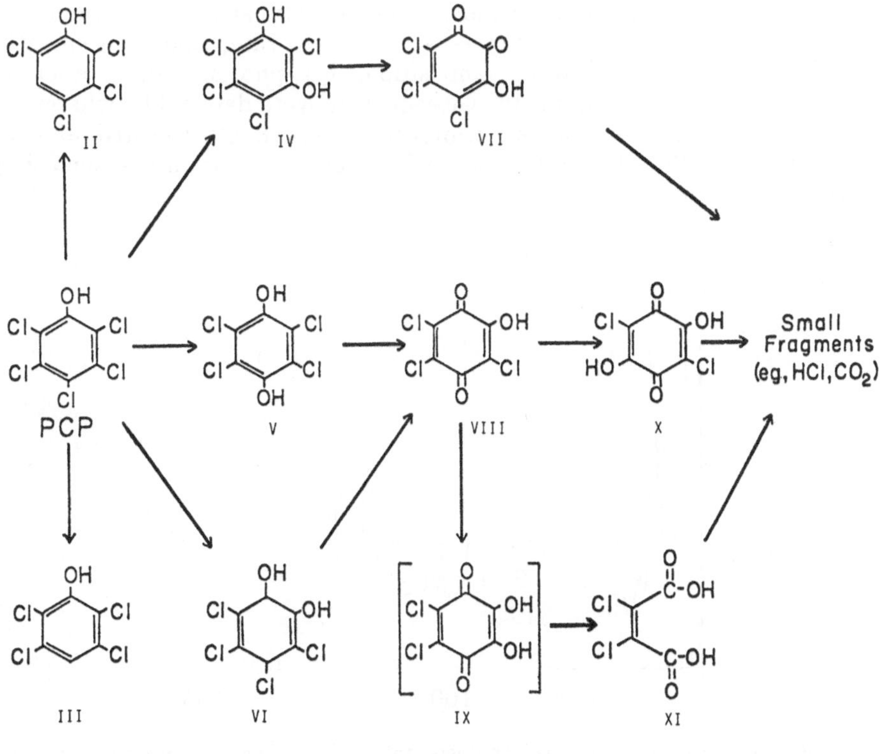

Figure 3. Proposed photolysis pathway for PCP.

TABLE 1. PCP in Selected North California Water

Location	Water	PCP (ppb)
Moss Landing	Near-shore seawater	< 1
Sacramento	River at sewage discharge site	< 1
Oroville	Drainage water	20
Oroville area	Drinking water wells (7)	< 1-800 (av. 227)

(Kuwahara *et al.,* 1966a, 1966b). In the highly dilute solutions usually represented by natural waters, however, further oxidation and hydration of the simple quinones would be preferred. The model for these reactions may be the perphthalic acid oxidation of tetrachloro-*o*-benzoquinone, reported by Karrer and Testa (1949), which produces tetrachloromuconic acid by ring cleavage and eventually leads to dichloromaleic acid.

Although its ultraviolet absorption maximum lies far below the sunlight region of the spectrum (λ_{max} = 265 nm), dichloromaleic acid still absorbs sufficient energy to react — presumably with water — at a slower rate, and consequently it accumulates during PCP photodecomposition. No photoproducts other than hydrogen chloride and carbon dioxide subsequently were detectable; whatever the degradation pathways of the numerous intermediate products, the striking feature of PCP photolysis is that it produces primarily — and perhaps only — such simple, gaseous final products.

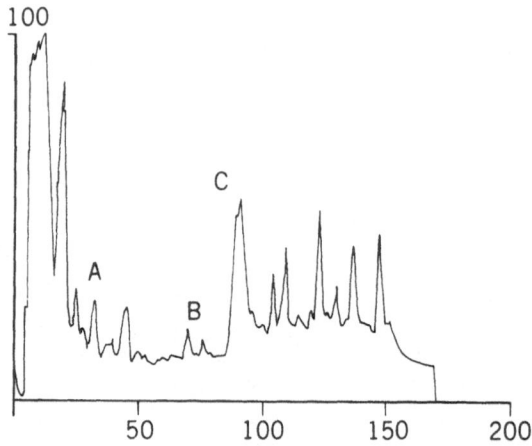

Figure 4. Typical GLC-mass spectral analysis (total-ion monitor) of an extract of natural water from near Oroville, California. (A) Dichlorophenol, (B) Tetrachlorophenol, (C) PCP.

In an attempt to verify this photochemical degradation in the field, water samples from several Northern California locations were analyzed by combined GLC-mass spectrometry. PCP was detected from several sources (Table 1) occasionally accompanied by 2,3,4,6-tetrachlorophenol (Fig. 4), but no dichloromaleic acid (as anhydride or methyl ester) ever was observed. However, the fact the PCP solutions at the appropriate pH obviously decompose in sunlight and that no reagents other than water and oxygen are required makes a strong case for the likelihood of essentially total PCP destruction in many aquatic environments.

References

Bevenue, A., and H. Beckman, 1967. Pentachlorophenol: A discussion of its properties and its occurrence as a residue in human and animal tissues. *Residue Rev.,* **19:** 83-134.

Crosby, D.G., K.W. Moilanen, M. Nakagawa, and A.S. Wong, 1972. Photonucleophilic reactions of pesticides. In, *Environmental Toxicology of Pesticides,* pp. 423-431 (F. Matsumura, G.M. Boush, and T. Misato, Eds.). Academic Press, New York. pp. 423-431.

Crosby, D.G., A.S. Wong, J.R. Plimmer, and E.A. Woolson, 1971. Photodecomposition of chlorinated dibenzo-*p*-dioxins. *Science,* **173:** 748-749.

Crosby, D.G., and A.S. Wong, 1973a. Photodecomposition of *p*-chlorophenoxyacetic acid. *J. Agr. Food Chem.,* **21:** 1049-1052.

Crosby, D.G., and A.S. Wong, 1973b. Photodecomposition of 2,4,5-trichlorophenoxyacetic acid (2,4,5-T) in water. *J. Agr. Food Chem.,* **21:** 1052-1054.

Crosby, D.G., and A.S. Wong, 1976. Photochemical generation of chlorinated dioxins. *Chemosphere,* **5:** 327-332.

Hamadmad, N., 1967. Photolysis of pentachloronitrobenzene, 2,3,5,6-tetrachloronitrobenzene, and pentachlorophenol. Ph.D. Dissertation, Univ. of California, Davis, California, 93 p.

Karrer, P., and E. Testa, 1949. Oxydatives Abbau des Tetrachloro-*o*-benzothinons mit Perphtalsäure. Untersuchung der Oxydationsprodukte. *Helv. Chim. Acta,* **32:** 1019-1028.

Kuwahara, M., N. Kato, and K. Munakata, 1966a. The photochemical reaction of pentachlorophenol. Part I. The structure of the yellow compound. *Agr. Biol. Chem., Japan,* **30:** 232-238.

Kuwahara, M., N. Kato, and K. Munakata, 1966b. The photochemical reaction of pentachlorophenol. Part II. The chemical structure of minor products. *Agr. Biol. Chem., Japan,* **30:** 239-245.

Omura, K., and T. Matsura, 1969. Photolysis of halogeno-phenols in aqueous alkali and cyanide. *Chem. Commun.,* 1394-1395.

von Rümker, R., W. Lawless, and A.F. Meiners, 1974. *Production, Distribution, Use and Environmental Impact Potential of Selected Pesticides.* Office of Pesticide Programs, Environmental Protection Agency, Washington, D.C. 439p.

Degradation of Pentachlorophenol in Soil, and by Soil Microorganisms

DONALD D. KAUFMAN

Abstract—Pentachlorophenol (PCP) readily degrades in the environment by chemical, microbiological and photochemical processes. Degradation in soil is affected by numerous chemical, physical and biological factors. PCP degrades more rapidly in flooded or anaerobic soil than in aerobic moist soil. Several pathways of degradation have been identified. The degradation of PCP in soil is primarily by reductive dehalogenation to simpler tetra-, tri-, and dichlorophenols. Methylation of PCP to pentachloroanisole and liberation of $^{14}CO_2$ from ^{14}C-PCP have also been observed. Isolated soil microorganisms metabolize ^{14}C-PCP by methylation, dehalogenation, and ring cleavage with ultimate conversion of the ^{14}C to $^{14}CO_2$ and normal ^{14}C-cell constituents.

Introduction

Pentachlorophenol (PCP) and its salts have probably had more varied uses than any other pesticide. It has been widely used as a pre- and post- plant herbicide in several crops, but its most extensive use as a herbicide has been in rice paddy fields. It also is used extensively as a wood preservative against wood rotting and staining microorganisms, and for the control of powder post beetles, wood-boring insects, and termites. Other uses have been as a biocide in cellulosic products, starches, adhesives, proteins, leather, oils, paints, rubber, rug shampoos, textiles, food processing plants, and the control of bacteria, fungi, insects, mollusks, and other

DONALD D. KAUFMAN • Pesticide Degradation Laboratory, Beltsville Agricultural Research Center — West, U.S. Department of Agriculture, Beltsville, Maryland 20705, U.S.A.

nuisance biota (Bevenue and Beckman, 1967). Pentachlorobenzyl alcohol and the acetate and barium salts of pentachlorophenol have been used as fungicides against *Pyricularia oryzae* which causes rice blast (Ishida, 1967; Kuwatsuka and Igarashi, 1975).

Worldwide production of PCP has been about 200 million pounds per year (Detrick, 1977). The occurrence of PCP as a residue in human and animal tissues (Bevenue and Beckman, 1967; Dougherty, this symposium) and its presence in 85% or more of the urine samples of people exposed non-occupationally to PCP (Dougherty, this symposium; Kutz *et al.,* this symposium) suggest a rather wide distribution of this compound. Whether such ubiquitous distribution of PCP is due to the patterns of usage of PCP alone has been questioned (Detrick, 1977). Other possible sources of PCP which have been suggested include degradation of hexachlorobenzene (Koss *et al.,* 1976) and pentachloronitrobenzene (Murthy and Kaufman, 1977). Regardless of the possible sources of environmental contamination by PCP, it is indeed present in the environment, and knowledge of its environmental fate and behavior is essential. The purpose of this presentation is to review available information regarding its degradation in soil and by soil microorganisms.

Degradation in Soil

Early information regarding the degradation and persistence of PCP in soil is based on determining the toxicity and biocidal activity of soil at various times after PCP treatment. Loustalot and Ferrer (1950) examined the persistence of PCP in soil treated with 0, 30, 60, and 90 lbs/A (acre) of the herbicide and stored for 0, ½, 1, and 2 months before being bioassayed with corn and cucumbers. The data obtained demonstrated that, in general, the toxicity of the herbicide decreased with the passage of time. There was no appreciable inactivation of PCP after two months when applied to air-dried soil. The material persisted somewhat longer in soil with medium moisture content than in saturated soil. Toxicity appeared to persist longer when applied to a heavy clay soil than when applied to a sandy or sandy-clay mixture (Loustalot and Ferrer, 1950). Young and Carroll (1951) also observed that PCP decomposed more rapidly in soils containing a higher content of organic matter. The rate of dissipation was greater when soils were near saturation point and when the soil temperature approached the optimum for microbiological activity.

In the investigations of Loustalot and Ferrer (1950), PCP persistence varied from 15 to more than 60 days and was dependent upon soil conditions and application rate. Young and Carroll (1951) observed PCP to persist in detectable quantities for as long as 49 days. When applied at 7.5 lb/A, Alban and McCombs (1949) observed residual effects of Na-PCP for as long as 12 weeks after application of 7.2 lb/A to muck soils. Taylor (1950) observed no residual toxicity in soil after 100 days from 25 lb/A applications of Na-PCP, but did find residual activity after 100 days from 5 lb/A applications of PCP in 50 gallons of Stoddard solvent. Thus, formulation, as well as soil type, organic matter and moisture contents, and temperature will affect PCP persistence in soil.

TABLE 1. Degradation Products of Pentachlorophenol

Chemical name	Source[a]	Reference
2,5-dichloro-3-hydroxy-6-pentachlorophenoxy-p-benzoquinone	P	Munakata and Kuwahara, 1969
3,4,5-trichloro-6-(2′,3′,4′,5′-tetrachloro-6′-hydroxyphenoxy)-o-benzoquinone	P	Munakata and Kuwahara, 1969
2,5-dichloro-3-hydroxy-6-(2′,4′,5′,6′-tetrachloro-3′-hydroxyphenoxy)-p-benzoquinone	P	Munakata and Kuwahara, 1969
3,5-dichloro-4-(2′,3′,5′,6′-tetrachloro-4-hydroxy)-6-(3,4,5,6-tetrachloro-2-hydroxyphenoxy)-o-benzoquinone	P	Munakata and Kuwahara, 1969
1,2,3,4,6,7,8,9-octachlorodibenzo-p-dioxin	P	Crosby et al., 1972
2,4,5,6-tetrachlororesorcinol	P	Munakata and Kuwahara, 1969
Chloranilic acid	P	Munakata and Kuwahara, 1969
2,3,4,5-tetrachlorophenol	S	Kuwatsuka and Igarashi, 1975; Ide et al., 1972; Igarashi and Kuwatsuka, 1973; Murthy et al., 1977
2,3,5,6-tetrachlorophenol	S	Kuwatsuka and Igarashi, 1975; Ide et al., 1972; Murthy et al., 1977
2,3,4,6-tetrachlorophenol	S	Kuwatsuka and Igarashi, 1975; Ide et al., 1972
2,4,5-trichlorophenol	S	Kuwatsuka and Igarashi, 1975; Ide et al., 1972
2,3,6-trichlorophenol	S	Kuwatsuka and Igarashi, 1975; Murthy et al., 1977
2,3,4-trichlorophenol	S	Kuwatsuka and Igarashi, 1975
2,3,5-trichlorophenol	S	Kuwatsuka and Igarashi, 1975; Ide et al., 1972; Igarashi and Kuwatsuka, 1973
2,4,6-trichlorophenol	S	Kuwatsuka and Igarashi, 1975
3,4-dichlorophenol	S	Kuwatsuka and Igarashi, 1975; Ide et al., 1972

TABLE 1. Continued

Chemical name	Source[a]	Reference
3,5-dichlorophenol	S	Kuwatsuka and Igarashi, 1975; Ide et al., 1972
2,3,4,5-tetrachloroanisole	S	Ide et al., 1972
2,3,5,6-tetrachloroanisole	S	Ide et al., 1972
2,3,4,6-tetrachloroanisole	S	Ide et al., 1972; Engel et al., 1966
2,3,5-trichloroanisole	S	Ide et al., 1972
2,4,5-trichloroanisole	S	Ide et al., 1972
3,4-dichloroanisole	S	Ide et al., 1972
3,5-dichloroanisole	S	Ide et al., 1972
3-chloroanisole	S	Ide et al., 1972
pentachloroanisole	S,M	Ide et al., 1972; Kuwatsuka and Igarashi, 1975; Suzuki and Nose, 1971; Cserjesi and Johnson, 1972; Igarashi and Kuwatsuka, 1971; Murthy et al., 1977
tetrachlorocatechol	M	Suzuki, 1977
tetrachlorohydroquinone	M	Suzuki, 1977
tetrachlorohydroquinone	M	Suzuki, 1977
tetrachlorohydroquinone, dimethyl ether	M	Suzuki and Nose, 1971
tetrachlorobenzoquinone	M	Reiner et al., this symposium
2,6-dichlorohydroquinone	M	Reiner et al., this symposium
$^{14}CO_2$	S,M	Chu and Kirsch, 1972; Kirsch and Etzel, 1973; Suzuki, 1977
Cl$^-$	M	Watanabe, 1973b; Suzuki, 1977
tetrachloromuconic acid		Lyr, 1962
β-hydroxytrichloromuconic acid		Lyr, 1962

[a]M = isolated microorganisms; P = photodegradation; S = soil.

The effect of numerous factors on the degradation of PCP in soil was recently discussed by Tsunoda (1965), Kuwatsuka (1972) and Aso and Sakamoto (1962). The rate of PCP dissipation from soil was closely related to temperature, aeration and organic matter. The degradation of PCP was partially dependent on cation exchange capacity and soil pH, whereas soil texture, clay content, degree of base saturation, and free iron oxides did not influence the rate of PCP degradation. At higher temperatures PCP disappeared more rapidly from soils under anaerobic (flooded) conditions than under aerobic conditions. A more rapid degradation of PCP in paddy soil under flooded conditions than in previously aerobic soils placed under anaerobic conditions indicates the importance of an established anaerobic population. The half-life of PCP at an initial soil concentration of 100 ppm was 10-40 days at 30 °C under flooded conditions, whereas almost 100% was detected in aerobic soil after two months. PCP degraded more rapidly in high organic matter soils than in soils with low organic matter content. In subsequent investigations Kuwatsuka and Igarashi (1975) further indicated a slight correlation of PCP degradation with free iron content and phosphate adsorption coefficient, but little correlation with available phosphorus content.

The effect of flooding or anaerobic conditions on PCP degradation in 11 different rice paddy soils was examined by Kuwatsuka and Igarashi (1975). The half-life of PCP varied from 10 to 70 days (average 30 days) in the arable soils under flooded conditions, and from 20 to 120 days (average 50 days) under upland conditions. The period of time required to degrade 90% of the PCP present was 30 days or more under flooded conditions, and 50 days or more under upland conditions. Almost 100% of the PCP remained even after 50 days when applied to a forest subsoil which contained only trace amounts of organic matter, and submitted to both water conditions. Similar results were obtained by Watanabe (1973a) and Ide et al. (1972) who reported that PCP disappeared rapidly in mature paddy fields, but very slowly in a newly reclaimed field and in soil collected from an immature paddy field. In other investigations the rate of PCP degradation varied among farm soils and was influenced by the presence of certain fungicides (Suzuki and Nose, 1970); repeated applications of PCP to paddy fields resulted in accelerated degradation (Watanabe and Hayashi, 1972a,b) and no PCP degradation occurred in submerged soils which had been sterilized (Kuwatsuka, 1972; Igarashi and Kuwatsuka, 1972). These results suggest that PCP degradation is related to microbial activity in the soil.

Numerous degradation products have been isolated from PCP treated soil. Ide et al. (1972) examined PCP degradation in sterile and nonsterile paddy soil. PCP was stable in sterilized soil, but was degraded in nonsterile soil. Decomposition products isolated from soil included 2,3,4,5-, 2,3,5,6-, and 2,3,4,6-tetrachlorophenols; 2,4,5- and 2,3,5-trichlorophenols; 3,4- and 3,5-dichlorophenols; and 3-chlorophenol (Table 1). Time course investigations were also conducted with the three tetrachlorophenols. Of these, 2,3,4,6-tetrachlorophenol was the most rapidly degraded. It was concluded that the chlorines *ortho* and *para* to the hydroxyl group underwent reductive dechlorination more easily than the *meta* chlorine.

Similar results were obtained by Kuwatsuka and Igarashi (1975) who also identified pentachloroanisole as a PCP degradation product. Pentachloroanisole formation and its subsequent degradation back to PCP was reported by Igarashi and Kuwatsuka (1971, 1972, 1973). Demethylation and methylation of phenolic groups in biological systems is well known (Williams, 1959). Ide *et al.* (1972) also found 2,3,4,5-, 2,3,5,6-, and 2,3,4,6-tetrachloroanisoles, 2,3,5- and 2,4,5-trichloroanisoles, 3,4- and 3,5-dichloroanisoles and 3-chloroanisole as methylated products of PCP in incubated soil, but not 2,3,6-, 2,5,6- and 2,4,6-trichloroanisoles. Based on the results obtained in these investigations Igarashi and Kuwatsuka (1972) and Matsunaka and Kuwatsuka (1975) propose a soil degradation pathway in which the initial degradative mechanism in soil is either methylation or reductive dechlorination (Fig. 1).

Photolytic degradation of PCP influences the fate of PCP in soil, as observed in paddy fields. The photodegradation products were isolated and identified by Munakata and Kuwahara (1969) (Table 1). 1,2,3,4,6,7,8,9-octachlorodibenzo-*p*-dioxin was identified later as a photodegradation product of PCP by Crosby *et al.* (1972). Others have also noted that photochemical decomposition is at least partly responsible for PCP dissipation (Kuwahara *et al.*, 1966a,b).

In considering the biodegradation of PCP in soils and other environments it is also important to consider which products are actually degradation products as opposed to contaminants in the original formulation. Typical commercial PCP contains a variety of substances such as tetrachlorophenols, trichlorophenols, hexachlorobenzene, chlorinated dibenzo-*p*-dioxins and chlorinated dibenzofurans

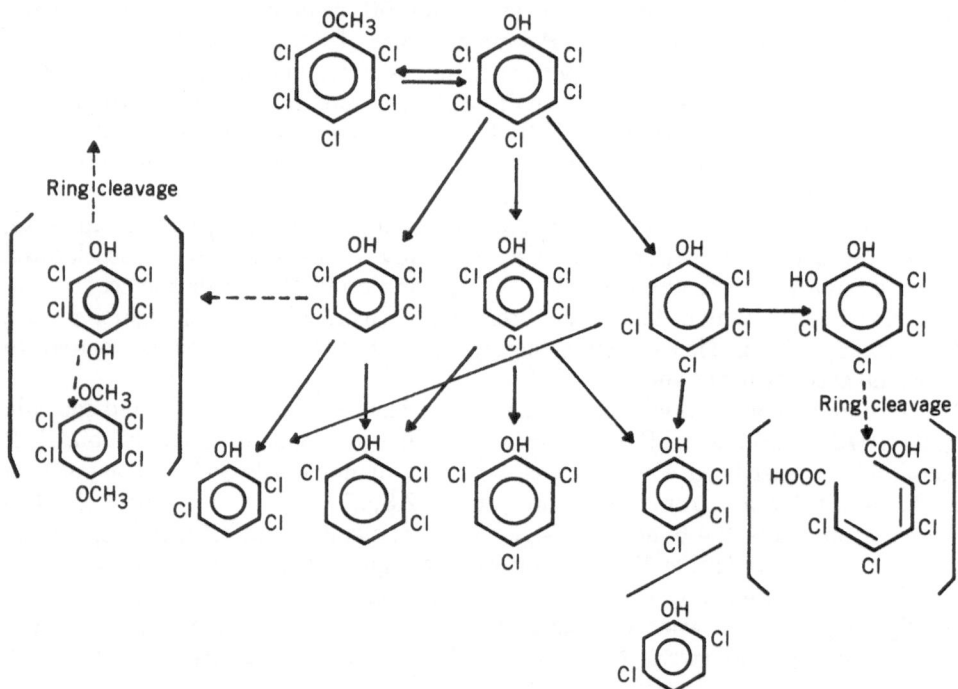

Figure 1. Proposed pathway of PCP degradation in soil.

(Johnson *et al.*, 1973; Schwetz *et al.*, this symposium). The degree of contamination seems to vary with the commercial preparations examined (Johnson *et al.*, 1973; Nilsson *et al.*, this symposium; Dougherty, this symposium).

Investigations which have involved the use of commercial formulations (other than analytically pure ^{14}C-PCP formulations) have reported a wide variety of tetra- and trichlorophenols as PCP degradation products. The predominant tetrachlorophenol in technical PCP samples is the 2,3,4,6-tetra isomer (Anon., 1977). It seems reasonable, therefore, to question which compounds are actually degradation products and which may actually be contaminants. From a chemical synthesis point of view placement of *meta* substituents is somewhat difficult. Similarly, from a biochemical point of view, abstraction of a substituent from a *meta* position is most difficult. Soil microorganisms have traditionally shown a preference for removing substituents which are either *ortho-* or *para* to reactive sites, or both. For these reasons it seems logical to question whether tetra- or trichlorophenols with only a missing *meta* substituent are actually degradation products or contaminants in the original formulation. The results of the few investigations with ^{14}C-PCP would support pathways, either in soil or with isolated microorganisms, which involve dehalogenation in either the *ortho* or *para* positions, or both. These observations, however, would not apply to the formation of methyl esters of any of the tetra- or trichlorophenols in soil. Methylation is a common phenomenon in the soil metabolism of phenolic moieties.

In recent investigations Murthy *et al.* (1977) examined the degradation of ^{14}C-PCP in both aerobic and anaerobic moist soil with and without cellulose amendments. The anaerobic soil was aerated with nitrogen. In anaerobic soil PCP reduced soil respiration in the presence of cellulose. Losses by volatilization accounted for only 0.5% of the PCP added to soil. No [14]CO_2 was detected. Organic solvent extractable radioactivity was the same from all anaerobic treatments. Gas chromatographic analysis of the soil extracts showed the presence of the methylether of PCP (0.7%). 2,3,5,6- and 2,3,4,5-tetrachlorophenols and 2,3,6-trichlorophenol were also detected as degradation products by gas chromatography after methylation. Total [14]C- recoveries from unamended and cellulose amended soils were 95.7 and 93.5%, respectively. Similar results were obtained in aerobic soils with regard to product formation. 2,3,4,5- and 2,3,5,6-tetrachlorophenol and 2,3,6-trichlorophenol were detected in extracts from aerobic soils. The principal product (51.5%) however, was the methylether of PCP, or pentachloroanisole.

Further degradation of the pentachloroanisole was examined in both aerobic and anaerobic soils. In aerobic soils only 5.6% of the pentachloroanisole was reduced back to PCP in 24 days, whereas in anaerobic soil 42.1% was reduced to PCP. These results indicate that while some interconversion of pentachloroanisole and PCP occurs in both aerobic and anaerobic soil, the reactions involved in the degradation of these compounds are reductive in anaerobic soils, and both reductive and oxidative in aerobic soils.

The results of the preceding investigation indicate that degradation of PCP in soil occurs primarily by reductive dehalogenation. This degradation pathway results in the formation of progressively simpler chlorophenols, i.e., tetrachlorophenols, trichlorophenols, and dichlorophenols. Investigations conducted with ^{14}C-PCP indicate that this proceeds through a logical sequence with the initial dehalogenation reactions occurring at either *ortho* or *para* positions. Methylation of the phenol group also appears to be a common reaction in PCP degradation as well as its tetra- and trichlorophenolic products. That dehydrohalogenated products are prominent in soil is interesting in view of the products formed by microorganisms isolated from soil. PCP metabolism by isolated soil microorganisms appears to occur by oxidative mechanisms (subsequent section). Thus, some discrepancies appear to exist between the degradative pathways observed in soil and those observed in isolated microbial cultures. Some oxidative metabolism of PCP degradation products must occur in soil, however. Suzuki and Nose (1970) reported the evolution of $^{14}CO_2$ from soil treated with ^{14}C-PCP.

Microbial Degradation of PCP

Early information regarding the microbial degradation of PCP emanates largely from its use as a wood preservative. Duncan and Deverall (1964) recovered less PCP from *Trichoderma*-treated wood and found that twice as much PCP was needed to control this organism in wood. They concluded that *Trichoderma* was able to degrade PCP. Stain-causing fungi growing in wood frequently tolerate higher concentrations of PCP than the wood-deteriorating Basidiomycetes. This increased tolerance by organisms such as *Cepholoascus fragrans* Hanawa and *Trichoderma* spp. is sometimes associated with the capacity of these fungi to degrade PCP. Leutritz (1965) questioned the biodegradability of this phenol in wood and suggested that other reasons for the dissipation of PCP from wood should be considered. The work of Lyr (1963) and others (Duncan and Deverall, 1964; Cserjesi, 1967; Ingols and Stevenson 1963), however, indicates that wood-rotting and -staining fungi do degrade PCP. Lyr (1963) associated the ability of basidiomycete fungi to degrade PCP with the presence of phenol oxidases. He observed that the wood-rotting basidiomycete *Tramates versicolor* was able to degrade PCP by secreting laccase into the culture medium. Although no metabolites were extracted or identified, changes in UV spectra, the liberation of chloride ion, and the appearance of colored products all point to a degradation of the parent molecule. Engel *et al.* (1966) described the isolation of a tetrachloroanisole from wood shavings used in chicken bedding. This compound which was the source of a musty taste in the eggs and broilers could have been derived from PCP degradation in treated wood.

Cserjesi (1967), however, isolated several *Trichoderma* spp. which gave negative response to phenol oxidase tests, but caused significant degradation of PCP. Investigations with other chlorinated phenols revealed that increased chlorination of the nucleus increased stability to oxidation and enzymatic degradation by the polyphenol oxidases, laccase and tyrosinase. Rich and Horsfall (1954) observed a good correlation between color formation and detoxification of 42 phenols exposed to mycelial extracts of *Stemphylium sarcinaeforme* containing a laccase type of polyphenol oxidase. All but two of the compounds producing color reactions were nontoxic to spores of the fungi, whereas those not visibly changed, retained their toxicity.

Cserjesi (1967) and Cserjesi and Johnson (1972) examined the degradation of PCP by cultures of the wood rotting fungi, *Cephaloascus fragrans, Trichoderma virgatum,* and *Pencillium* sp. *C. fragrans* appeared to have an adapted tolerance to PCP, whereas *T. virgatum* methylated PCP to form pentachloroanisole. The formation of pentachloroanisole, however, did not account for the total loss of pentachlorophenol from the culture medium, which suggested that this reaction was either an initial step or parallel reaction in the degradation process.

Microbial degradation of PCP has also been observed in soil perfusion systems and with isolated bacterial cultures. Watanabe (1973a,b) examined PCP degradation in soil perfused with 40 ppm of PCP and observed the typical soil enrichment type phenomena. After an eight day lag period during which essentially no degradation occurred, chloride ion liberation was initiated, and was essentially complete within three weeks. Subsequent additions of PCP were degraded more rapidly with no lag period. A species of *Pseudomonas* was subsequently isolated which was capable of utilizing PCP as a sole source of carbon with complete liberation of the chloride ion. The dechlorination process corresponded approximately with PCP disappearance.

The effect of medium composition on PCP degradation by *Pseudomonas* sp. was examined. Yeast extracts accelerated degradation, whereas glucose at 100 ppm suppressed degradation. The substitution of ammonium sulfate for sodium nitrate as a nitrogen source also suppressed degradation. PCP degradation and microbial growth at 40 ppm were greater than at 100 ppm. Neither degradation nor growth occurred at 200 ppm PCP, which suggests some inherent PCP toxicity to the organism. The organism was incapable of anaerobic metabolism of PCP. A loss of UV absorption occurred with metabolism of PCP in culture solutions. Pentachloroanisole and the dimethylether of tetrachlorohydroquinone were identified as PCP degradation products.

Suzuki and Nose (1971), Kamisako *et al.* (1975), and Suzuki (1975, 1977) have also examined PCP metabolism by isolated *Pseudomonas* spp. Suzuki and Nose (1971) observed that up to 47% of the ^{14}C-PCP was degraded to ^{14}CO$_2$ within one hour, and that a small percent of the radioactivity was incorporated into amino acids. They also detected tetrahydroquinone as an intermediate product in PCP degradation. In subsequent work, Suzuki (1977) demonstrated that a *Pseudomonas* sp. degraded ^{14}C-PCP rapidly with the release of approximately 50% of the ^{14}C as ^{14}CO$_2$. The results of amino acid analysis of the bacterial cells indicated that the ^{14}C derived from ^{14}C-PCP was incorporated rapidly into the cell constituents, and that the pattern of ^{14}C-amino acids in the cell constituents was not significantly different from 15-minute and 24-hour incubation periods. Tetrachlorocatechol and tetrachlorohydroquinone were identified as PCP degradation products.

The *Pseudomonas* sp. investigated by Kamisako *et al.* (1975) degraded PCP, liberating the theoretical amount of chlorine atoms and producing CO$_2$ up to 55-70% of the added PCP. This bacterium also rapidly metabolized several other chlorophenols, benzoic acid and phthalic acid, but only slightly affected 2,3,4,5-tetrachlorophenol. Chu and Kirsch (1972) observed that ^{14}C-PCP was metabolized with evolution of ^{14}CO$_2$ by a saprophytic coryneform bacterium isolated from wastewater. In other investigations (Chu and Kirsch, 1973) they observed that this culture was capable of utilizing PCP as a single source of carbon and energy for growth, but exhibited pronounced substrate selectivity for various other halogenated phenols. The organism did not seem to distinguish between bromo- and chlorophenols. 2,3,4,6-Tetra and 2,4,6-trichlorophenol were satisfactory alternate growth substrates. Several additional multihalogenated phenols were removed from solution by respiring cells, but did not support growth. The remaining halophenols, including phenol and the three monochlorophenols were neither incorporated, nor did they exhibit a high degree of toxicity. PCP metabolism was shown to be highly responsive to enzyme induction with PCP as the inducer. Partial induction of the PCP-degrading system occurred when 2,4,6-trichlorophenol was employed as an inducer.

Kirsch and Etzel (1973) examined the PCP-oxidative capacity of a mixed population of soil microorganisms growing in a fill-and-draw, completely mixed aerator with daily increments of dilute nutrient broth and PCP. They observed that the PCP-oxidative capacity reached a maximum of 68% in 25 days, remained stable for approximately 17 days and then began to diminish to a negligible level during the next 14 days. The reason for the sharp decrease in PCP-oxidizing capacity was not known. It is conceivable, however, that this reduction may have resulted from either an excessive chloride ion concentration ultimately achieved through the fill-and-draw system, or a gradual but significant change in the composition of the dominant microbial flora of the mixed population, or both.

The evolution of $^{14}CO_2$ from ^{14}C-PCP treated cultures offers conclusive proof that ring cleavage does occur (Suzuki, 1977). It is generally accepted that the conversion of phenols to *ortho* or *para*-dihydroxyphenol derivatives occurs prior to ring cleavage in the metabolism of aromatic compounds by microorganisms (Evans, 1963). The isolation and identification of tetrachlorophenol and tetrachlorohydroquinone (Suzuki, 1977) would suggest that these products are intermediate metabolites prior to ring cleavage in PCP degradation (Fig. 1). Lyr (1962) hypothesized that tetrachloromuconic acid and β-hydroxytrichloromuconic acid are probably produced by the ring cleavages of tetrachlorocatechol and tetrachlorohydroquinone, respectively, during the reaction of PCP with hydrogen peroxide. This supposition would support the results of Suzuki (1977). Evidence obtained by Reiner *et al.* (this symposium) suggests the probable role of 2,6-dichlorohydroquinone and tetrachlorohydroquinone or tetrachlorobenzoquinone as intermediates in the metabolism of PCP.

Conclusions

A review of the literature indicates that PCP is degraded in the soil environment. Several pathways of degradation have been identified. Methylation yields the methylether of PCP and other phenolic products in soil. These products, however, are ultimately dealkylated and returned to the chlorinated phenol degradation pathways. Reductive dehalogenation appears to be the most significant PCP degradation pathway in soil. These pathways ultimately lead to ring cleavage and complete degradation of the PCP moiety. Numerous mono-, di-, tri-, and tetrachlorinated phenols have been reported as soil degradation products of PCP. A question arises as to which products isolated from PCP-treated systems are indeed degradation products or are merely contaminants in the original PCP materials applied.

Investigations conducted with ^{14}C-PCP have clearly demonstrated that very specific degradation pathways exist in soil and isolated microbial systems. These pathways involve either reductive or oxidative dehalogenation either *ortho* or *para* to the hydroxy group to yield either the 2,3,5,6- or 2,3,4,5-tetrachlorophenols, or tetrachlorocatechol or tetrachlorohydroquinone, respectively. Further degradation ultimately results in ring cleavage, ^{14}CO$_2$ evolution, and additional chloride ion liberation.

References

Alban, E.K., and L. McCombs, 1949. Pre- and post-emergence weed control with vegetable crops, 1947-1949. *Proc. North Central Weed Control Conf.,* **1949**: 81-86.

Anon., 1977. Pentachlorophenol — A wood preservative. Memorandum for the Office of Pesticide Programs, Environmental Protection Agency. American Wood Preservers Institute, McLean, Virginia. 40 pp.

Aso, S., and K. Sakamoto, 1972. Studies on the behavior of pentachlorophenol (PCP) under paddy field conditions. I. Differences of the behavior of PCP among several soil types. *J. Sci. Soil Manure, Japan,* **43**: 119-122.

Bevenue, A., and H. Beckman, 1967. Pentachlorophenol: A discussion of its properties and its occurrence as a residue in human and animals tissues. *Residue Rev.,* **19**: 83-134.

Chu, J. and E.J. Kirsch, 1972. Metabolism of pentachlorophenol by an axenic bacterial culture. *Appl. Microbiol.,* **23**: 1033-1035.

Chu, J., and E.J. Kirsch, 1973. Utilization of halophenols by a pentachlorophenol metabolizing bacterium. *Developments in Industrial Microbiology,* **14**: 264-273.

Crosby, D.G., K.W. Moilanen, M. Nakagawa, and A.S. Wong, 1972. photonucleophilic reactions of pesticides. In, *Environmental Toxicology of Pesticides* pp. 423-431 (F. Matsumura, G.M. Boush, and T. Misato, Eds.), Academic Press, New York.

Cserjesi, A.J., 1967. The adaptation of fungi to pentachlorophenol and its biodegradation. *Can. J. Microbiol.,* **13**: 1243-1249.

Cserjesi, A.J., and E.L. Johnson, 1972. Methylation of pentachlorophenol by *Trichoderma virgatum. Can. J. Microbiol.,* **18**: 45-49.

Detrick, R.S., 1977. Pentachlorophenol: Possible sources of human exposure. *Forest Prod. J.,* **27**: 13-16.

Duncan, C.G., and F.J. Deverall, 1964. Degradation of wood preservatives by fungi. *Appl. Microbiol.,* **12**: 57-62.

Engel, C., A.P. DeGroot, and C. Weurman, 1966. Tetrachloroanisol: A source of musty taste in eggs and broilers. *Science,* **154**: 270-271.

Evans, W.C., 1963. Microbial degradation of aromatic compounds. *J. Gen. Microbiol.,* **32**: 177-184.

Ide, A., Y. Niki, F. Sakamoto, I. Watanabe, and H. Watanabe, 1972. Decomposition of pentachlorophenol in paddy soil. *Agric. Biol. Chem.,* **36**: 1937-1944.

Igarashi, M., and S. Kuwatsuka, 1971. Degradation of PCP in soil. *Soc. Sci. Soil and Manure Japan,* Abstr., p. 24.

Igarashi, M., and S. Kuwatsuka, 1973. Degradation of PCP methylether in soils. *Soc. Sci. Soil and Manure, Japan,* Abstr., p. 19.

Ingols, R.S., and P.C. Stevenson, 1963. Biodegradation of the carbon-chlorine bond. *Res. Engr.* **18**: 4-8.

Johnson, R.L., P.J. Gehring, R.J. Kociba, and B.A. Schwetz, 1973. Chlorinated dibenzodioxins and pentachlorophenol. *Environ. Hlth. Persp.*, **5**: 171-175.

Kamisako, T., T. Sasaki, T. Iwamoto, M. Inaaka, and M. Kawashita, 1975. Degradation of pentachlorophenol (PCP) by a soil microorganisms. *Annual Meeting, Agr. Chem. Soc., Japan*, Abstr., p. 206.

Kirsch, E.J., and J.E. Etzel, 1973. Microbial decomposition of pentachlorophenol. *J. Water Pollut. Contr. Fed.*, **45**: 359-364.

Koss, G., W. Koransky, and K. Steinbach, 1976. Studies on the toxicology of hexachlorobenzene. II. Identification and determination of metabolites. *Arch. Toxicol.*, **35**: 107-114.

Kuwahara, M., N. Kato, and K. Munakata, 1966a. The photochemical reaction of pentachlorophenol. Part I. The structure of the yellow compound. *Agr. Biol. Chem., Japan*, **30**: 232-238.

Kuwahara, M., N. Kato, and K. Munakata, 1966b. The photochemical reaction of pentachlorophenol. Part II. The chemical structure of minor products. *Agr. Biol. Chem., Japan*, **30**: 239-245.

Kuwatsuka, S., 1972. Degradation of several herbicides in soils under different conditions. In, *Environmental Toxicology of Pesticides*, pp. 385-400 (F. Matsumura, G.M. Boush, and T. Misato, Eds.), Academic Press, New York.

Kuwatsuka, S., and M. Igarashi, 1975. Degradation of PCP in soils. II. The relationship between the degradation of PCP and the properties of soils, and the identification of the degradation products of PCP. *Soil Sci. Plant Nutr.*, **21**: 405-414.

Kuwatsuka, S., and Y. Niki, 1976. Fate and behavior of herbicides in soil environments with special emphasis on the fate of principal paddy herbicides in flooded soils. *Rev. Plant Protection Res.*, **9**: 143-163.

Leutritz, J., Jr., 1965. Biodegradability of pentachlorophenol as a possible source of depletion from treated wood. *Forest Prod. J.*, **15**: 269-272.

Loustalot, A.J., and R. Ferrer, 1950. The effect of some environmental factors on the persistence of sodium pentachlorophenate in the soil. *Proc. Amer. Hort. Sci.*, **56**: 294-298.

Lyr, H., 1962. Uber den oxydativen abbau chloriester phenole. *Holztechnologie*, **3**: 201-208.

Lyr, H., 1963. Enzymatische detoxification chlorierter phenole. *Phytopathol. Z.*, **47**: 73-83.

Matsunaka, S., and S. Kuwatsuka, 1975. Environmental problems related to herbicidal use in Japan. *Environ. Qual. Safety*, **4**: 149-159.

Munakata, K., and M. Kuwahara, 1969. Photochemical degradation products of pentachlorophenol. *Residue Rev.*, **25**: 13-23.

Murthy, N.B.K., and D.D. Kaufman, 1977. Degradation of pentachloronitrobenzene (PCNB) in anaerobic soils. *J. Agric. Food Chem.*, Submitted for publication.

Murthy, N.B.K., D.D. Kaufman, and G.F. Fries, 1977. Aerobic and anaerobic soil degradation of pentachlorophenol. *J. Environ. Sci. Hlth.*, Submitted for publ.

Rich, S., and J.G. Horsfall, 1954. Relation of polyphenol oxidases to fungitoxicity. *Proc. Natl. Acad. Sci., (U.S.)*, **40**: 139-145.

Suzuki, T., 1975. Metabolism of PCP by PCP degrading bacteria. *Agr. Chem. Soc., Japan*, Abstr., p. 44.

Suzuki, T., 1977. Metabolism of pentachlorophenol by a soil microbe. *J. Environ. Sci. Hlth.*, **12B**: 113-127.

Suzuki, T., and K. Nose, 1970. Decomposition of pentachlorophenol in farm soil. (Part I). Some factors relating to PCP decomposition. *Noyaku Seisan Gijutsu, Japan*, **22**: 27-30.

Suzuki, T., and K. Nose, 1971. Decomposition of pentachlorophenol in farm soil. (Part 2). PCP metabolism by a microorganism isolated from soil. *Noyaku Seisan Gijutsu, Japan*, **26**: 21-24.

Taylor, C.E., 1950. Chemical control of weeds in vegetable crops in Illinois. *Ill. Agr. Expt. Sta. Bull.*, H-436.

Tsunoda, H., 1965. Pentachlorophenol (PCP) derivatives as weed killers. 2. Movement and decomposition of PCP derivatives in soil. *J. Sci. Soil Manure, Japan,* **36**: 200-202.

Watanabe, I., 1973a. Degradation of pentachlorophenol (PCP) by soil microorganisms. *Soil and Microorganisms, Japan,* **14**: 1-7.

Watanabe, I., 1973b. Isolation of pentachlorophenol decomposing bacteria from soil. *Soil Sci. Plant Nutr.,* **19**: 109-116.

Watanabe, I., and S. Hayashi, 1972a. Degradation of PCP (pentachlorophenol in soil. I. Microbial depletion of PCP under dark and submerged conditions. *J. Sci. Soil Manure, Japan,* **43**: 119-122.

Watanabe, I., and S. Hayashi, 1972b. The participation of microorganisms in the disappearance of PCP in submerged soil. (Part I) Degradation of PCP in soils. *J. Sci. Soil and Manure, Japan,* **43**: 119-122.

Williams, R.T., 1959. Detoxification Mechanisms. 2nd Edn., John Wiley and Sons, Inc. New York. 796 pp.

Young, H.C., and J.C. Carroll, 1951. The decomposition of pentachlorophenol when applied as a residual pre-emergence herbicide. *Agron. J.* **43**: 504-507.

The Fate of Pentachlorophenol in an Aquatic Ecosystem

RICHARD H. PIERCE, Jr. and DILNA M. VICTOR

Abstract—Extensive fish kills occurred in a freshwater lake in December 1974 and again in December 1976 due to the accidental release of wood-treating wastes containing pentachlorophenol (PCP) in fuel oil. Samples of water, sediment, leaf litter, and fish collected from the lake were analyzed to determine the persistence and distribution of PCP and major PCP-degradation products in the aquatic environment.

PCP was found to persist in the water and in fish for over six months following the spill. Sediment and leaf litter samples retained high concentrations of PCP throughout the two-year period of investigation.

The major degradation products observed were pentachloroanisole (PCP-OCH$_3$) and the 2,3,5,6- and 2,3,4,5-tetrachlorophenol (TCP) isomers. These products were also found to persist in the sediment and fish. The methyl ethers (anisoles) of both TCP isomers were observed in some samples but the small amounts were difficult to quantitate. Tetrachlorophenol appeared to have been formed by photodegradation in the fuel oil solution before entering the lake, while PCP-OCH$_3$ appeared to have been formed within the aquatic environment.

Variations observed in the concentrations of PCP and PCP-degradation products throughout the study indicate that periodic influx occurred from the contaminated water shed area. The persistence of these chemicals in sediment also provided a source of continuous contamination of the aquatic environment.

RICHARD H. PIERCE, Jr. and DILNA M. VICTOR • Institute of Environmental Science, University of Southern Mississippi, Box 5215, Hattiesburg, Mississippi 39401, U.S.A.

Introduction

In December 1974 water containing pentachlorophenol (PCP) in fuel oil overflowed the banks of a wood-treatment company's waste water holding pond and entered a sixty-acre freshwater lake near Hattiesburg, Mississippi. The resulting fish kill was described as extensive to total (Mississippi Air and Water Pollution Control Commission, 1975). Another extensive fish kill was observed in the lake in December 1976.

We analyzed samples of sediment, water, leaf litter and fish collected at periodic intervals up to seventeen months after the first spill to assess the distribution and persistence of PCP in the lake (Pierce *et al.,* 1977). These studies indicated that the levels of PCP in water and fish returned to background levels by ten months after the spill. However, PCP persisted in leaf litter and sediments for at least seventeen months following the spill.

While the above investigations were in progress, we improved the methods for analysis of PCP and developed methods for analysis of PCP-degradation products. These techniques were employed during the second year of investigation (July 1976 — July 1977). As the second major PCP spill occurred during December 1976, we were able to study the ecosystem before and after this spill. The results of this phase of our investigation are presented here together with a discussion of our earlier findings.

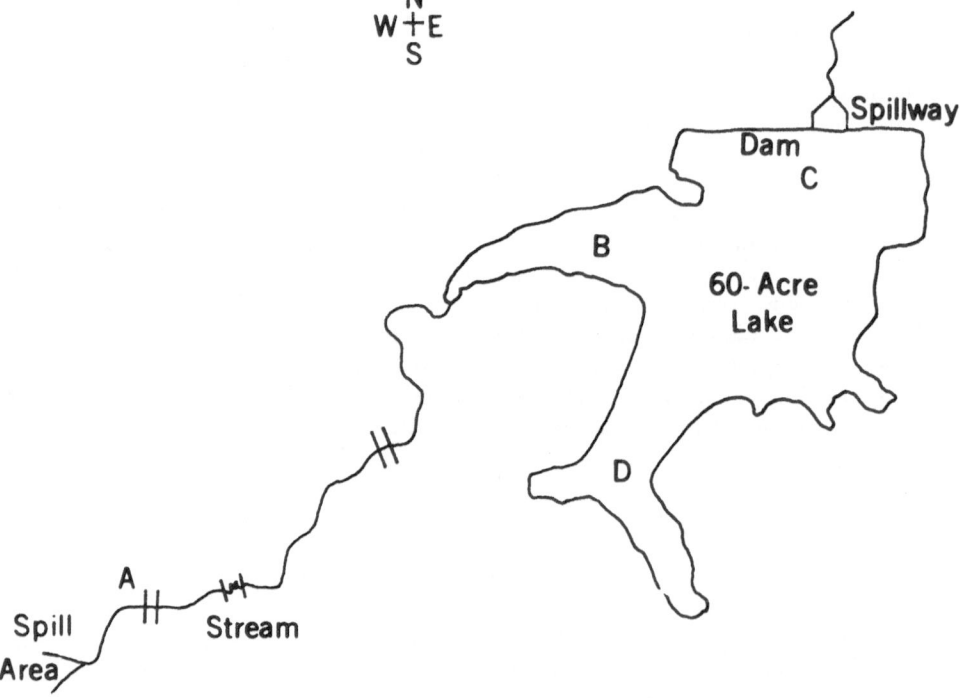

Figure 1. Sites of sample collection during the year 1976-1977.

Collection of Samples

The samples of water, sediment and fish were collected from four sites (Fig. 1) on a quarterly basis. Water samples were collected in 4 l jugs with aluminum-lined caps. Dissolved and particulate PCP were obtained by filtering the water through medium porosity glass-fiber filter pads. Fish were collected by gill net to obtain large specimens so that various organs could be subjected to chemical analysis individually. Sediment samples were collected with an Eckman dredge to provide at least 1 Kg of wet sediment from each site.

Analytical Procedures

Fish tissue was analyzed by lysing 1 g (wet weight) in 2 ml of 50% H_2SO_4, adding 2 ml acetonitrile and extracting with 2 × 5 ml hexane. The hexane solution (hexane-I) was then washed with distilled water and the base-soluble components (phenols) were extracted with 2 × 5 ml hexane (hexane-II). Both hexane-I (containing base-insoluble components) and hexane-II (containing base-soluble components) were analyzed by GC-EC before and after methylation with diazomethane.

Water samples were analyzed in triplicate by acidifying filtered 1-liter samples to pH 2 and extracting with 2 × 25 ml hexane. The hexane extract was separated into hexane-I and hexane-II and analyzed as described above for fish samples. Sediment samples were air-dried, washed with 0.1 N HCl and extracted with a solution of acetone/hexane (60/40, v/v) under reflux for 20 hours. The acetone was removed by washing with water and the hexane solution recovered and treated as described above. Leaf litter and particulates were analyzed in like manner.

The base-insoluble components (hexane-I solution) from selected samples were further fractionated by elution through an alumina micro-column (Buser, 1975) in an attempt to isolate and identify dioxins and dibenzofurans. Control samples of water, sediment, leaf litter, and fish were concurrently collected from an isolated 5-acre pond which received no industrial or agricultural drainage. Levels of PCP and PCP-degradation products in the control pond are considered to represent background concentrations in this area.

The extraction efficiency for PCP from six or more replicate spiked samples ± standard deviation was found to be 95 ± 10% for water, 113 ± 25% for sediments, and 90 ± 8% for fish.

Samples were analyzed with a Varian model 2700 gas chromatograph with Sc³H electron capture detectors. Two 3mm × 2m stainless steel columns were used: a non-polar 3% SP-2100 on 80/100 Supelcoport and a polar 10% SP-1000 on 80/100 chromosorb W.A.W. Injector temperature was 175 °C, column 200 °C, and detector 250 °C. The carrier gas was N_2 at a flow rate of 25 ml/min. Compound identity was verified from representative samples utilizing a Hewlett-Packard 5933 gas chromatography-mass spectrometry (GC-MS) Data System. All solvents were reagent grade, redistilled in a glass distillation apparatus and tested for purity by concentrating 100 fold, methylating and analyzing by GC-EC.

The Fate of PCP After the First Spill (December 1974)

The results, discussed by Pierce *et al.* (1977), indicate that the PCP content of the water was 10 ppb two months after the spill (February 1975), approached background levels (< 1 ppb) by October 1975, increased again to about 15 ppb in February 1976, and steadily declined through June 1976. PCP concentrations in fish collected two months after the spill were 2,500 ppb in the whole body (air-dried), decreased to background levels (< 50 ppb) by 10 months after the spill (October 1975), increased again in February 1976 and decreased through June 1976. Sediments and leaf litter contained large concentrations (an average of 100 ppb for sediments and 4,500 ppb for leaf litter) throughout the year, thus providing a source for chronic pollution of the aquatic ecosystem. The increase in PCP concentration observed in February 1976 followed a period of heavy rainfall, indicating that PCP was leached from the contaminated water shed area or that another small spill had occurred from the holding pond.

Studies on the Levels of PCP and Its Degradation Products in Relation to the Second Spill (December 1976)

Samples collected from the control pond in July 1976 and again in February 1977 contained PCP concentrations of 0.3 ppb in water, 3 ppb in sediment, and 7 ppb in fish muscle tissue (Table 1) indicating a slight background concentration of PCP in supposedly non-contaminated environmental samples.

TABLE 1. PCP and PCP-Degradation Products in Control Samples, ng PCP/g sample (ppb)

Sample	Number	PCP	PCP-OCH$_3$	2,3,5,6-TCP
water	(3)	0.28[a]	< .01[b]	0.07
sediment	(3)	3.3	0.1	1.00
fish-muscle	(2)	7.0	< 1[b]	< 1[b]

[a]Average value
[b]Lower limit of detection relative to background noise.

**TABLE 2. Pentachlorophenol
Concentrations (ppb) in Water (Dissolved)
and Sediment**

Date	Sites			
	A	B	C[a]	D
Aug. 11, 1976				
water[b]	11	0.1	0.1	0.1
sediment[c]	857	429	520	142
Oct. 22, 1976				
water	dry	<0.3	<0.3	1
sediment	166	994	389	212
Jan. 5, 1977				
water	81	24	25	16
sediment	277	239	150	170
Feb. 22, 1977				
water	147	N.A.[d]	29	N.A.
sediment	N.A.	1518	N.A.	N.A.
Apr. 27, 1977				
water	16	5	5	5
sediment	4	250	238	132

[a]Anaerobic sediment
[b]Average of triplicate values, ng/ml
[c]Average of duplicate values, ng/g air-dried sediment
[d]N.A., not analyzed

Pentachlorophenol in the Contaminated Lake

The concentration of PCP in samples of water and sediment for a 5 month period before the December 1976 spill and for a 4 month period after the spill are shown in Table 2. PCP concentration in water appeared to be near background levels (< 0.3 ppb) in October 1976 and higher (average, 40 ppb) in January 1977 immediately after the spill. PCP levels in April 1977 were markedly lower than in January 1977. Suspended particulates generally contained less than 10% of the PCP in the water column. Although particulate PCP was not as abundant as dissolved PCP, it may be important for the transport of PCP to the sediment. The sediment contained a high concentration (460 ppb average) of PCP in August and

October 1976, whereas the sediment samples collected in January 1977 immediately after the spill showed a lower PCP content (210 ppb average). The reason for the relatively lower PCP content in the sediment samples is unknown. Lake sediment showed an increase near the mouth of the stream in February 1977 (1,500 ppb) but in April the overall concentration in sediment was close to the January level. This apparent decline could be due to the degradation of PCP in the sediment or to the influx of uncontaminated sediment from soil erosion.

Fish contained only background levels of PCP in October 1976 (Table 3), but had very high concentrations in January 1977 immediately after the spill. Concentrations were somewhat lower by April 1977 but were still well above background levels. Fish collected shortly after the spill in January 1977 contained PCP in concentrations of 200,000 ppb in liver tissue, 40,000 ppb in gills, and 12,000 ppb in muscle. These values represent concentration factors over the PCP content in the water of 500 for muscle, 1,500 for gills, and 8,000 for liver. These data are similar to the results obtained by exposing the fish, *Lepomis macrochirus*, to 0.1 ppm PCP in water under controlled laboratory conditions (Pruitt *et al.*, 1977).

TABLE 3. Concentrations of Pentachlorophenol (ng/g wet weight; ppb)
in the Tissues of Fish

Date	Muscle	Gills	Liver
Oct. 11, 1976			
sunfish-1	5	N.A.[a]	26
sunfish-2	4	N.A.	150
Jan. 6, 1977			
sunfish-1	9,400	48,400	130,000
sunfish-2	6,400	N.A.	N.A.
bass-1	7,000	42,000	200,000
bass-2	17,000	N.A.	140,000
bass-3	16,000	N.A.	325,000
catfish-1	19,000	N.A.	214,000
Apr. 27, 1977			
sunfish-1	900	N.A.	14,600
sunfish-2	1,000	N.A.	14,900
catfish-1	8,200	N.A.	50,600
catfish-2	1,500	N.A.	20,200

[a]N.A., not analyzed.

Pentachlorophenol-Degradation Products in the Contaminated Lake

The major degradation products observed in the contaminated lake were pentachloroanisole (PCP-OCH₃), 2,3,5,6-tetrachlorophenol, and 2,3,4,5-TCP. Although the 2,3,4,5-TCP isomer was present in concentrations equal to or greater than the 2,3,5,6-TCP isomer in some samples, it exhibited a low response to GC-EC and was difficult to quantitate in many instances. Studies on the photodegradation of PCP have shown the major products to include TCP (Crosby, 1972). Pentachloroanisole has been found to be produced from PCP by soil bacteria (Cserjesi and Johnson, 1972). The concentrations of 2,3,5,6-TCP in samples of the

TABLE 4. 2,3,5,6-Tetrachlorophenol
Concentrations (ppb) in Water
(Dissolved) and Sediment

Date	Sites			
	A	B	C[a]	D
Aug. 11, 1976				
water[b]	0.21	0.07	0.08	0.10
sediment[c]	235	196	130	28
Oct. 22, 1976				
water	dry	0.03	0.06	0.06
sediment[d]	13	64	67	55
Jan. 5, 1977				
water	1.53	0.85	0.97	0.60
sediment[d]	12	97	27	24
Feb. 22, 1977				
water	1.62	N.A.[e]	0.25	N.A.
sediment[d]	N.A.	339	N.A.	N.A.
Apr. 27, 1977				
water	2.0	0.72	0.94	1.0
sediment[d]	3.8	71	63	24

[a]Anaerobic sediment
[b]Average of triplicate values
[c]Single composite sample
[d]Average of duplicate values
[e]N.A., not analyzed

industrial waste-holding pond oil slick and water, collected after the spill, were 13% and 9% of the PCP content, respectively. This is about five to ten times greater than the relative concentration of 2,3,5,6-TCP to PCP in a sample of commercial PCP (1 to 2%) indicating that TCP was formed in the holding pond, probably by photodegradation as suggested by Crosby (1972). Due to the high concentration of oil, PCP-OCH$_3$ could not be distinguished in the industrial holding pond; however, its presence was not anticipated.

Varying quantities of the methyl ether (anisole) of both TCP isomers were also observed but proved difficult to quantitate due to low concentrations and interference from naturally-occurring substances. The 2,3,4,6-TCP isomer was not observed but may be present in small quantities. Trichlorophenol eluted with the solvent front under the chromatographic conditions used, thus its concentration was not determined. Chloranil, polychlorinated dibenzo-*p*-dioxins and polychlorinated dibenzofurans have been detected in sediment and holding pond oil samples in ppb quantities utilizing the procedure described by Buser (1975). However, our present analytical system does not provide sufficient separation of isomers for adequate quantitation. The gas chromatographic retention of the methylated TCP isomers relative to methylated PCP was observed to be 0.5 for 2,3,5,6-TCP and 0.80 for 2,3,4,5-TCP on 3% SP-2100, and 0.46 for 2,3,5,6-TCP and 1.21 for 2,3,4,5-TCP on 10% SP-1000. The relative response factor for peak height was 1.0 for 2,3,5,6-TCP and 0.25 for 2,3,4,5-TCP on 3% SP-2100; and 1.5 for 2,3,5,6-TCP and 0.25 for 2,3,4,5-TCP on 10% SP-1000.

The concentration of 2,3,5,6-TCP in the control pond samples was 0.07 ppb in water, 1 ppb in sediment, and less than 1 ppb in fish muscle (Table 1). The TCP levels in the contaminated lake are shown in Table 4. Water samples contained background levels in October 1976, while the TCP content increased to 1 ppb after the spill in January 1977, and remained near 1 ppb through April 1977. Sediment contained relatively high concentrations of TCP in October 1976. The changes in TCP concentration followed the pattern of PCP indicating that a large portion of the TCP was present with PCP as the waste entered the lake, probably formed by photodegradation in the holding pond.

The concentration of PCP-OCH$_3$ in water and sediment is shown in Table 5. Near background concentrations were observed in the water in October 1976. The concentration remained relatively low probably due to the low solubility of PCP-OCH$_3$ in water. Sediment contained PCP-OCH$_3$ concentrations above background in August and October 1976. The concentration increased slightly after the spill in January 1977, decreased in February 1977, and increased again through April 1977, suggesting that PCP-OCH$_3$ may be produced in the sediment.

Fish muscle and liver contained about the same concentration of PCP-OCH$_3$ as PCP in October 1976 (Table 6). The concentration of PCP-OCH$_3$ increased in January 1977 immediately after the spill and remained relatively high through April. The high concentration of PCP-OCH$_3$ in fish obtained from water containing very low concentrations of the compound suggests an extremely high partitioning of PCP-OCH$_3$ from water to fish. It is also possible that the PCP-OCH$_3$ in fish was obtained from food or from PCP converted to PCP-OCH$_3$ by flora in the fish gut.

Fish muscle and liver tissue contained low concentrations of TCP in October

TABLE 5. Pentachloroanisole
Concentrations (ppb) in Water
(Dissolved) and Sediment

Date	Sites			
	A	B	C[a]	D
Aug. 11, 1976				
water[b]	0.06	0.05	0.07	0.05
sediment[b]	24	11	19.4	25.7
Oct. 22, 1976				
water[b]	dry	0.05	0.02	0.02
sediment[c]	7.8	2.2	13.0	3.0
Jan. 5, 1977				
water[b]	1.94	0.10	0.06	0.08
sediment[c]	14.0	17.0	17.0	15.0
Feb. 24, 1977				
water[c]	0.14	N.A.[d]	0.07	N.A.
sediment[c]	N.A.	1.5	N.A.	N.A.
Apr. 27, 1977				
water[c]	0.03	0.04	0.03	0.03
sediment[c]	0.2	67.0	80.0	25.0

[a]Anaerobic sediment
[b]Analysis of single composite sample
[c]Average for two samples
[d]N.A., not analyzed

1976 (Table 6). The concentration increased to an average of 200 ppb (muscle) and 4,000 ppb (liver) after the spill in January 1977 and decreased to about 40 ppb (muscle) and 700 pbb (liver) by April 1977. Thus, it appears that fish rapidly accumulated TCP from the water immediately after the spill and retained it in a manner similar to PCP.

Although identification of the 2,3,4,5-TCP isomer was not accomplished for many samples, 2,3,4,5-TCP concentrations in April 1977 samples were the following: sediment, 15 ppb; fish muscle, 6 ppb; and fish liver, 30 ppb; which indicate a similar, yet somewhat lower concentration than the 2,3,5,6-TCP isomer in the same samples.

TABLE 6. PCP-Degradation Products in Fish
(ng/g wet weight; ppb)

Date	Muscle		Liver	
	PCP-OCH$_3$	2,3,5,6-TCP	PCP-OCH$_3$	2,3,5,6-TCP
Oct. 11, 1976				
sunfish-1	4	< 1[a]	10	30
sunfish-2	2	< 1[a]	12	50
Jan. 6, 1977				
sunfish-1	94	95	530	950
sunfish-2	32	60	N.A.	N.A.
bass-1	N.A.[b]	N.A.	N.A.	N.A.
bass-2	250	300	500	1,600
bass-3	90	130	700	8,200
catfish-1	164	219	1,200	8,500
Apr. 27, 1977				
sunfish-1	30	27	190	250
sunfish-2	28	22	115	150
catfish-1	100	82	140	1,400
catfish-2	177	41	575	940

[a]Lower limit of detection
[b]N.A., not analyzed

General Discussion and Conclusions

In addition to acute PCP poisoning (Bevenue and Beckman, 1967; Cote, 1972; Cardwell *et al.,* 1976), there is concern for contamination and biological magnification of PCP in aquatic organisms resulting from chronic exposure (Rudling, 1970; Stark, 1969; Buhler *et al.,* 1973; Zitko *et al.,* 1974; Kobayashi *et al.,* 1976). The problem is magnified by the persistence of PCP and PCP-impurities and degradation products, many of which are also highly toxic, such as tetrachlorophenol (TCP), tetrachlorobenzo-*p*-quinone (chloranil), tetrachloro-resorcinol, polychlorinated dibenzo-*p*-dioxins (PCDD) and polychlorinated dibenzo-furans (PCBF) (Munakata and Kuwahara, 1969; Crosby *et al.,* 1973; Rappe and Nilsson, 1972; Buser, 1976; Crosby and Wong, 1976).

Our investigations revealed that PCP persisted in the water column and in fish for six to ten months following the first major PCP spill and remained well above background levels in the last set of samples collected four months after the second spill. Sediment and leaf litter retained high concentrations of PCP and PCP-degradation products throughout the study, thus providing a source for chronic pollution of the aquatic ecosystem. Interpretation of the data is complicated by the chronic influx of PCP from the contaminated water shed area and the possible periodic release of small amounts of PCP-containing waste from the industrial holding pond.

The major degradation products observed in the lake weᵢe PCP-OCH₃, 2,3,5,6-TCP and 2,3,4,5-TCP. The concentrations of TCP in the various samples were low but they followed a pattern similar to that of PCP, indicating that TCP was present in the oil solution when it entered the lake. The concentration of TCP relative to PCP in the industrial waste-holding pond indicates that additional TCP may have been formed by photodegradation in the holding pond.

The low concentrations of PCP-OCH₃ in the water compared with the high concentration in fish and sediments probably reflects the low solubility of PCP-OCH₃ in water. Since PCP-OCH₃ is not present in technical-grade PCP, these data show that it is formed in the aquatic environment, probably in the sediment.

Fish were observed to accumulate PCP and PCP-degradation products rapidly from the water. The concentration in fish decreased as the concentration in the water decreased, but required six to ten months to reach background concentrations. Fish liver tissue had the highest concentration followed by gills and muscle.

ACKNOWLEDGMENTS

This investigation is supported by Grant No. R-803-82-0010 from the U.S. Environmental Protection Agency. The authors are grateful to Dr. C.A. McDaniel at the U.S. Department of Agriculture, Gulfport, Mississippi, for providing GC-MS analysis of some of our samples, and to the project director, Dr. N. Lee Wolfe, of the U.S. EPA Environmental Research Laboratory, Athens, Georgia.

References

Bevenue, A., and H. Beckman, 1967. Pentachlorophenol: A discussion of its properties and its occurrence as a residue in human and animal tissues. *Residue Rev.,* **19:** 83-134.

Buhler, D.R., M.E. Rasmusson, and H.S. Nakane, 1973. Occurrence of hexachlorophene and pentachlorophenol in sewage and water. *Environ. Sci. Tech.,* **7:** 929-934.

Buser, H.R., 1975. Analysis of polychlorinated dibenzo-*p*-dioxins and dibenzofurans in chlorinated phenols by mass fragmentography. *J. Chromatogr.,* **107:** 295-310.

Buser, H.R., 1976. High-resolution gas chromatography of polychlorinated dibenzo-*p*-dioxins and dibenzofurans. *Anal. Chem.,* **48:** 1553-1557.

Cardwell, R.D., D.G. Foreman, T.R. Payne, and D.J. Wilbur, 1976. *Acute Toxicity of Selected Toxicants to Six Species of Fish.* U.S. EPA-600/3-76-008, pp. 1-117, United States Environmental Protection Agency, Washington, D.C.

Cote, R.P., 1972. A literature review of the toxicity of pentachlorophenol and pentachlorophenates. Environmental Protection Service manuscript report, 72-2, pp. 1-14, Halifax, Nova Scotia, Canada.

Crosby, D.G., 1972. Photodegradation of pesticides in water. *Amer. Chem. Soc., Advances in Chemistry Ser.,* 111: 173-188.

Crosby, D.G., K.W. Moilanen, and A.S. Wong, 1973. Environmental generation and degradation of dibenzodioxins and dibenzofurans. *Environ. Hlth. Persp.,* 5: 259-266.

Crosby, D.G., and A.S. Wong, 1976. Photochemical generation of chlorinated dioxins. *Chemosphere,* 5: 327-332.

Cserjesi, A.J., and E.L. Johnson, 1972. Methylation of pentachlorophenol by *Trichoderma virgatum. Can. J. Microbiol.,* 18: 45-49.

Kobayashi, K., S. Kimura, and H. Akitake, 1976. Studies on the metabolism of chlorophenols in fish — VII. Sulfate conjugation of phenol and PCP by fish livers. *Bull. Japan. Soc. Fish.,* 42: 171-177.

Mississippi Air and Water Pollution Control Commission, 1975. Court hearing, January 14, 1975. Jackson, Mississippi.

Munakata, K., and M. Kuwahara, 1969. Photochemical degradation products of pentachlorophenol. *Residue Rev.,* 25: 13-23.

Pierce, R.H., Jr., C.R. Brent, H.P. Williams, and S.G. Reeves, 1977. Pentachlorophenol distribution in a freshwater ecosystem. *Bull. Environ. Contam. Toxicol.,* 18: 251-258.

Pruitt, G.W., B.J. Grantham, and R.H. Pierce, Jr., 1977. Accumulation and elimination of pentachlorophenol in bluegill, *Lepomis macrochirus. Trans. Amer. Fish. Soc.,* in press.

Rappe, C., and C.A. Nilsson, 1972. An artifact in the gas chromatographic determination of impurities in pentachlorophenol. *J. Chromatogr.,* 67: 247-253.

Rudling, L., 1970. Determination of PCP in organic tissues and water. *Water Res.,* 4: 533-537.

Stark, A., 1969. Analysis of pentachlorophenol in soil, water and fish. *J. Agric. Food Chem.,* 17: 871-873.

Zitko, V., O. Hutzinger, and P.M.K. Choi, 1974. Determination of pentachlorophenol and chlorobiphenyls in biological samples. *Bull. Environ. Contam. Toxicol.,* 12: 649-653.

The Environmental Fate of ^{14}C-Pentachlorophenol in Laboratory Model Ecosystems

PO-YUNG LU, ROBERT L. METCALF, and LARRY K. COLE

Abstract—The fate of ^{14}C-pentachlorophenol (PCP) was evaluated in three model ecosystems: aquatic, terrestrial-aquatic, and terrestrial. The principal degradative products appeared to be tetrachlorohydroquinone, pentachlorophenyl acetate, and conjugates. In the terrestrial-aquatic model ecosystem, the bioaccumulation factors for PCP were: alga 5, daphnia 205, snail 21, mosquito 26, and fish 132. The parent PCP constituted 11.1% of the total ^{14}C in alga, 12.2% in snail, 33.3% in mosquito, 55.5% in daphnia, and 51.2% in fish. Tetrachlorohydroquinone constituted 5.4% of the total extractable ^{14}C in alga and 10.5% in snail. None was detected in other organisms.

In the terrestrial model ecosystem, the vole at the top of the food chain contained 0.5% of the total dosage applied to the corn-soil interface, and 6.4% of this was intact PCP.

Similar model ecosystem experiments with ^{14}C-hexachlorobenzene showed that this compound is slowly converted to PCP.

Introduction

Pentachlorophenol was first introduced as a wood preservative in 1936 and has also been used as a fungicide, herbicide, defoliant and insecticide. Sodium pentachlorophenate has been used widely as a molluscacide and wood perservative, and pentachlorophenyl laurate has been used as a mothproofing agent. The

PO-YUNG LU, ROBERT L. METCALF and LARRY K. COLE • Department of Entomology and Institute for Environmental Studies, University of Illinois, Urbana, Illinois 61801, U.S.A.

production of PCP in the U.S. was 46×10^6 lb. in 1969, and more than 25×10^6 lb. of this was used as a wood preservative (Fowler and Shepard, 1971).

Hundreds of millions of pounds of PCP have been applied world wide, yet there is little information on the environmental fate and impact of this chemical. Such studies are of particular importance because of the relative environmental stability of PCP and its high toxicity to nearly all forms of life as an inhibitor of oxidative phosphorylation. More recently, PCP and dioxin impurities have been implicated in the death of farm animals chewing on PCP treated wood (Chem. Eng. News, 1977).

TABLE 1. R_f Values of Potential
Metabolites of PCP

Compound	R_f value[a]
Pentachlorobenzene	0.84
Pentachloroanisole	0.76
1,2,4,5-Tetrachlorobenzene	0.76
1,2,4-Trichlorobenzene	0.76
1,2,3,4-Tetrachlorobenzene	0.74
1,2,3-Trichlorobenzene	0.73
1,3,5-Trichlorobenzene	0.72
Pentachlorophenyl acetate	0.66
Pentachlorophenol	0.53
Chloranil	0.52
2,4,6-Trichlorophenol	0.52
2,3,5,6-Tetrachlorophenol	0.51
2,5-Dichlorophenol	0.44
3,5-Dichlorophenol	0.44
2,3-Dichlorophenol	0.41
3,4-Dichlorophenol	0.41
2,4,5-Trichlorophenol	0.41
2,3,4,5-Tetrachlorophenol	0.41
2,4-Dichlorophenol	0.38
Tetrachlorohydroquinone	0.28
4,6-Dichlororesorcinol	0.22
4-Chlororesorcinol	0.17
Chlorohydroquinone	0.16
2,3-Dichloro-4-hydroxyphenol	0.09
Chloranilic acid	0.05

[a]TLC with hexane : acetone : acetic acid = $80 : 20 : 2$ by volume on Silica Gel GF-254, 0.25 mm thickness.

Perhaps the widest scale usage of PCP has been in Japan where its use as a fungicide and herbicide has resulted in contamination of almost all surface waters to 0.01-0.1 ppb (Goto, 1971).

To provide information to fill some of these lacunae we have evaluated the distribution and fate of ¹⁴C-PCP in three types of laboratory model ecosystems, the aquatic model ecosystem (Lu and Metcalf, 1975), the terrestrial-aquatic model ecosystem (Metcalf *et al.*, 1971), and the terrestrial model ecosystem (Cole *et al.*, 1976). These model ecosystem studies with pentachlorophenol are considered from the perspective of similar model ecosystem evaluations conducted in this laboratory with some 200 environmental pollutants: pesticides, plasticizers, heat transfer agents, veterinary drugs and food additives, rubber chemical, flame retardants, fuel additives, carcinogens, and heavy metals (Lu and Metcalf, 1975; Lu *et al.*, 1975; Metcalf and Sanborn, 1975; Metcalf, 1977).

The model aquatic ecosystem simulates the direct application of pentachlorophenol to water or its discharge through leaching into drainage systems. The closed system facilitates the direct detection of radiolabeled organic vapors and ¹⁴CO₂. The terrestrial-aquatic ecosystem has a farm portion and a lake and models the transfer of pesticides from crops into an aquatic ecosystem. Both aquatic and terrestrial-aquatic systems have identical food chain organisms: alga, *Oedogonium cardiacum;* snail, *Physa* spp.; daphnia, *Daphnia magna;* mosquito larva, *Culex pipiens quinquefasciatus;* and mosquito fish, *Gambusia affinis.* These were carefully chosen to model food chain transfer and lead to quantitative calculations of ecological magnification (E.M., or concentration of the parent compound in the organisms/concentration of the parent compound in the water) and biodegradability index (B.I., or concentration of the polar metabolites/concentration of the non-polar metabolites). Identification of the degradation products was made

TABLE 2. Fate of ¹⁴C-Pentachlorophenol in Model Aquatic Ecosystem

		PCP equivalents — ppm				
	H_2O	*Oedogonium* (alga)	*Daphnia* (daphnia)	*Physa* (snail)	*Culex* (mosquito)	*Gambusia* (fish)
Total ¹⁴C	0.00794	0.0107	1.1725	1.1377	0.1733	1.0764
Pentachlorophenyl acetate (R_f 0.75)[a]	0.00021	—	—	—	—	—
Pentachlorophenol (R_f 0.41)	0.00271	0.0043	0.4484	0.3289	0.0456	0.8043
Unknown I (R_f 0.34)	0.00038	—	—	—	—	—
Unknown II (R_f 0.26)	—	—	—	0.0272	—	—
Tetrachlorohydro- quinone (R_f 0.22)	0.00004	—	—	0.0332	—	—
Unknown III (R_f 0.16)	0.00009	—	—	0.1629	—	—
Conjugates (R_f 0.0)	0.00426	0.0064	0.7241	0.5855	0.1277	0.2721
Unextractable	0.00025					

[a]TLC with hexane : acetone : acetic acid = 80 : 20 : 2 by volume.

by comparison of the R_f values of the authentic standards (Table 1) upon
cochromatography with several solvent systems.

The terrestrial model ecosystem simulates a crop-soil interaction, typically corn
and silty clay loam and has a compliment of organisms: earthworm, *Lumbricus
terrestris;* slug, *Limax maximus;* pillbug, *Armadillidium vulgare;* saltmarsh
caterpillar, *Estigmene acrea;* and prairie vole, *Microtus ochregaster* which provide
for food chain transfer. The terrestrial system is equipped with traps for organic
vapors and $^{14}CO_2$ and can be leached to determine runoff.

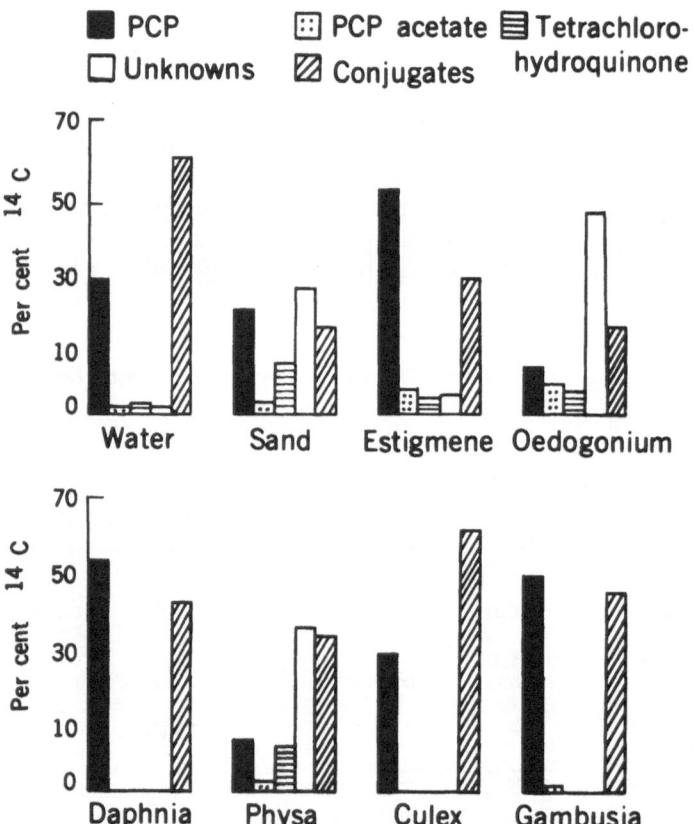

Figure 1. Relative detoxification capacity of key components of model terrestrial-aquatic
ecosystem following treatment with ^{14}C-pentachlorophenol.

TABLE 3. Fate of ¹⁴C-Pentachlorophenol in Model Terrestrial-aquatic Ecosystem

	PCP equivalents — ppm						
	H_2O	Sand	Oedogonium (alga)	Daphnia (daphnia)	Physa (snail)	Culex (mosquito)	Gambusia (fish)
Total ¹⁴C	0.05235	0.0186	0.8061	6.2441	2.9576	1.3355	4.3765
Pentachlorophenyl acetate (R_f 0.86)	0.00028	0.0005	0.0673	—	0.0438	—	0.0197
Pentachlorophenol (R_f 0.64)	0.01693	0.0048	0.0893	3.4692	0.3619	0.4451	2.2408
2,3,4,5-tetrachlorophenol (R_f 0.56)	0.00109	0.0010	0.0771	—	—	—	—
Unknown I (R_f 0.49)	—	—	0.0575	—	—	—	—
Unknown II (R_f 0.42)	—	—	0.0477	—	—	—	—
Tetrachlorohydroquinone (R_f 0.33)	0.00009	0.0023	0.0440	—	0.8122	—	—
Unknown III (R_f 0.22)	0.00010	0.0011	0.1700	—	0.3129	—	—
Unknown IV (R_f 0.15)	0.00008	0.0013	—	—	0.2798	—	—
Unknown V (R_f 0.10)	—	0.0020	—	—	—	—	—
Unknown VI (R_f 0.05)	0.00012	—	0.0808	—	—	—	—
Conjugates (R_f 0.0)	0.02220	0.0056	0.1724	2.7749	1.1470	0.8904	2.1160
Unextractable	0.01146						

aTLC with hexane : acetone : acetic acid = 80 : 20 : 2 by volume.

Laboratory Model Ecosystem Studies

Model Aquatic Ecosystem

No volatile products or $^{14}CO_2$ were detected in the traps over the three-day experimental period. ^{14}C-Pentachlorophenol and its metabolites were very rapidly accumulated in the food chain organisms and substantial amounts of intact parent compound were stored as shown in Table 2. The E.M. (Ecological Magnification) value for PCP in fish was 296 and the parent compound represented 74% of the total extractable ^{14}C. The other organisms had lower quantities as shown by lower E.M. values: alga 1.5, mosquito 16, snail 121, and daphnia 165. Conjugation at the phenolic OH was the most important means of degradation found among the organisms. The B.I. (Biodegradability Index) values for the food chain members are: alga 1.48, daphnia 1.61, mosquito 2.80, snail 1.06, and fish 0.34. However, pentachlorophenyl acetate comprised 2.6% and tetrachlorohydroquinone comprised 2.9% of the total extractable ^{14}C in the water extract and 1% of tetrachlorohydroquinone in the snail.

Model Terrestrial-aquatic Ecosystem

In this system, PCP was applied to the sand of the model ecosystem at approximately 1.0 kg/h by injection of 20 μl of acetone solution at 50 locations equidistant from the *Sorghum* seeds. The salt marsh caterpillar, *Estigmene acrea* feeding on the *Sorghum* leaves was the dispersing agent. Intact PCP was recovered as 54.1% of the total radioactivity in the caterpillar and its excreta and 25.8% in sand. Tetrachlorohydroquinone represented 2.9% of the total radioactivity in caterpillar and 12.4% in sand. Pentachlorophenyl acetate represented 5.5% of the total radioactivity in caterpillar and 2.6% in sand, as shown in Figure 1.

In the food chain organisms, the parent compound constituted 11.1% of the total radioactivity in alga (E.M. 5), 12.5% in snail (E.M. 21), 33.3% in mosquito (E.M. 26), 55.5% in daphnia (E.M. 204), and 51.2% in fish (E.M. 132). Tetrachlorohydroquinone constituted 5.4% of the total extractable ^{14}C in alga and 10.5% in snail. None was detected in the other organisms as shown in Table 3. Pentachlorophenyl acetate comprised 8.0% of the total extractable ^{14}C in alga, 1.5% in snail, and 1% in fish. None was detected in daphnia and mosquito. The B.I. values are: alga 0.27, daphnia 0.22, snail 0.66, mosquito 2.0, and fish 0.93. The water contained a more complicated series of degradation products as shown in Figure 2. Although the majority of the radioactivity was found as PCP and conjugates, it was apparent that PCP was undergoing a series of reductive dechlorinations, presumably beginning with tetrachlorohydroquinone. The lower chlorinated phenols are the products with lower R_f values found in the water, the organisms, and in white sand, e.g., R_f 0.56, 2,3,4,5-tetrachlorophenol. This study confirmed the results from the 3-day aquatic system study. The phenolic OH of PCP provides a ready center for conjugation and this opportunity is further increased in tetrachlorohydroquinone and accounted for the high levels of conjugates found in water and in various organisms as shown in Figure 1.

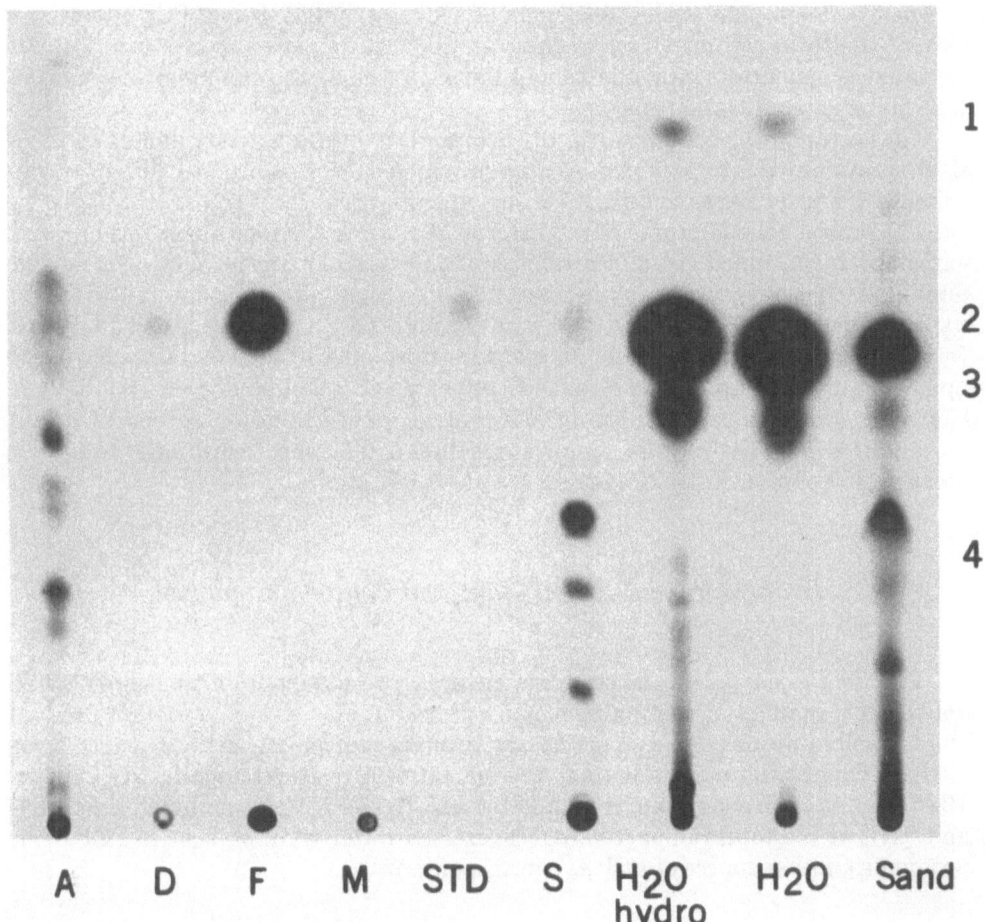

Figure 2. Radioautograph from TLC plate of extracts from components of model terrestrial-aquatic ecosystem treated with ^{14}C-PCP. A: alga, D: daphnia, F: fish, M: mosquito, STD: standard, S: snail, H_2O hydro: hydrolized water, and H_2O: unhydrolized water. 1: pentachlorophenyl acetate, 2: PCP, 3: 2,3,4,5-tetrachlorophenol, and 4: tetrachlorohydroquinone.

Model Terrestrial Ecosystem

The laboratory terrestrial ecosystem revealed the final distribution of ^{14}C-pentachlorophenol and degradation products at the end of the 20 day experimental period as 51% of the total applied radioactivity in the air, 48% in the soil, and only 1% in the animals of the ecosystem.

The drummer soil, a typical silty clay loam, was treated with 5 mg of radiolabeled PCP in acetone solution by injection of 20 μl at 50 locations, 1 cm deep equidistant from the corn seed. The initial soil concentration was 1.25 ppm and this declined to 0.634 ppm at the conclusion of the experiment. Intact PCP comprised 19% of the total radioactivity in the soil, 20% was a series of minor unknown degradation products, and the remainder was unextractable polar conjugates determined by combustion analysis.

The corn plants, *Zea mays* rapidly accumulated radioactivity and at 14 days after treatment the total was 6.30 ppm of which 16% was intact PCP, 40% was unknowns, and 44% was conjugates as shown in Figure 3.

The prairie vole, *Microtus ochregaster* at the top of the food chain was exposed for 5 days. It consumed virtually all the plant and animal material in the system and contained 0.5% of the total applied dose (25 μg), 7% of which was intact PCP, 22% was unknown metabolites and 71% was conjugates. At autopsy intact PCP was found in the tissues in the following order: uterus 0.135 ppm > intestines 0.045 ppm > brain 0.044 ppm > carcass 0.037 ppm > liver 0.034 ppm > skin 0.026 ppm. However, the total accumulation of radioactivity, including intact PCP and metabolites was: liver 1.484 ppm > intestines 1.071 ppm > brain 0.336 ppm > uterus 0.302 ppm > skin 0.204 ppm > carcass 0.147 ppm.

Hexachlorobenzene as a Source of Pentachlorophenol

PCP is a micropollutant that has entered the environment not only when deliberately applied by humans for many years but also as a degradation product from the ubiquitous contaminant hexachlorobenzene. ^{14}C-Hexachlorobenzene was evaluated in similar experiments in the terrestrial-aquatic (Metcalf and Sanborn, 1975) and aquatic ecosystem (Lu and Metcalf, 1975). It was concluded that 4.8% and 0.9% of the total radioactivity recovered in water were present as PCP, thus providing another source of PCP as a micropollutant.

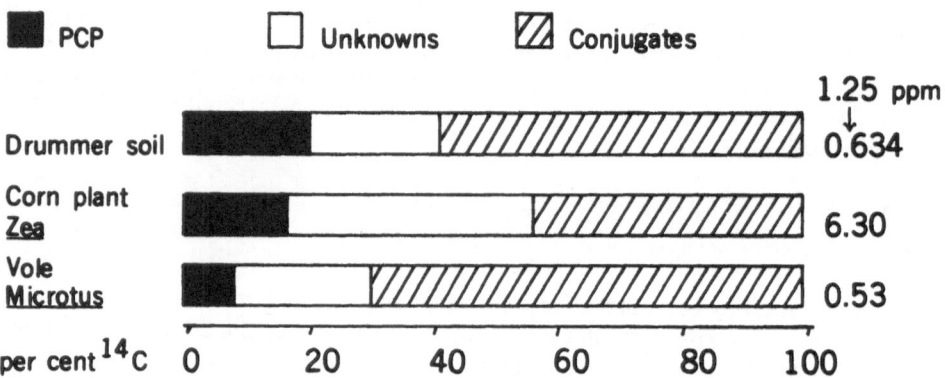

Figure 3. Distribution of radioactivities in the key components of the model terrestrial ecosystem.

Biological Degradation

The role of microsomal enzymes in the living organisms to cope with xenobiotics is to increase the polarity of the xenobiotics which will be discharged more effectively, e.g., hydroxylation. Therefore, the hydrophilicity of the compound is increased and lipophilicity is decreased. As discussed above, hexachlorobenzene is degraded to the relatively polar compound, pentachlorophenol. Similarly, pentachlorophenol is degraded to tetrachlorohydroquinone and pentachlorophenyl acetate as their metabolites in the organisms. The water solubilities and the octanol/water partition coefficients are summarized as follows:

	Water solubility	Octanol/water partition coefficient
hexachlorobenzene	0.006 ppm	13560
pentachlorophenol	20	6400
tetrachlorohydroquinone	21.5	3200
pentachlorophenyl acetate	20	2350

Thus the process of ecosystem degradation of PCP involves formation of more water soluble and less lipophilic degradation products which are more readily eliminated by the various organisms of the ecosystem.

Ecological Magnification

The accumulation of lipid-soluble, water-insoluble pesticides in living organisms is one of the most disturbing features of environmental pollution by pesticides. The laboratory model ecosystem is particularly suitable for determining "ecological magnification" since it is simulating the real world environmental conditions. Ecological magnification is a function of the partition coefficient in lipid/water and the stability of the pesticide and its metabolites in the animal. As shown in Figure 4, an effective approximation is obtained when the water solubility of the pesticide in parts per billion (ppb) is plotted as a log function against ecological magnification. There is clearly an inverse relationship with the least water soluble pesticides accumulating to the highest degree. This relationship is highly significant with a correlation coefficient of $r = -0.76$ and it is substantially predictable (Metcalf and Sanborn, 1975). Thus it is of great importance to know the water solubility of even the least soluble compounds. From this information it is possible to make a reliable estimate of the potentialities of new pesticides to accumulate in the tissues of fish and other aquatic organisms. Our study suggests a classification of pesticides as:
1. water solubility < 0.5 ppm, likely to be environmentally hazardous
2. water solubility > 50 ppm, likely to be environmentally nonhazardous
3. water solubility from 0.5 to 50 ppm, to be used with caution.

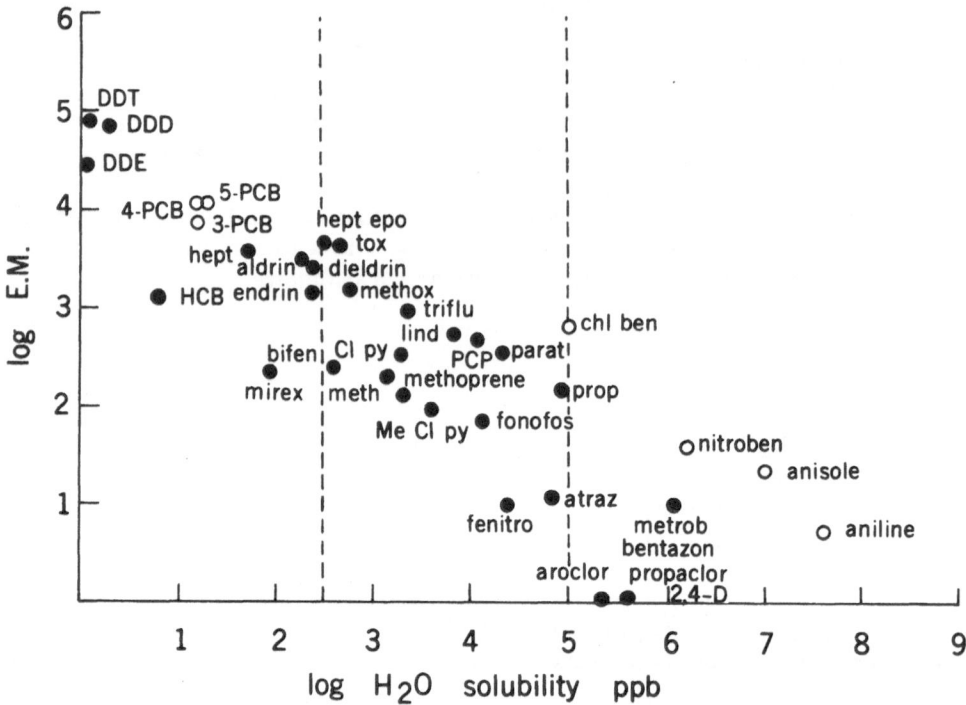

Figure 4: Correlation of log ecological magnification (E.M.) in the fish, *Gambusia,* and log water solubility of the pesticides.

The lines of demarcation between the three classes obviously are not sharp, and the ultimate hazard also depends upon lipid partitioning, the rapidity of pesticide degradation in living animals, use patterns, amounts applied, and the nature of toxic action. However, practical experience has already shown that most of the pesticides with water solubilities of < 0.5 ppm demonstrate bioaccumulation following field use and that most of those with water solubilities of > 50 ppm have not shown bioaccumulation. The large group of pesticides with water solubilities between 0.5 and 50 ppm represent those which may demonstrate bioaccumulation under some condition of use, e.g. in lakes or oceans with very cold water. Their use patterns should be judged accordingly.

The water solubility of PCP is 20-25 ppm at 25 °C (Bailey and White, 1965). Tetrachlorohydroquinone and pentachlorophenyl acetate have water solubilities of 21.5 and 20.0 ppm respectively. Considering the relatively high lipid/H$_2$O partition coefficient 6400 for PCP (Lu and Metcalf, 1975) 3200 for tetrachlorohydroquinone and 2350 for pentachlorophenyl acetate, its resistance to degradation, its known high toxicity to both plants and animals (Bevenue and Beckman, 1967), its teratogenicity to Sprague-Dawley rats and hamsters (Schwetz *et al.,* 1974; Hinkle, 1973), and its presence in the U.S. drinking water to levels as high as 1.7 ppb, we conclude that pentachlorophenol is an undesirable pollutant whose use patterns should be carefully regulated to avoid contamination of soil, water and food. The Acceptable Daily Intake (ADI), for PCP has been calculated as 0.003 mg/kg/day and a suggested no-adverse effect in drinking water is proposed as 0.021 mg/liter (Report of the Safe Drinking Water Committee, National Academy of Sciences, 1977).

ACKNOWLEDGMENTS

This work was supported in part by National Science Foundation RANN Program (Grant ESR 74-22760), U.S. Environmental Protection Agency (Development of a Terrestrial Model Ecosystem; Grant R-80-3249), and the University of Illinois Water Resources Center (Project B-050, Agreement 14-31-0001-3273).

References

Bailey, G.W., and J.L. White, 1965. Herbicides: a compilation of their chemical, physical and biological properties. *Residue Rev.,* **10:** 97-122.

Bevenue, A., and H. Beckman, 1967. Pentachlorophenol: A discussion of its properties and its occurrence as a residue in human and animal tissues. *Residue Rev.,* **19:** 83-134.

Chem. Eng. News, 1977. Michigan bans PCP wood preservative. P. 7, March 21, 1977.

Cole, L.K., R.L. Metcalf, and J.R. Sanborn, 1976. Environmental fate of insecticides in terrestrial model ecosystems. *Intern. J. Environ. Studies,* **10:** 7-14.

Fowler, L.D., and H.H. Shepard, 1971. *The Pesticide Review 1970,* U.S. Department of Agriculture, Agricultural Stabilization and Conservation Service, Washington, D.C.

Goto, M., 1971. Organochlorine compounds in the environment of Japan. International Symposium on Pesticide Terminal Residues. Pure and Applied Chemistry Supplement, 105-110, Butterworth's,London.

Hinkle, D.Y., 1973. Fetotoxic effects of pentachlorophenol in the golden syrian hamster. *Toxicol. Appl. Pharm.,* **25:** 455.

Lu, P.Y., and R.L. Metcalf, 1975. Environmental fate and biodegradability of benzene derivatives as studied in a model aquatic ecosystem. *Environ. Hlth. Persp.,* **10:** 269-284.

Lu, P.Y., R.L. Metcalf, R. Furman, R. Vogel and J. Hassett, 1975. Model ecosystem studies of lead and cadmium and of urban sewage sludge containing these elements. *J. Environ. Qual.,* **4:** 505-509.

Metcalf, R.L., 1977. Model ecosystem approach to insecticide degradation: a critique. *Ann. Rev. Entomol.,* **22:** 241-261.

Metcalf, R.L., and J.R. Sanborn, 1975. Pesticides and environmental quality in Illinois. *Ill. Nat. Hist. Surv. Bull.,* **31:** 381-436.

Metcalf, R.L., G.K. Sangha, and I.P. Kapoor, 1971. Model ecosystem for the evaluation of pesticide biodegradability and ecological magnification. *Environ. Sci. Tech.,* **5:** 709-712.

National Academy of Sciences, 1977. Report of Safe Drinking Water Committee Washington, D.C.

Schwetz, B.A., P.A. Keeler, and P.J. Gehring, 1974. The effect of purified and commercial grade pentachlorophenol on rat embryonal and fetal development. *Toxicol. Appl. Pharm.,* **28:** 151-161.

II

Pharmacology and Toxicology of Pentachlorophenol

A. Metabolism

Microbial Metabolism of Pentachlorophenol

ERIC A. REINER, JOSEPH CHU and EDWIN J. KIRSCH

Abstract—Microbial degradation of pentachlorophenol has been studied in mixed microbial communities and in an axenic bacterial culture. Using the sodium salt of ^{14}C-pentachlorophenol, bacteria have been shown to metabolize the compound with the formation of $^{14}CO_2$ and Cl-ions. Conditions for the biodestruction of pentachlorophenol in synthetic wastewater and authentic wood preservative wastewater were studied using an aerobic, continuous flow, high cell recycle system. The biomass formed in the presence of pentachlorophenol was capable of significant degradation of several other chlorophenols, however, monochlorophenols were not eliminated at a significant rate.

The characteristics of bacterial attack on pentachlorophenol were studied with a single bacterial species (KC-3) which was shown to use pentachlorophenol, 2,3,4,6-tetrachlorophenol and 2,4,6-trichlorophenol as growth substrates. Mutants of this isolate with impaired pentachlorophenol metabolism were obtained. Pentachlorophenol metabolites were extracted from the culture filtrate and were identified as chlorinated hydroquinones or benzoquinones. Evidence was obtained

---

For the sake of brevity, pentachlorophenol (PCP) was used in the text in place of sodium pentachlorophenate. Other abbreviations used are as follows:

COD: Chemical oxygen demand   PDC: Pentachlorophenol degrading capability
CHQ: Monochlorohydroquinone   TCHBQ: Trichlorohydroxybenzoquinone
DCHQ: 2,6-Dichlorohydroquinone   TeCBQ: Tetrachlorobenzoquinone
PCA: Pentachloroanisole   TeCHQ: Tetrachlorohydroxyquinone

ERIC A. REINER • 3M Company, Post Office Box 33331, Stop 58, St. Paul, Minnesota 55133, U.S.A.
JOSEPH P. CHU • General Motors Corporation, Warren, Michigan 48090, U.S.A.
EDWIN J. KIRSCH • School of Environmental and Civil Engineering, Purdue University, West Lafayette, Indiana 47907, U.S.A.

for the probable participation of 2,6-dichlorohydroquinone and tetrachloro-hydroquinone or tetrachlorobenzoquinone as intermediates in the catabolism of pentachlorophenol.

Introduction

The primary application of pentachlorophenol (PCP) in the United States is to inhibit mold and control wood-boring insects in the lumber and construction industry. As is typical of chlorinated pesticides, PCP has proven to be somewhat resistant to biodegradation. However, the ability to degrade the biocide has been demonstrated among bacterial populations and fungi, in both pure and mixed cultures (Chu and Kirsch, 1972; Cserjesi, 1967; Cserjesi and Johnson, 1972; Ide *et al.* 1972; Kirsch and Etzel, 1973; Suzuki, 1977; Suzuki and Nose, 1971). Most studies have concentrated on the conditions and rate of degradation, but little information exists on the microbial degradation pathway.

Ide *et al.* (1972) have identified conditions under which PCP is degraded through reductive dechlorinations in rice paddy soil. Dechlorination proceeded more rapidly in a reducing environment and was assumed to be the result of microbial action since no decomposition proceeded in sterilized soil. Using gas chromatography-mass spectrophotometry, all three tetrachlorophenol isomers were identified as degradation products, although 2,3,4,5-tetrachlorophenol was the predominant metabolite. When the three tetrachlorophenols were subjected to the same conditions as PCP, additional reductive dechlorinations occurred, causing the accumulation of less substituted chlorophenols.

Cserjesi and Johnson (1972) have described a methylation of pentachlorophenol to pentachloroanisole (PCA) by the fungus, *Trichoderma virgatum*. Three more fungi capable of this conversion were identified by Curtis *et al.* (1972) as *Aspergillus sydowi, Scopulariopsis brevicaulis* and a *Penicillium* species. Although PCA represents a metabolic modification of PCP, it is an unlikely intermediate in its degradation since PCA seems to be more resistant to biological and chemical degradation than PCP (Cserjesi and Johnson, 1972).

Suzuki (1977) has identified tetrachlorohydroquinone (TeCHQ) in yields of 0.2 to 0.5% based on the original PCP concentration from extracts of a PCP-degrading *Pseudomonas*. Smaller yields of tetrachlorocatechol were also obtained. Tetrachlorohydroquinone dimethyl ether and tetrachlorobenzoquinone (TeCBQ) have also been observed following PCP degradation by bacterial and fungal cultures, respectively (Konishi and Inoue, 1972; Suzuki and Nose, 1971).

TeCBQ and TeCHQ have likewise been found as PCP metabolites in mammalian systems. Following oral administration of PCP Tashiro *et al.* (1970)   recovered 0.2% as TeCBQ in rabbit urine. They also found TeCBQ in the internal organs of mice after intraperitoneal injection. Jakobson and Yllner (1971) reported the detection of TeCHQ in mouse urine following intraperitoneal injection of PCP. In *in vivo* and *in vitro* studies, Ahlborg (this symposium) found that rat liver microsomal enzymes rapidly dechlorinated PCP and formed TeCHQ and trichlorohydroquinone.

In our laboratories, we have been interested since the late 1960's in the microbial metabolism of halophenols and particularly of PCP. We were challenged initially to seek answers to two intriguing questions: 1) Could we unequivocally demonstrate the microbial decomposition of PCP in aerobic mixed bacterial culture systems? and 2) Is aerobic biological wastewater treatment a feasible process for purifying process wastes containing PCP? When it appeared that the answers to these questions were affirmative, we were prepared to ask and attempt to answer additional questions: 1) Does PCP qualify as an energy substrate for individual microbial cultures or is it subject to co-metabolic attack in mixed culture systems? 2) What are the limiting parameters in microbial biodegradation of PCP? and 3) What is the biochemical mechanism of PCP degradation in a bacterial culture? The following summary of the work in our laboratory provides some of the answers to these intriguing questions.

Enrichment of PCP-Degrading Bacteria

When we began our work in 1969, one could find occasional references in the literature to indicate that PCP disappearance and transformation in soil was subject to variations in environmental conditions such as moisture content, pH and temperature, suggesting possible microbial action. However, Kincannon et al. (1967) and Ingols et al. (1966) were unable to demonstrate degradation of PCP in aerated wastewater treatment systems despite attempts to adapt the biomass to PCP. We reinvestigated the metabolic transformation of PCP in a mixed culture biomass, by inoculating 50 grams of soil taken from a storage yard, where PCP treated wood poles were stored, into four liters of 0.1% nutrient broth. The culture was aerated for several days and an adaptation schedule was initiated in which increments of phenol (to encourage ring-splitting organisms) plus increments of PCP were mixed with nutrient broth and were batch fed daily to the aerator for 5 days. At 5 day intervals the levels of phenol and PCP were increased. After one month, the level of phenol in the fresh nutrient supply was reduced stepwise and PCP was increased stepwise until the PCP addition was 120 mg/4 l/day. The entire acclimation process took about 3 months. Unfortunately we did not monitor the adaptation process but spent this time preparing randomly labeled radioactive PCP. After acclimation, a sample of the cells was removed from the aerator, washed and transferred to Warburg flasks to which radioactive PCP was added. Carbon dioxide was trapped, and within 24 hours 14% of the PCP activity appeared as CO_2. Heat-treated (100°C) cells incubated for 48 hours released essentially no carbon dioxide (Kirsch and Etzel, 1973). This experimental observation was repeated many times and led to the conclusion that ultimate microbial degradation (mineralization) of PCP was obtainable.

Heterogeneous cell biomass, adapted to growth on PCP (40 mg/l/day) plus nutrient broth (500 mg/4 l/day) in long term, fill-and-draw reactors, was enriched for PCP degrading organisms (Kirsch and Etzel, 1973). Assays of the biomass with radioactive PCP as the sole oxidizable substrate indicated that in this study as much as 68% of the applied PCP could be respired in 24 hours. The true degree of enrichment could not be determined from this study since internal rates of PCP oxidation were not measured. After 20 days of increasing activity there was a 12 day plateau, followed by gradual loss of PCP-degrading activity. Subculture to fresh medium was seen to reestablish the PCP culture and suggested that growth conditions in the fill-and-draw system were eventually not supportive of the continuing competitive position of PCP-degrading organisms in the biomass. However, the ability of the bacteria to mineralize PCP in a highly complex chemical environment populated by heterogeneous flora was clearly established (Kirsch and Etzel, 1973).

Biological Treatment of PCP in Wastewater

The feasibility of biological treatment of PCP-containing process wastes was examined by using aerated continuous flow systems and a contrived, soluble, PCP-supplemented waste. The composition of the synthetic waste was described in a previous report (Etzel and Kirsch, 1974) and included substantial quantities of rapidly degradable organics (Chemical oxygen demand, COD-515 mg/l, reducing sugar 300 mg/l), nutrient salts and variable low levels of PCP. In early experiments, a conventional laboratory activated sludge reactor was employed. Cell recycle was achieved by sectioning the unit to obtain a quiescent area for sludge settling and permitting return sludge to be carried back to the aeration chamber by mixing currents. Adequate reduction of COD and high purity PCP was obtained in this system. The average destruction of PCP for 19 days of operation was 99.4% and the efficiency of treatment as measured by decrease of COD and reducing sugar in the effluent was 89 and 94%, respectively. On day 19 the high purity Eastman PCP normally used was replaced by a composite of 5 commercial PCP preparations which had been well blended. A very marked decrease of the biodegradability of PCP was observed. In fact, as the 25th day approached, it seemed probable that the PCP-degrading culture would be lost. This had not been our first experience with differences in treatability between commercial grade PCP and analytical grade PCP.

Long term operation of the conventional bench scale activated sludge units described above tended to give erratic treatability of PCP during periods when the sludge did not settle well. The biodestruction of an exotic organic chemical at low input concentration mixed with high input concentrations of easily degraded, energy-yielding substrates may often require artificial enrichment of bacterial species to keep a critical mass of specialized cells in the aerator. Operationally, this can be done by applying a high rate of cell recycle (long cell residence time) to prevent washout of desired organisms. Experimentally, a novel fiber wall aerator was used to artificially obtain long cell residence time at short hydraulic turnover time. The configuration of this reactor was described in a previous paper by Etzel and Kirsch (1974). It should be emphasized that this unit was not intended to model a recommended wastewater treatment system for PCP process wastes, but was a simple laboratory device to promote the physical separation of liquid turnover and cell turnover. In practice, a fixed film reactor such as a trickling filter should accomplish the same purpose.

Some typical experimental data obtained with the fiber wall aerator were described in the report cited above (Etzel and Kirsch, 1974). The destruction of reagent grade PCP was on the average 98.4% at an input concentration of 20 mg/l.

On the other hand, blended commercial PCP averaged 89.1% destruction at 20 mg/l input concentration. Apparently, reduction of the input concentration of commercial grade PCP led to better treatability than that observed previously. However, a significant inhibitory effect attributable to commercial PCP was observed in this study as well. The change in activity of the biomass, which is given by the PDC (PCP degrading capability) or the mg of PCP respired/gram of cells/hour, was initially 0.49 and finally 0.40 for analytical grade PCP. However, when commercial PCP was used a 10-fold decrease in biomass activity was observed.

At approximately this time, one of the commercial PCP suppliers (Dow Chemical Co.) developed a PCP production process which resulted in a much reduced level of chlorinated dioxins and other contaminants in their commercial product. Both the old and the improved commercial preparations were said to contain approximately 75% PCP. Comparison was made of the treatability of the improved commercial PCP with reagent grade PCP (Etzel and Kirsch, 1974). At 60 mg/l input concentration the new improved commercial preparation was as degradable as the reagent grade PCP and gave no indication of inhibition. In fact, the PDC values at the termination of the experiment indicated that substantial enrichment of the biomass with PCP-metabolizing organisms had occurred since the unit rate of utilization increased from 0.4 mg/g cells/hour to 3.2 mg/g cells/hour.

Finally, as reported in a previous paper (Etzel and Kirsch, 1974) authentic wood-preserving wastewater was obtained and supplied to the PCP degrading biomass. In this study, chlorophenols were measured by a modification of the aminoantipyrene method which did not distinguish between 2,3,5,6-tetrachlorophenol and PCP. Chlorophenol content was 23 mg/l in the input waste and was 96% destroyed, COD reduction was 86% and reducing sugar decreased by 85%. The biomass acclimated to PCP-containing wastewater was also assayed for its activity in eliminating other chlorophenols. Phenol substituents in the 2,4,6-positions were taken up most rapidly. Monochlorophenols seemed to be inhibitory, as was 2,3,4,5-tetrachlorophenol. Inhibition was suggested by an initial uptake of the phenol, cessation of uptake and a release of the compound. Unsubstituted phenol was initially not significantly eliminated, however, uptake was noticeable after several hours. Apparently the cells that utilized chlorophenols were not the same as those utilizing the unsubstituted phenol. Conversely, the cells which were capable of utilizing unsubstituted phenols were not induced to phenol attack by the presence of the substituted phenols. It seems reasonable to conclude that given appropriate environmental conditions, PCP-containing wastes are treatable via aerobic biological systems and that adaptation to PCP will probably provide adaptation to several other chlorophenols, particularly those with 2,4,6 substitutions.

Biodegradation of PCP by an Axenic Bacterial Culture

A single axenic bacterial culture obtained from the heterogeneous biomass was shown to attack PCP (Chu and Kirsch, 1972; Chu and Kirsch, 1973). After many futile attempts to isolate a PCP-degrading bacterium directly from the aerator biomass, a continuous-flow enrichment culture was started with very low concentration of PCP as the sole organic carbon source. The effluent liquor was tested for loss of PCP and when this was observed, the effluent was plated on a complex organic medium supplemented with 100 mg/l PCP. The first bloom of fast-growing colonies was tested but did not yield PCP-degrading isolates. A large number of small yellow raised colonies accounting for about 50% of the total count appeared after prolonged incubation with a complex organic medium containing PCP. Several were selected and tested for PCP destruction, and all of these yellow colonies showed PCP utilization. One was selected for purification and subsequent study. This culture was designated KC-3 and resembled a saprophytic soil corynebacterium. Washed, PCP-adapted cells were incubated with radioactive PCP, and a variety of metabolic parameters were measured with time. PCP uptake was rapid and was followed by a rapid discharge of chloride and a slower discharge of carbon dioxide. Oxygen uptake was rapid and excessive, indicating that this was not an accurate parameter of biodegradation of PCP. Approximately 67% of the PCP was respired, as shown by carbon dioxide accumulation, and 33% was assimilated in a resting cell preparation (Chu and Kirsch, 1973). This is fairly typical for common energy-yielding substrates. The organism was also shown to reproduce in a mineral salts medium with PCP, 2,3,4,6-tetrachlorophenol or 2,4,6-trichlorophenol as sole carbon sources. None of the other phenols tested supported growth even though evidence was obtained that almost all of the chlorophenols could be transformed to some extent by the culture. This would suggest that attack on all but the three growth supporting phenols involved co-metabolism.

The inducibility of the PCP oxidizing system was demonstrated in an experiment comparing the pattern of halophenol uptake by PCP-preadapted cells with cells which were preadapted to 2,4,6-trichlorophenol. These data were reported in detail by Chu and Kirsch (1973). Elimination of all six chlorophenols tested occurred without a discernable lag when PCP adapted cells were used. However, the elimination rates varied depending upon structure of the compound added. Growth-producing chlorophenols were most rapidly utilized. For cells adapted to 2,4,6-trichlorophenol, a clearly defined lag of approximately 2 hours was required before substantial PCP uptake occurred. All of the other substrates tested except 2,4,6-trichlorophenol required varying amounts of time before the uptake rate became maximal. The substrate, 2,3,5-trichlorophenol was not eliminated and apparently was unable to act as in inducer for its own metabolism despite the fact that PCP-induced cells were able to transform this compound.

Microbial Metabolism of PCP

Having defined the bacteriology of the PCP degrading culture, the mechanism of PCP degradation was investigated. Attempts to find intermediates accumulating in the culture liquor of KC-3 were made but were not productive. Furthermore, whenever the surfaces of cells were disrupted or even altered by detergents or other active agents, PCP-oxidizing activity quickly disappeared. Broken cell preparations did not give useful information on the mechanism of PCP catabolism since we were unable to stabilize the enzyme activity. The method finally used was to study metabolites accumulating in KC-3 mutant cultures.

About 70 mutants with impaired PCP metabolism were isolated and a few of them were selected for intensive study. The following is a description of some work with two of these mutants. ER-47 and ER-7, contrary to the parent culture did not grow in a minimal medium with PCP and salts and grew poorly in a cellobiose-PCP synthetic medium. In liquid medium containing PCP and a supplemental energy source, such as glucose or cellobiose, both of these cultures produced chromogenic metabolites. Culture ER-7 released fewer atoms of chloride per molecule of PCP than did ER-47, implying that the enzyme block in the PCP decomposition pathway in ER-7 preceeded the block in ER-47. In addition, the UV absorption characteristics of the culture broths differed slightly from each other; both differed markedly from that of the parent strain. This is clearly shown in Figure 1. In the parent culture, PCP absorption at 318 nm decreased with time and no new peaks developed. In ER-47 cultures 2 peaks were observed, PCP at 318 nm and a second peak at a wavelength approximating 293 nm. In ER-7 cultures a metabolite developed with an absorbance peak at approximately 287 nm. This metabolite also seemed to be subsequently metabolized or transformed as incubation progressed. Having established differences in metabolite production between these cultures, let us consider separately each of the mutant strains and its products. The data which follow suggest that ER-47 converts PCP primarily to 2,6-dichlorohydroquinone.

The relative amounts of PCP and chlorinated benzoquinone or hydroquinone intermediates in the culture fluid were analyzed by first extracting clarified culture broth at a neutral pH with chloroform. Ultra-violet spectra of the extracts were made and respective compounds were measured at their absorbance maxima. Figure 2 illustrates an experiment in which the pattern of PCP transformation by ER-47 is

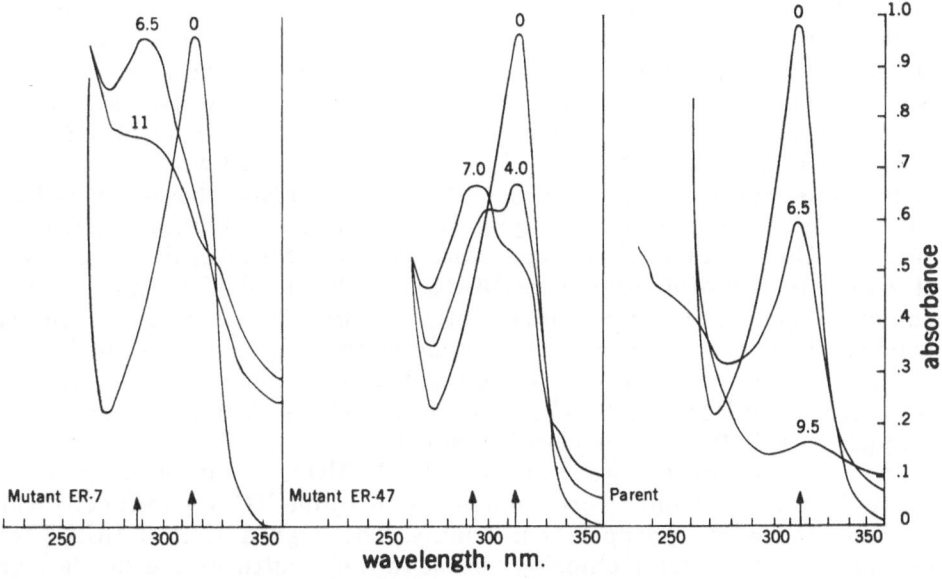

Figure 1. Ultra-violet absorption spectra of culture filtrates of KC-3 Parent, ER-47 Mutant and ER-7 Mutant. (Numbers above peaks indicate time in hours).

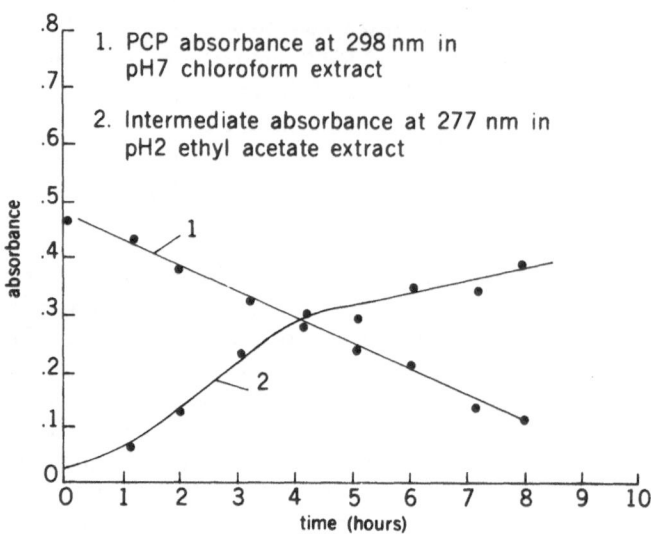

Figure 2. Kinetics of PCP degradation and formation of an intermediate product by the mutant ER-47.

shown using the above methodology. PCP concentration was reduced smoothly over the 8-hour incubation period and a metabolite absorbing at 277 nm in ethyl acetate was produced. Mutant ER-47 was next grown in a 14 liter Microferm fermentor in a glucose, PCP, salts medium. The cells were separated by continuous flow centrifugation and the clarified broth was extracted, first to remove residual PCP and then to recover the metabolites. Water was removed from the metabolite extract with sodium sulfate. The extract was evaporated to dryness with a stream of N_2 resuspended in acetone and purified by temperature-programmed preparative gas chromatography using a methyl silicone grease, OV-1 coated on an 80/100 mesh silanized diatomaceous earth (gas-chrom Q) and developed with nitrogen gas. Samples were collected in pre-weighed U-shaped tubes. ER-47 showed a very large metabolite peak in gas chromatography. This material was isolated, purified and cast into a KBr pellet for infrared spectroscopy. A comparison of the IR spectra of authentic 2,6 -dichlorohyroquinone (DCHQ) and the metabolite can be made by examining Figure 3. The spectra are essentially identical. It should be noted that the preparative recovery of DCHQ by gas chromatography was 23% based on the PCP applied. However, the actual yield estimated by UV absorption indicated that the conversion of PCP to 2,6-DCHQ was closer to 70%. A trace of monochloro-hydroquinone (CHQ) also accumulated in the culture.

In order to establish more clearly that 2,6-DCHQ is an intermediate of PCP degradation, experiments were done in which viable KC-3 parent cells were supplemented with appropriate nutrients. The degradation of DCHQ was determined by measuring chloride release in these cultures and the data are presented in Fig. 4. A PCP-adapted KC-3 parent culture rapidly released chloride from DCHQ without a lag. Unadapted cells were delayed but were induced by

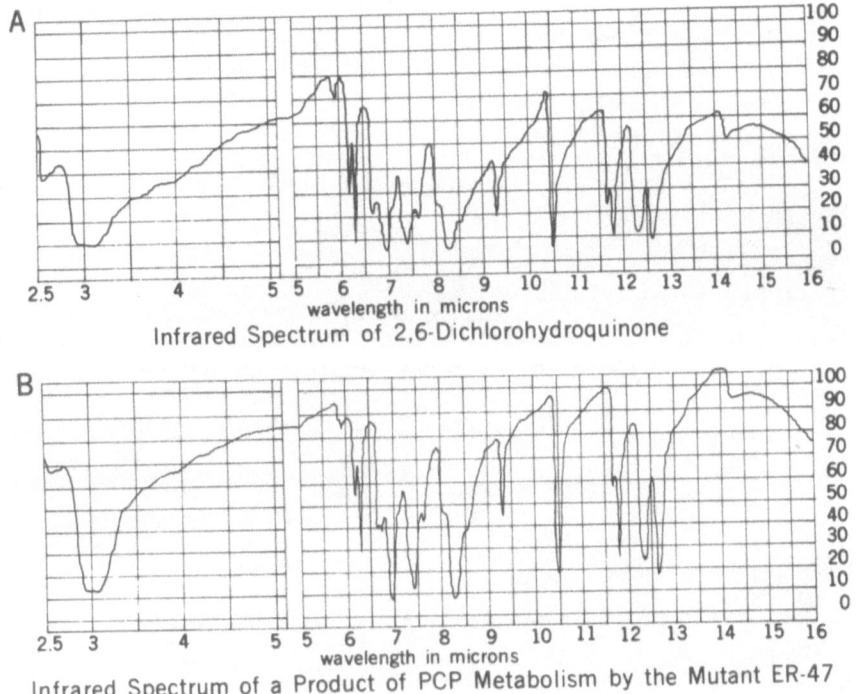

Infrared Spectrum of 2,6-Dichlorohydroquinone

Infrared Spectrum of a Product of PCP Metabolism by the Mutant ER-47

Figure 3. Identification of 2,6-dichlorohydroquinone as a metabolite of PCP metabolism in cultures of the mutant ER-47.

      A. Synthesized from Eastman 2,6-DCBQ by reduction with sodium sulfite, Recrystallized from benzene. KBr pellet, 1.5 mg/ 70 mg KBr; fast scan.

      B. Collected at 4.9 min from OV-1, 250°, 50 ml/min N, 0.85 mg/70 mg KBr. Slow scan; product identified as 2,6-dichlorohydroquinone.

exogenous DCHQ to rapid oxidation of the metabolite within an hour. A boiled KC-3 control showed that spontaneous release of chloride was very slow and that the metabolite was fairly stable during the experiment. Figure 4 shows that the oxidized form of the metabolite, 2,6-dichlorobenzoquinone is not significantly metabolized, perhaps it was not incorporated into the cell. Chlorohydroquinone (CHQ) did undergo slow dechlorination which seemed to be a result of microbial attack. Again it was not at all clear what the metabolic significance of CHQ was, however, since a trace of CHQ was also shown to accumulate in the ER-47 culture it cannot be overlooked. The slow rate at which CHQ was attacked in comparison with 2,6-DCHQ suggested that the former compound may not be a true intermediate in the degradation of PCP by KC-3.

    The data obtained with the mutant ER-7 culture are somewhat more difficult to interpret. This mutant was shown to accumulate more than one metabolite in significant concentrations. All of the metabolites isolated were shown to contain

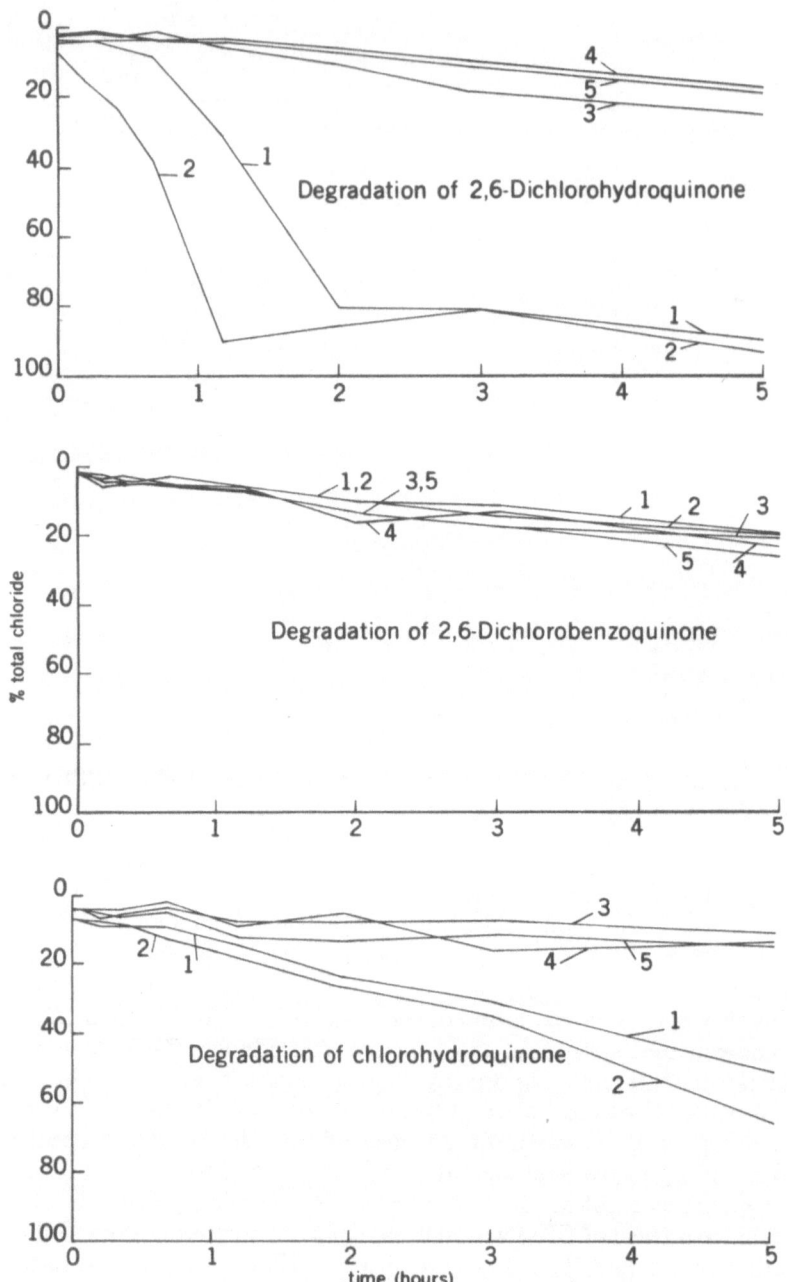

Figure 4. Degradation of 2,6-dichlorohydroquinone, 2,6-dichlorobenzoquinone and chlorohydroquinone by cultures - 1: unadapted KC-3; 2: PCP-adapted KC-3; 3: boiled KC-3; 4: unadapted ER-47; 5: PCP-adapted ER-47.

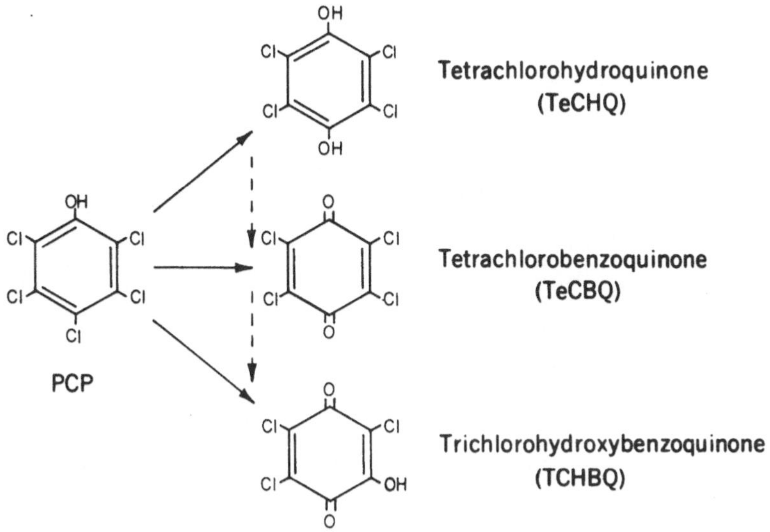

Figure 5. Products of PCP metabolism in cultures of the mutant ER-7.

either three or four atoms of chlorine. The three metabolites isolated, identified, and to be discussed are illustrated in Figure 5.

The UV spectrum of an acid, ethyl acetate extract of ER-7 culture filtrate after metabolism of PCP suggested that the sample contained trichlorohydroxy-benzoquinone (TCHBQ). This compound has been isolated from the culture filtrate in good yield (approximately 25%). However, kinetic studies suggested that it was not the earliest quinone produced in PCP degradation, and in fact it may not be a true intermediate at all. The identity of TCHBQ was established by its chlorine content and infrared spectrum. Since an authentic sample of TCHBQ was unavailable, the spectrum of the unknown metabolite was compared with that published by Slifkin *et al.* (1967). The spectra seemed identical, therefore, the identity of one metabolite was designated as TCHBQ. In addition to TCHBQ, tetrachlorobenzoquinone (TeCBQ) and tetrachlorohydroquinone (TeCHQ) were obtained in less than 10% yields after extraction and preparative gas chromatography. Comparative spectra of authentic TeCHQ and one of the ER-7 metabolites are shown in Figure 6. It appears that there is excellent agreement in these spectra to verify the identity of TeCHQ. TeCBQ was also shown to be present and was identified in the same manner. A question can be raised as to whether all of these metabolites are true intermediates in PCP degradation or whether spontaneous oxidation in the samples might lead to the formation of some of these materials. It was observed rather early in these studies that the absorption spectra and visible color of stored, cell-free mutant culture liquor changed and intensified with time. When cells were present the changes also occurred, but the products formed were somewhat different. In addition, it had been noted that a fresh solution of TeCHQ buffered at a pH of 7 or higher underwent a rapid transition of

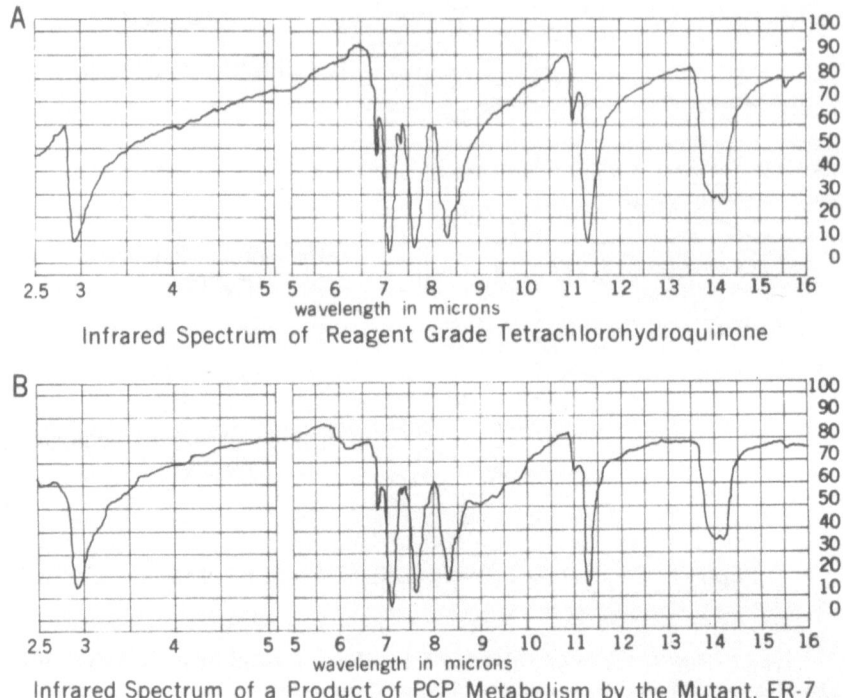

Infrared Spectrum of a Product of PCP Metabolism by the Mutant, ER-7

Figure 6. Identification of tetrachlorohydroquinone as a metabolite of PCP metabolism in cultures of the mutant, ER-7.

    A. KBr pellet, 3.5 mg/100 mg KBr, slow scan.
    B. Purified by GLC on OV-210, at 250°, 50 ml/min N_2. Recrystallization from chloroform. KBr pellet, slow scan.

color changes starting from colorless, proceeding through intense yellow, to green-gray and finally to purple. The rapidity of this transition was increased with increasing pH. These observations, coupled with the finding of Hancock *et al.* (1962) that TeCBQ undergoes spontaneous reaction to form TCHBQ, led us to consider the possibility that spurious reactions in the culture fluid not directly associated with PCP metabolism, might lead to the formation of TCHBQ and its occurrence in large quantities.

    Figure 7 shows data describing the spontaneous conversion of TeCHQ through an intermediate step involving TeCBQ. The identity of the reaction products was determined by extracting the reaction mixture at selected times, reisolation and purification of the extracts by preparative gas chromatography and analysis of the purified products by infrared spectroscopy. Clearly, TeCHQ was converted spontaneously within a few minutes to TeCBQ, which was subsequently transformed to the more stable TCHBQ. In the presence of KC-3 cells, we have observed a retardation of this reaction but not complete cessation. TCHBQ has been

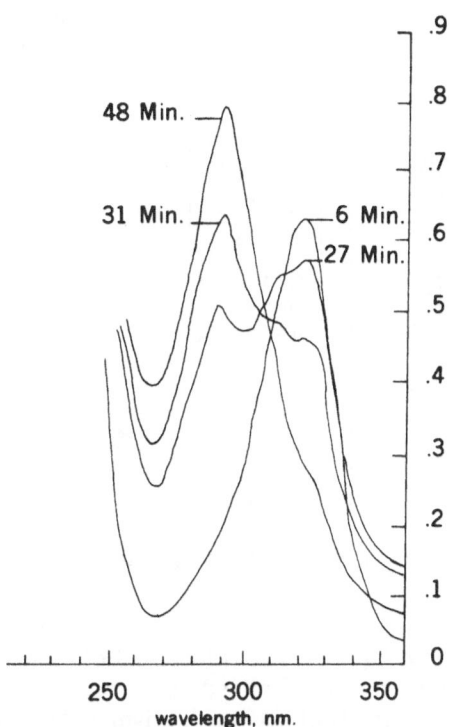

Figure 7. The spontaneous conversion of tetrachlorohydroquinone to tetrachlorobenzoquinone and trichlorohydroxybenzoquinone at pH 7.0.

shown to undergo metabolic attack by KC-3 but was neither completely dechlorinated nor was the ring ruptured. Apparently another product was formed which was not metabolically active. Contrary to this, TeCBQ was readily metabolized by the KC-3 parent culture and has been shown to yield substantial amounts of 2,6-DCHQ when applied to the ER-47 mutant, which accumulated 2,6-DCHQ from PCP metabolism. The application of TeCHQ to the ER-47 mutant resulted in the formation of measurable amounts of DCHQ but it appeared that the rapid spontaneous transformation of TeCHQ to TCHBQ competed vigorously with metabolic activity either in the mutant or in the parent culture, leading to difficulty in interpretation of the data.

Figure 8 hypothesizes a partial pathway for the metabolism of PCP by culture KC-3 in which TeCHQ or TeCBQ and 2,6-DCHQ are given as essential intermediates. Further studies must be completed before this pathway can be firmly established. It is essential, for example, that the enzymes responsible for this sequence of reactions be isolated and characterized.

In summary, we have shown that PCP and some of the related halophenols can be extensively decomposed by an aerated heterogeneous microflora. A single axenic bacterial isolate from this biomass has been shown to use pentachlorophenol,

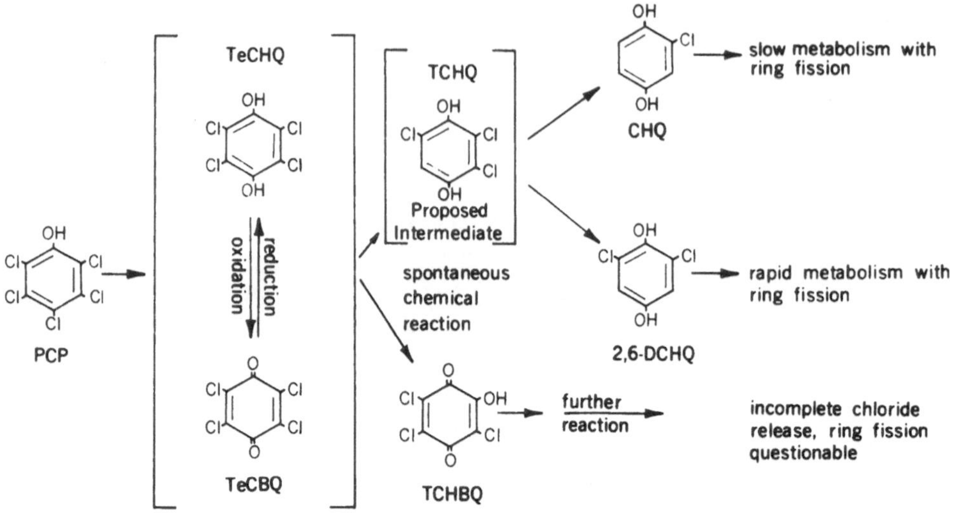

Figure 8. Hypothetical pathway for the biodegradation of pentachlorophenol by the bacterial culture, KC-3.

2,3,4,6-tetrachlorophenol and 2,4,6-trichlorophenol as growth substrates. These and other chlorophenols are metabolized to various degrees by the culture. It appears reasonable to assume that the mechanism of breakdown of PCP involves the conversion of PCP to partially dechlorinated hydroquinone intermediates which then undergo ring breakage. The critical intermediates appear to be 2,6-dichlorohydroquinone and 2,3,5,6-tetrachlorohydroquinone or 2,3,5,6-tetrachlorobenzoquinone.

References

Chu, J.P., and E.J. Kirsch, 1972. Metabolism of pentachlorophenol by an axenic bacterial culture. *Appl. Microbiol.,* **23**: 1033-1035.

Chu, J.P., and E.J. Kirsch, 1973. Utilization of halophenols by a pentachlorophenol metabolizing bacterium. *Dev. Ind. Microbiol.,* **14**: 264-273.

Cserjesi, A.J., 1967. The adaptation of fungi to pentachlorophenol and its biodegradation. *Can. J. Microbiol.,* **13**: 1243-1249.

Cserjesi, A.J., and E.L. Johnson, 1972. Methylation of pentachlorophenol by *Trichoderma virgatum. Can. J. Microbiol.,* **18**: 45-49.

Curtis, R.F., D.G. Land, and N.M. Griffiths, 1972. 2,3,4,6-Tetrachloroanisole association with musty taint in chickens and microbiological formation. *Nature,* **235**: 223-224.

Etzel, J.E., and E.J. Kirsch, 1974. Biological treatment of contrived and industrial wastewater containing pentachlorophenol. *Dev. Ind. Microbiol.,* **16**: 287-295.

Hancock, J.W., C.E. Morrell, and D. Rhum, 1962. Trichlorohydroxyquinone. *Tetrahedron Lett.,* **22**: 987-989.

Ide, A., Y. Niki, F. Sakamoto, I. Watanabe, and H. Watanabe, 1972. Decomposition of pentachlorophenol in paddy soil. *Agr. Biol. Chem.,* **36**: 1937-1944.

Ingols, R., P. Gaffney, and P. Stevenson, 1966. Biological activity of halophenols. *J. Water Poll. Contr. Fed.,* **38:** 629-635.

Jakobson, I., and S. Yllner, 1971. Metabolism of ¹⁴C-pentachlorophenol in the mouse. *Acta Pharmacol. Toxicol.,* **29:** 513-518.

Kincannon, D.F., J.A. Heidman, and A.F. Gaudy, 1967. Metabolic response of activated sludge to sodium pentachlorophenol. *Proceedings Purdue Industrial Waste Conference,* **22:** 561-574.

Kirsch, E.J., and J.E. Etzel, 1973. Microbial decomposition of pentachlorophenol. *J. Water Poll. Contr. Fed.,* **45:** 359-364.

Konishi, K., and Y. Inoue, 1972. Detoxification mechanism of pentachlorophenol by the laccase of *Coriolus versicolor. Japan. Wood Res. Soc. J.,* **18:** 463-469.

Slifkin, M.A., R.A. Sumner, and J.G. Heathcote, 1967. The interactions of chloranil in aqueous solvents. I. The absorption spectrum of chloranil in 50% aqueous ethanol. *Spectrochimica Acta,* **23A:** 1751-1756.

Suzuki, T., 1977. Metabolism of pentachlorophenol by a soil microbe. *J. Environ. Sci. Hlth.,* **12B:** 113-127.

Suzuki, T., and K. Nose, 1971. Decomposition of pentachlorophenol in farm soil. II. PCP metabolism by a microorganism isolated from soil. *Pesticide Prodn. Technique Japan,* **26:** 21-24.

Tashiro, S., T. Sasamoto, T. Ajkawa, S. Tokunaga, E. Taniguchi, and M. Eto, 1970. Metabolism of pentachlorophenol in mammals. *J. Agr. Chem. Soc. Japan,* **44:** 124-129.

Fate of Pentachlorophenol in the Blue Crab,

Callinectes sapidus

AJAY K. BOSE and H. FUJIWARA

Abstract—The fate of PCP in blue crabs was studied using GC-CI-MS and negative CI-MS. Negative CI-MS has already been reported by Dougherty and Piotrowska (1976) as a sensitive method for the detection of PCP. We have observed that CI-MS (NH₃) is also a convenient and sensitive analytical method, especially in conjunction with gas chromatography and "multiple ion detection" techniques, for studies on PCP and related compounds and metabolites.

Essentially pure sodium pentachlorophenate (purity checked by negative CI-MS) was injected into the hemolymph of crabs at sub-lethal doses and observations made at various time intervals up to 4 days. It was observed that within a few hours the level of PCP became very low in the hemolymph while the level in the hepatopancreatic tissues became high and remained so for the period of observation. Negative CI-MS of the extract of hepatopancreatic tissues with organic solvents showed some free PCP as well as ion clusters (chloro compounds) starting at m/e 227 and 320. Much of the PCP in the hepatopancreatic tissues was in the form of esters since hydrolysis with 6 N KOH produced strong PCP peaks in negative CI-MS.

We have noticed significant changes in the lipid profile of the hepatopancreatic tissues within a few hours of the injection of PCP although the lipids of the hemolymph showed little change. Determination of the structure of the altered lipid components is in progress.

AJAY K. BOSE and H. FUJIWARA • Department of Chemistry and Chemical Engineering, Stevens Institute of Technology, Hoboken, New Jersey 07030, U.S.A.

Introduction

The metabolism of PCP by bacteria, clams, fish and mammals has been well documented. Little is known about the fate of PCP in crustaceans. Kobayashi *et al.* (1976) studied the sulfate conjugation of phenol and PCP by the hepatopancreas of the spiny lobster, *Panulirus japonicus*. The sulfate conjugation of this tissue was relatively lower than that of the liver of goldfish. In collaboration with K. Ranga Rao at The University of West Florida, we have studied the fate of PCP in the blue crab, *Callinectes sapidus*.

In view of the extremely low levels of metabolites to be expected from the injection of about 0.5-3 mg of PCP per 100 g body weight of crabs, the analytical methods selected had to be particularly sensitive to PCP and expected metabolites. For repetitive determinations of metabolites from small samples of biological fluids or tissues it was also desirable to reduce the need for time consuming or laborious separation procedures. From these considerations mass spectrometry appeared to be the most promising analytical method. Dougherty and Piotrowska (1976) have reported negative chemical ionization (NCI) mass spectrometry (MS) to be a sensitive method for the detection of PCP. In connection with several biomedical programs in our laboratory (Bose *et al.*, 1976, 1977; Fujiwara *et al.*, 1977) we have been using CI-MS (NH$_3$) as well as NCI-MS as convenient and sensitive tools. The present study on the metabolic fate of PCP in crabs was carried out with the help of CI-MS methods. We present here a preliminary report on our observations.

Instruments and Materials

Chromatography was performed with a Perkin Elmer model 900 gas chromatograph using a 6 ft. 3% JXR column and a flame ionization detector. The column temperature was programmed from 90-240 °C.

For mass spectral work a chemical ionization mass spectrometer ("Biospect" of Scientific Research Instruments Inc., Baltimore, Maryland) was employed. A Varian 1440 gas chromatograph fitted with a 6 ft. SE-30 (5%) column was directly connected to the Biospect. Helium was used as the carrier gas for GC.

Samples of hemolymph and hepatopancreatic tissues from crabs injected with different quantities of Na-PCP were supplied by K. Ranga Rao, University of West Florida. Sodium pentachlorophenate (0.5 to 3.0 mg/100 g body wt.) was injected into blue crabs weighing 162 to 275 g. The crabs were sacrificed 6, 48, 72 and 96 hours after the injection of Na-PCP. The hemolymph was first withdrawn with the aid of a syringe. The carapace of each crab was cut open and the hepatopancreas was removed. The samples were stored in dry ice during shipment by air from Pensacola, Florida to Newark, New Jersey and subsequent transfer to our laboratory.

Analytical Methods

Hemolymph or hepatopancreatic tissue samples from crabs were heated under reflux with 6 N alcoholic KOH solution under a blanket of nitrogen for 1 hour. The reaction mixture was cooled, acidified with HCl, and extracted with CHCl$_3$. A part of the CHCl$_3$ extract was used for CI-MS studies, while another part was treated with diazomethane to convert free fatty acids to their methyl esters which were analyzed by gas chromatography to study the lipid profile.

For detecting the presence of non-covalently bound PCP and phenolic metabolites, the biological sample was treated with a 1 : 9 mixture of chloroform and ethanol and a few drops of H$_2$SO$_4$ (conc.). After shaking or stirring for a few minutes the organic layer was separated from the protein and other solids that precipitated. The organic solution was evaporated under a stream of N$_2$. The residue was submitted to mass spectral examination.

For quantitative estimation of PCP a standard method is to methylate PCP to PCA and then to analyze by GC. This method will give erroneous results if PCA is present as a metabolite. For this reason we have selected the benzoate derivative of PCP for quantitative CI-MS analysis. Treatment of a sample containing PCP with benzoyl chloride and (t-Bu)$_3$N leads to the formation of PCP benzoate in quantitative yield. Since pentadeuterated benzoic acid is readily available, pentadeuterated benzoate of PCP would be an easily available internal standard for mass spectral analysis using "multiple ion detection" accessories. This analytical approach would also be applicable to other phenolic metabolites.

Fate of PCP in Blue Crabs

The hemolymph of crabs injected with PCP showed no free PCP by negative chemical ionization spectrometry. Hepatopancreatic tissues were acidified and extracted with organic solvents as described earlier. These extracts were first examined by the NCI technique using methane as the reagent gas to test for

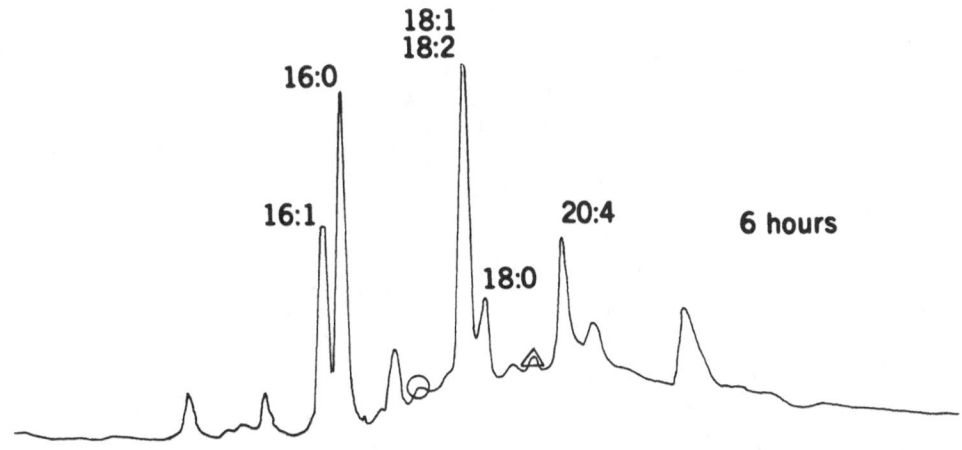

Figure 1. Lipids from the hepatopancreatic tissues of blue crabs 6 hours after injection of a sublethal dose of Na-PCP. GC on 6 ft 3% JXR, 90°-240°C, flame ionization detector.

unbound PCP. Higher levels of PCP were detectable after saponification indicating that much of the PCP was covalently bound in the ester form. Conjugates of PCP, such as the glucuronide, are known to occur in other animals (Kobayashi, this symposium). The NCI spectrum showed several chlorine containing ion clusters. The two most prominent ion clusters were due to PCP and pentachloroanisole (PCA). There were two ion clusters at lower mass ranges that have not been identified yet. A trace amount of a cluster corresponding to tetrachloro-dibenzodioxin (TCDD) was also observed. Recently Hass *et al.* (1977) have reported

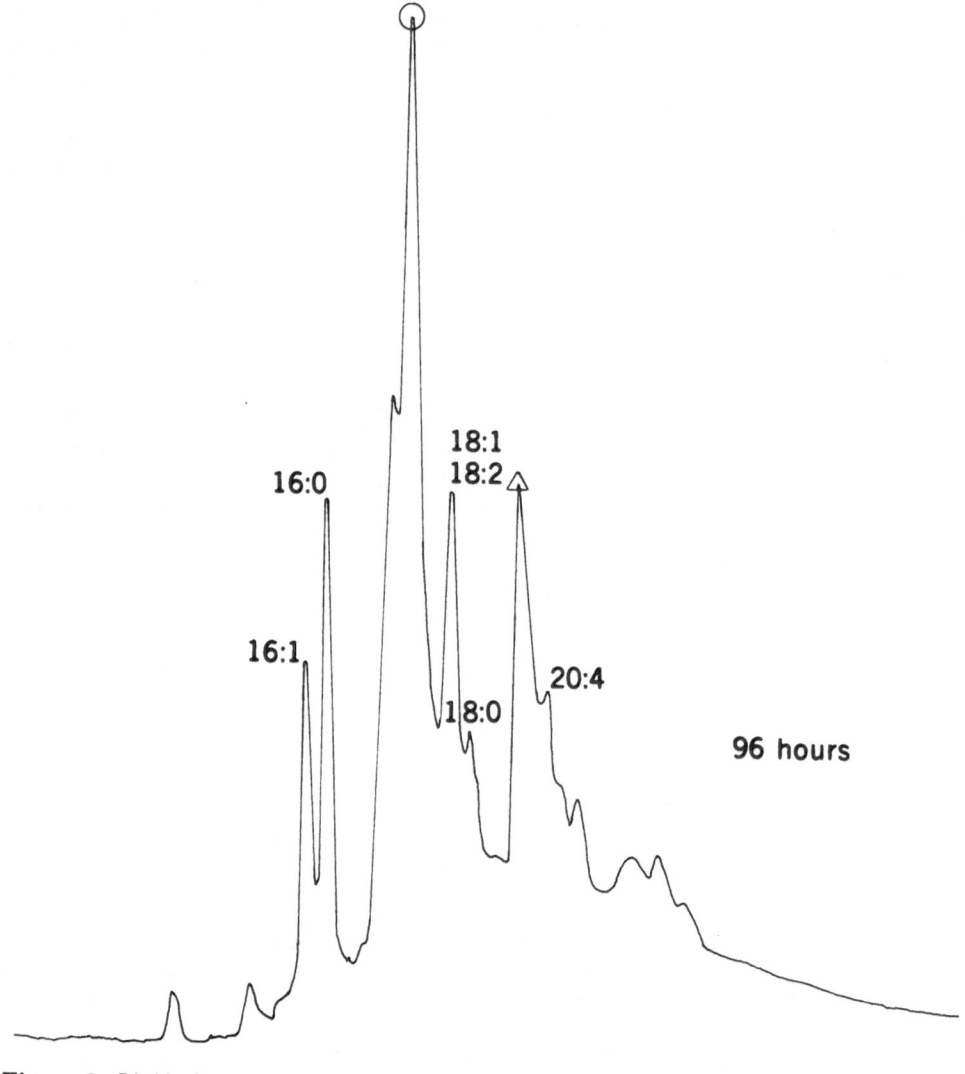

Figure 2. Lipids from the hepatopancreatic tissues of blue crabs 96 hours after injection of a sublethal dose of Na-PCP. GC on 6 ft 3% JXR, 90°-240°C, flame ionization detector.

the presence of TCDD among metabolites produced by cows exposed to PCP on a Michigan farm. The sample of PCP used in our studies did not show any TCDD by NCI-MS analysis. It is possible that TCDD present in PCP was below our detection level but it had been accumulated by hepatopancreatic tissues to a detectable level.

Lipid Profiles of Hemolymph and Hepatopancreas of Blue Crabs

Lipids obtained by the saponification of hemolymph samples from crabs sacrificed at different intervals after Na-PCP injection showed essentially similar profiles when examined by gas chromatography. However, the lipid profile as seen after the saponification of hepatopancreatic tissues changed considerably with time after injection of Na-PCP. At the end of 6 hours after injection of Na-PCP the major lipid peaks were due to the fatty acids 16:1, 16:0, 18:1 (and/or 18:2), 18:0 and 20:4. At the end of 96 hours there was an appreciable increase in the amount of unsaturated C-16 acid while the relative amounts of 18:1 (or 18:2) and 18:0 did not change much. Two peaks appearing between the 16:0 and 18:1 peaks were bigger than any other peak in the GC. Yet another change in the lipid pattern concerned a peak appearing just before the 20:4 peak. This peak increased in size with increase in time after injection of Na-PCP into crabs (Figs. 1 and 2).

Some of these samples were examined by GC-CI-MS techniques. In one experiment NH_3 was used as the reagent gas and positive ions were monitored. In a complementary run the NCI spectra were monitored using CH_4 as the reagent gas. The identity of C_{16}, C_{18}, and C_{20} acids was confirmed. The new peaks were found to be free of chlorine. Further work will be necessary for the identification of these lipid components.

ACKNOWLEDGMENTS

The authors are thankful to Professor K. Ranga Rao for encouraging this study and for his valuable suggestions. We are grateful to Mr. Philip J. Conklin and Dr. Angela C. Cantelmo, University of West Florida, for collecting hemolymph and tissue samples from blue crabs. The CI-MS facilities used for this research were established with a grant from the Schering-Plough Foundation.

References

Bose, A.K., H. Fujiwara, E.J. Lazaro, and C. Spillert, 1976. Biomedical studies using ammonia as reagent gas in CI mass spectroscopy. Abstract of a paper presented at the 170th ACS National Meeting, San Francisco, California, August 1976.

Bose, A.K., H. Fujiwara, B.N. Pramanik, and M.S. Manhas, 1977. Negative ion chemical ionization mass spectroscopy — a convenient tool for biomedical studies. Abstract of a paper presented at the Joint Conference of ACS and CIC, Montreal, Canada, May 1977.

Dougherty, R.C., and K. Piotrowska, 1976. Multiresidue screening by negative chemical ionization mass spectrometry of organic polychlorides. *J. Ass. Offic. Anal. Chem.,* **59**: 1023-1027.

Hass, J.R., M.D. Friesen, and M.K. Hoffmann, 1977. Negative ion chemical ionization mass
    spectrometry of chlorinated dibenzo-*p*-dioxins. Workshop on Negative Ion Chemical
    Ionization Mass Spectrometry, National Institute of Environmental Health Sciences,
    Research Triangle Park, North Carolina.
Fujiwara, H., A.K. Bose, and B.N. Pramanik, 1977. Positive and negative chemical
    ionization mass spectrometry for biomedical studies. Abstract of a paper presented at
    the 26th IUPAC, Tokyo, Japan, September 1977.
Kobayashi, K., S. Kimura, and H. Akitake, 1976. Studies on the metabolism of
    chlorophenols in fish. VII. Sulfate conjugation of phenol and PCP by fish livers. *Bull.
    Japan. Soc. Sci. Fish.*, **42:** 171-177.

Metabolism of Pentachlorophenol in Fishes

KUNIO KOBAYASHI

Abstract—PCP was rapidly absorbed by goldfish from media and accumulated in various organs, especially gall bladder. PCP in gall bladder rapidly increased with exposure time and displayed a further increase even after fish had been transferred to running water, whereas a decrease was observed in all other organs.

PCP absorbed by goldfish was quickly excreted into surrounding water, mostly in a conjugated-form accompanied with a small amount of free-form. The conjugate was isolated and identified as pentachlorophenylsulfate. The PCP conjugated by short-necked clam in media was also identified as the sulfate. The presence of the enzyme system responsible for the sulfate conjugation of phenols was confirmed in the liver-soluble fractions of various species of fishes. This enzyme activity for chlorophenols, however, abruptly decreased with increasing the Cl-atom number, resulting in the lowest activity for PCP.

Most of the PCP found in gall bladder was a conjugate. The conjugate was identified as pentachlorophenyl-β-glucuronide. It was revealed that fish dispose of PCP by both the sulfate and glucuronide conjugations, contrary to the conclusion of Brodie and Maickel (1962).

Introduction

Pentachlorophenol (PCP) has been widely used as a herbicide, fungicide or bactericide. In Japan, PCP was produced in quantities from 1960 to 1971 as shown

KUNIO KOBAYASHI • Department of Fisheries, Faculty of Agriculture, Kyushu University, Hakozaki, Fukuoka, Japan 812.

in Figure 1, and used mostly in paddyfields as a herbicide until its use was restricted in 1971 by the Japanese Government because of its high toxicity to fish.

At that time, the river and coastal waters near paddyfields were liable to be contaminated with PCP which flowed from the paddyfields due to heavy rainfalls. Although a number of studies were conducted on the toxic effects of PCP on aquatic organisms, no information on the metabolism of PCP in aquatic organisms was available at the time we initiated these studies.

In a test for the toxicity of PCP to short-necked clam, *Tapes philippinarum,* which is commercially important along the coastal area in Ariake sea, Kyushu Island and also more sensitive (0.1 ppm is lethal) than common fish and shrimp (Tomiyama and Kawabe, 1962; Tomiyama *et al.,* 1962b), we found that the concentration of PCP in the media holding the shellfish was remarkably decreased

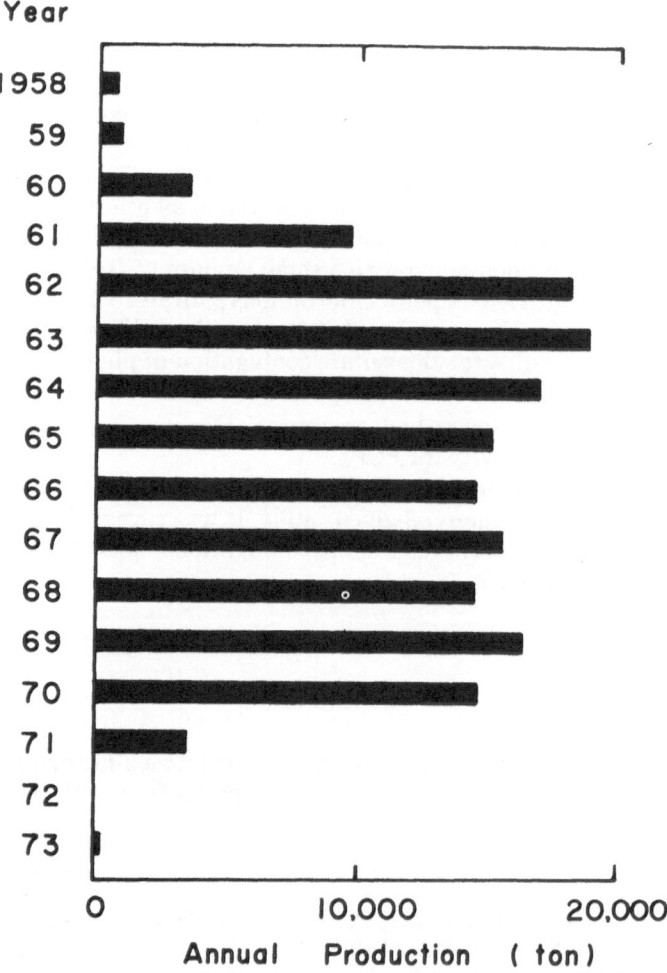

Figure 1. Annual production of PCP in Japan (1958-1973, as PCP-Na). (Japan Plant Protection Association, 1959-1974. Noyaku Yoran)

with exposure time, while the amount of PCP accumulated in the shellfish was smaller than the loss of PCP in the media (Tomiyama *et al.,*1962a). This suggested the presence of some mechanism of PCP detoxification in the shellfish and led us to study the metabolism of PCP in fish and shellfish.

Complying with the request of the conveners of this symposium, the present paper reviews the absorption, excretion and detoxification of PCP in fish and shellfish, for the most part, summarizing the work done in our laboratory.

In several of our investigations we used the goldfish, *Carassius auratus,* because of its high tolerance and ability to detoxify PCP among common fishes.

Absorption of PCP

When goldfish were exposed to PCP media (0.1, 0.2 and 0.4 ppm), the amount of PCP accumulated by the fish rapidly increased with both exposure time and PCP concentration in media, until a level of approximately 100 μg PCP/g body weight was reached when fish died from toxic effects of PCP, as shown in Figure 2 (Kobayashi and Akitake, 1975a). The concentration factor after 120-hr exposure in 0.1 ppm was approximately 1000, although that of phenol was only ca. 2 (Kobayashi and Akitake, 1975c). However, no surviving fish accumulated more

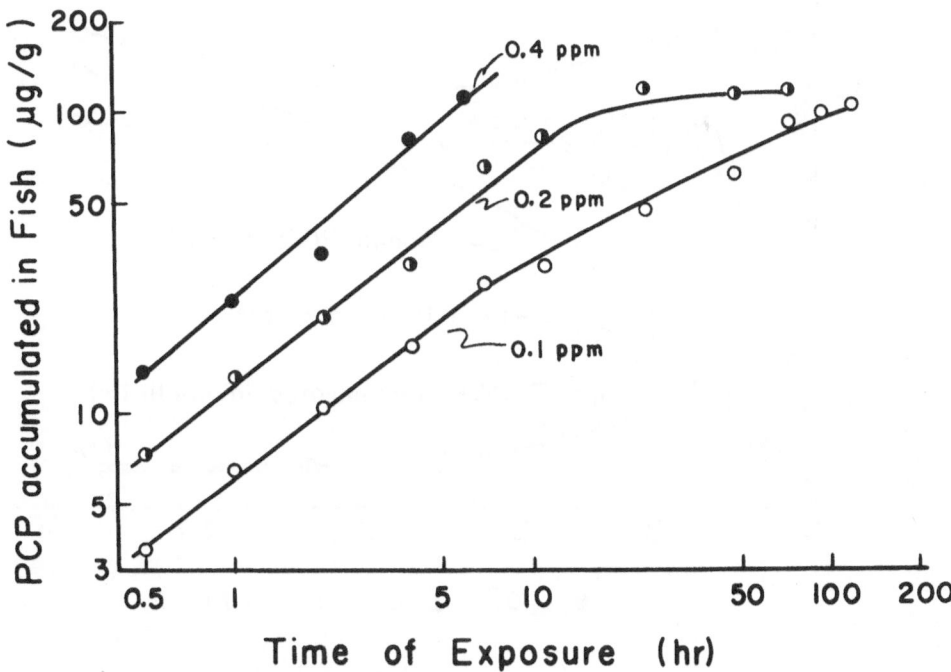

Figure 2. Accumulation of PCP in goldfish surviving in PCP-media (0.1, 0.2 and 0.4 ppm). (Kobayashi and Akitake, 1975a)

than 114 μg PCP/g and the range in dead fish was 82 to 116 μg/g (Kobayashi and
Akitake, 1975a).

Biochemical Change of PCP in Media During Exposure

During exposure of several fishes to PCP, the amount of PCP accumulated in
the fish bodies was always less than that lost from the PCP-medium. However, we
found that this loss of PCP is caused, not by decomposition, but by transformation
of the PCP by the fish to some bound-form. This bound-PCP in the medium is not
directly detectable by the 4-aminoantipyrine colorimetry, but it is detectable after
hydrolysis by heating in acidic solution which releases free-PCP (Akitake and
Kobayashi, 1975).

We also found the formation of a bound-PCP by short-necked clam similar to
that by fish mentioned above. Figure 3 shows the change with time in amounts of

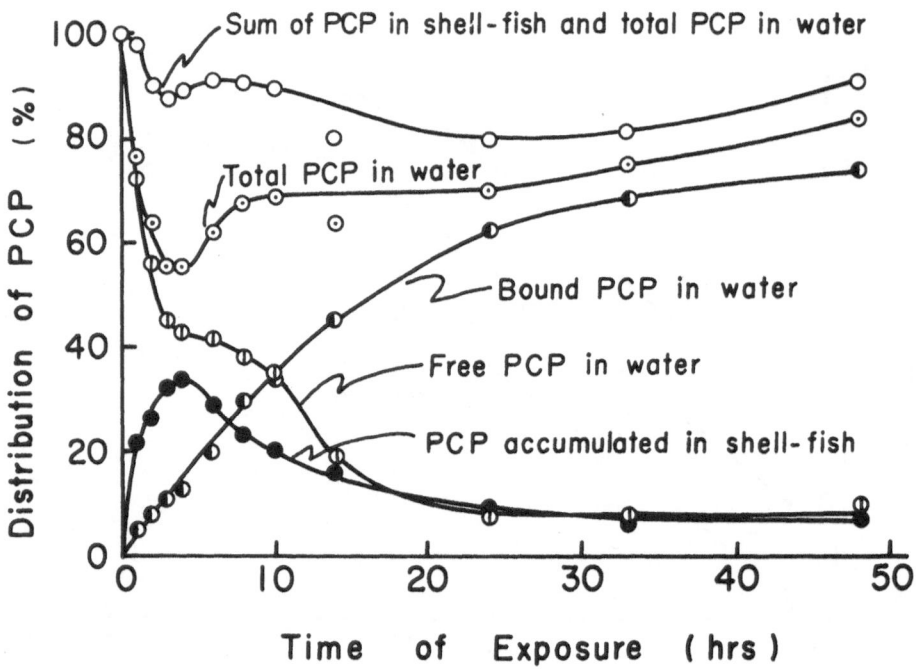

Figure 3. Change in amounts of free- and bound-PCP in the medium, and PCP
accumulated in short-necked clam from the medium, during exposure (Kobayashi *et al.*,
1970a). Amounts are expressed as percent of the initial amount of PCP in the medium.

free- and bound-PCP in the medium and PCP accumulated in the shellfish, during exposure to PCP-medium (Kobayashi *et al.,* 1970a). In the experiment, it was also confirmed that the excreta of the shellfish and microorganisms in the medium played no role in the formation of the bound-PCP (Kobayashi *et al.,* 1970a).

Excretion of Absorbed PCP

When goldfish were transferred to running water after 24-hr exposure to PCP media (0.1 and 0.2 ppm), the PCP concentration in the fish fell to half within 10 hr and ca. 20% of the initial values at 20-hr, as shown in Figure 4 (Kobayashi and Akitake, 1975a). PCP will not remain at high concentrations in fish and shellfish

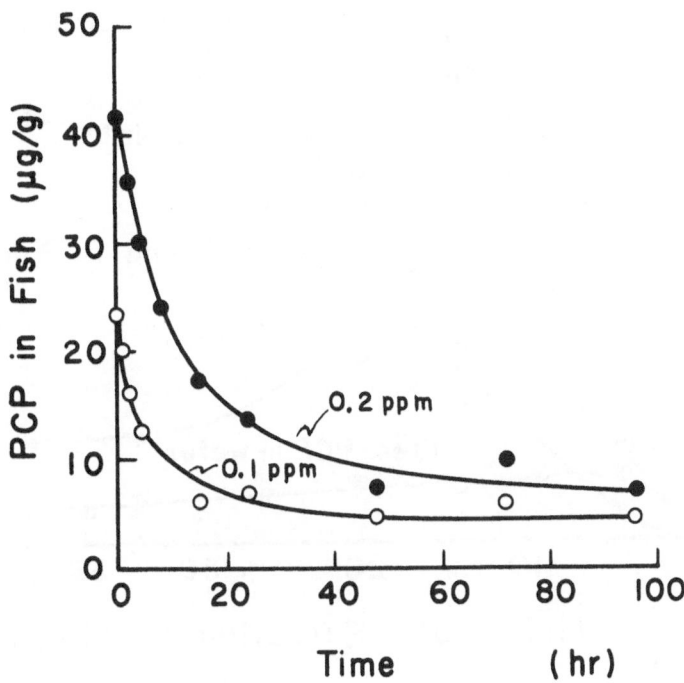

Figure 4. Retention of PCP in goldfish during culture in running water after 24-hr exposure to PCP-media (0.1 and 0.2 ppm). (Kobayashi and Akitake, 1975a)

for a long time without causing death. Most of the PCP excreted by the fish was in a bound-form and only 10% was found as free-PCP, as shown in Figure 5 (Akitake and Kobayashi, 1975). In the case of phenol, however, approximately 75% of the phenol accumulated in goldfish was excreted within 1 hr mostly in free-form accompanied with a small amount of a bound-form (Kobayashi and Akitake, 1975c).

Turnover of Absorbed PCP in Various Organs

Figure 6 shows the turnover of absorbed ^{14}C-labeled PCP (^{14}C-PCP) in various organs of goldfish, during exposure to PCP-medium and excretion after transfer to

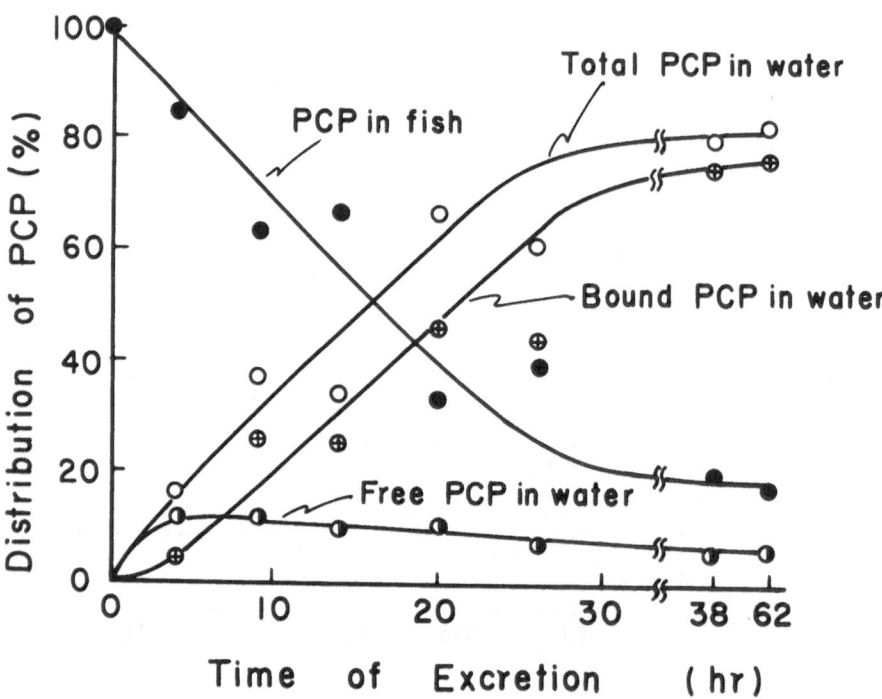

Figure 5. Change in amounts of PCP in goldfish and in free- and bound-PCP excreted in PCP-free water (Akitake and Kobayashi, 1975). Amounts are expressed as percent of the initial amount of PCP (24.8 μg/g-body weight) in the fish.

running water (Kobayashi and Akitake, 1975b). PCP absorbed by the fish from the medium was accumulated in various organs, especially the gall bladder. Though the PCP concentration in the gall bladder was the lowest among the organs examined at 1-hr exposure, it rapidly increased with time and displayed a further increase even after fish had been transferred to running water, whereas a decrease was observed in all other organs. Although the weight of gall bladder was only 0.28% of the whole body weight, the concentration of PCP in the gall bladder eventually reached a level of 1077 μg/g, corresponding to a concentration factor of 5400, and this accounted for 41% of the total PCP detected in fish after 24-hr culture in running water.

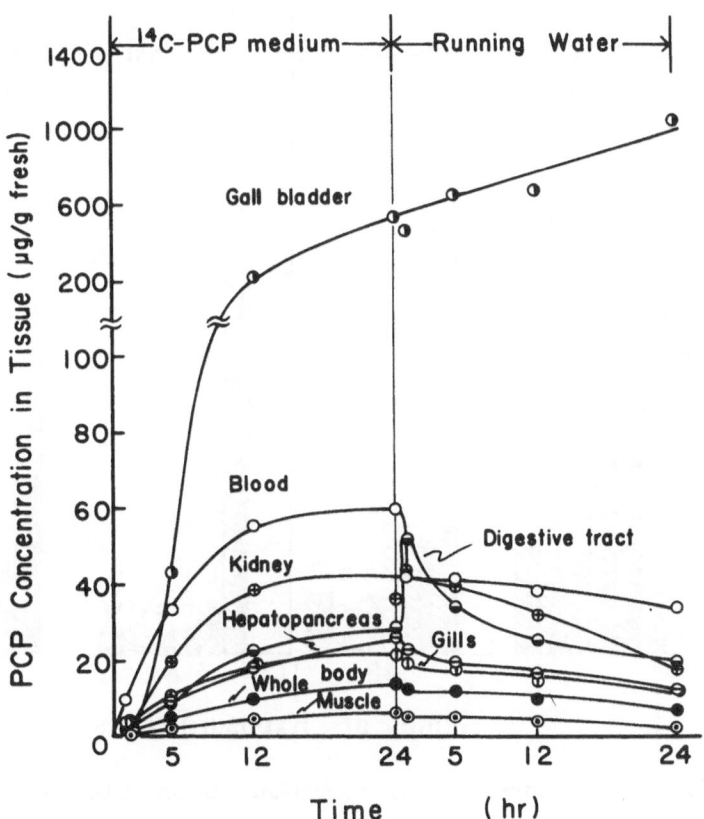

Figure 6. Change in concentration of PCP in various tissues of goldfish during exposure to PCP-medium and culture in running water after 24-hr exposure (Kobayashi and Akitake, 1975b)

It was presumed that a large proportion of the PCP found in the gall bladder would be transferred from the hepatopancreas after detoxification by conjugation or decomposition. Some decomposition of ^{14}C-PCP in the fish was confirmed by a slight increase of the specific activity found in the gall bladder. However, the decomposition of PCP in fish is of little consequence in detoxification of PCP, when compared to the formation of conjugates (Kobayashi and Akitake, 1975b).

^{14}C-PCP in the medium was rapidly absorbed by short-necked clam and distributed into all organs examined, especially the Bojanus' organ, and then quickly eliminated after transfer to PCP-free water, as shown in Figure 7 (Kobayashi *et al.*, 1969). It was also revealed that most of the PCP accumulated in the organs was undecomposed, as observed in goldfish.

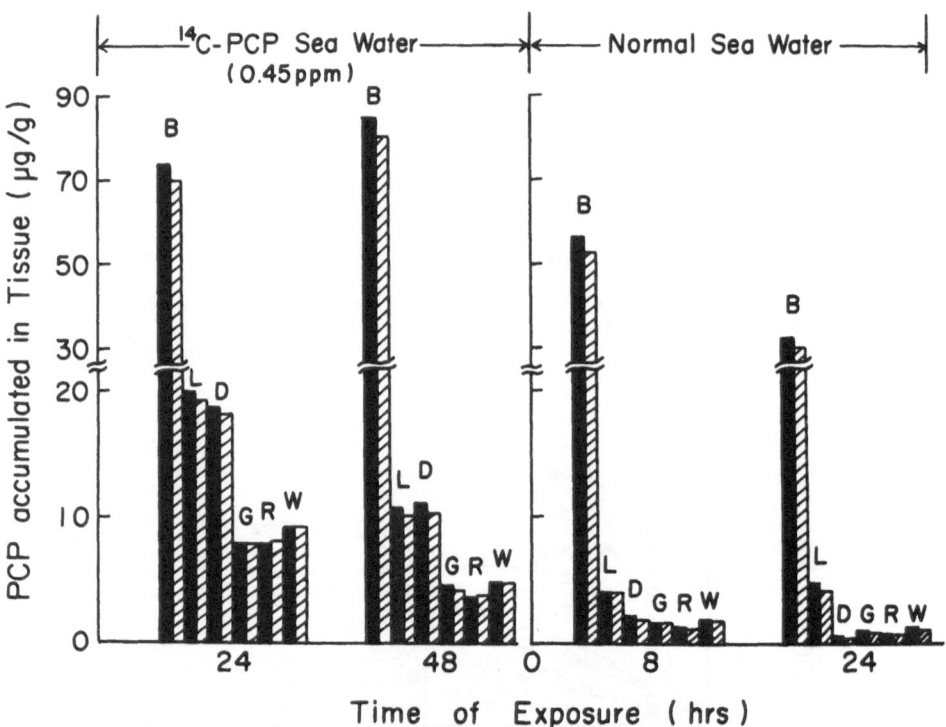

Figure 7. Change in concentration of PCP accumulated in various tissues of short-necked clam (Kobayashi *et al.*, 1969).

Solid column indicates PCP concentration calculated from values for radioactivities of tissues, based on the value for the specific activity of ^{14}C-PCP used in the experiment. Cross-hatched column indicates PCP concentration determined by the 4-aminoantipyrine method. Names of tissues were abbreviated as follows: B, Bojanus' organ; L, liver; D, digestive tract; G, gills; R, remainder; W, whole body.

Detoxification of PCP

Maickel *et al.* (1958) have reported that goldfish and perch were completely incapable of forming either glucuronides or sulfates of phenols, which are well-known as typical detoxification forms of phenolic compounds in mammals. However, both the bound-PCP excreted into surrounding water by goldfish and the characteristic accumulation of PCP in the gall bladder of the fish indicate that fish may dispose of PCP by active elimination such as conjugation and decomposition.

A study was made of the isolation and identification of both the bound-PCP excreted into surrounding water and the PCP accumulated in gall bladder. On the sulfate conjugation, further studies have been made *in vitro* by liver slices from several fish and shellfish, and also by various liver cell fractions of fish, using ^{35}S-labeled K_2SO_4 (^{35}S-K_2SO_4).

1. Sulfate Conjugation

(1) Isolation and identification of a conjugated PCP excreted by goldfish: After 15-hr exposure to PCP-medium (0.5 ppm, 560 liters), 250 goldfish (av. 35 g) were transferred to PCP-free water (560 liters) and cultured for 24 hr. A conjugate containing 96.6 mg PCP was excreted in the PCP-free water during the 24 hr culture period. The conjugated PCP was isolated by treating the medium with activated charcoal columns, followed by elution with an ammonia-acetone mixture, and finally by passing the concentrated eluate through a Sephadex G-10 column.

The isolated conjugate was identified as pentachlorophenylsulfate by precipitation with $BaCl_2$, by extractability with xylene before and after steam-distillation, by column and thin-layer chromatography, by UV-absorption spectra and by determination of the molar ratio of PCP to SO_4, by co-chromatography with sodium pentachlorophenate (PCP-Na) and synthetic potassium pentachloro-phenylsulfate (PCP-SO_3K) (Akitake and Kobayashi, 1975). Glucuronide was not detected in the conjugate.

The conjugated phenol excreted by goldfish into the culture medium was also identified as phenylsulfate, although it was accompanied by a large amount of free-phenol (Kobayashi *et al.*, 1975).

(2) Isolation and identification of a conjugated-PCP excreted by the short-necked clam: By a method similar to that used for goldfish described above, the PCP conjugate biosynthesized by the shellfish was isolated from the culture medium and identified as pentachlorophenylsulfate which was identical to that found in goldfish (Kobayashi *et al.*, 1970b).

(3) Sulfate conjugation of phenol and PCP by fish livers: To reconfirm the sulfate conjugation of phenol and PCP *in vivo,* a study has been made *in vitro* with livers of goldfish; carp, *Cyprinus carpio;* rainbow trout, *Salmo gairdneri irideus;* spiny lobster, *Panulirus japonicus;* and short-necked clam, using ^{35}S-K_2SO_4. The sulfate conjugation of phenol and PCP by albino rat liver was also determined for comparative purposes (Kobayashi *et al.*, 1976a).

All the liver slices of the test animals exhibited sulfate conjugation activity with phenol, the activity being 1-30% of that in rat liver (approx. 1.8 μmole/g dry liver/hr) as shown in Figure 8 (Kobayashi *et al.*, 1976a). However, the liver homogenates prepared by a Waring blender had activities which were markedly lower than those shown by the liver slices (Table 1).

On the other hand, PCP added to the incubation mixture containing liver slices not only failed to conjugate with sulfate, but at concentrations higher than 1.4×10^{-6} M, it inhibited even the sulfate conjugation of phenol, as shown in Figure 9 (Kobayashi *et al.*, 1976a).

(4) Sulfate conjugation of chlorophenols by various liver cell fractions of goldfish: A goldfish liver homogenate prepared by a Potter's homogenizer was subjected to fractionation using an ultracentrifuge. Among the nuclear, mitochondrial, lysosomal, microsomal and soluble fractions, only the last fraction displayed the sulfate conjugation activity for phenol (630 nmole/g liver/hr). The soluble fraction also displayed the sulfate conjugation activity for all the tested

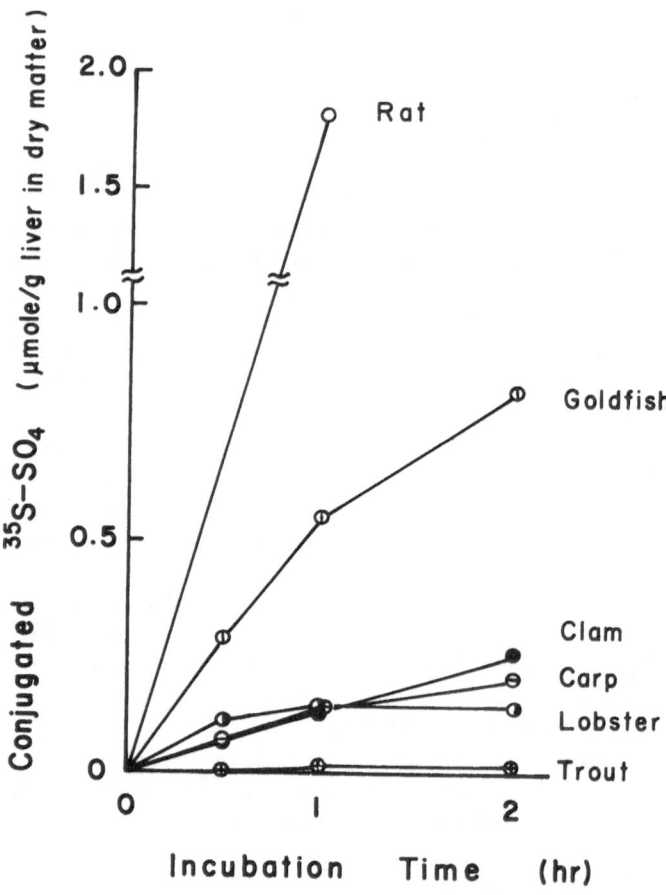

Figure 8. Sulfate conjugation of phenol by liver slices of test animals (Kobayashi *et al.* 1976a).

TABLE 1. Effect of homogenization on the sulfate conjugation of phenol by livers

Animal	Liver preparation	Time		Conjugated ^{35}S–SO$_4$ (nmole/g liver*)
		Homogenization (sec)	Incubation (hr)	
Goldfish	Slice		2	476
	Homogenate	10	2	38
		30	2	14
		120	2	2
Albino rat	Slice		1	1778
	Homogenate	30	1	813
		120	1	7
	Homogenate with 4 times of medium	30	1	63
		120	1	13

*in dry matter.
(Kobayashi *et al.*, 1976a)

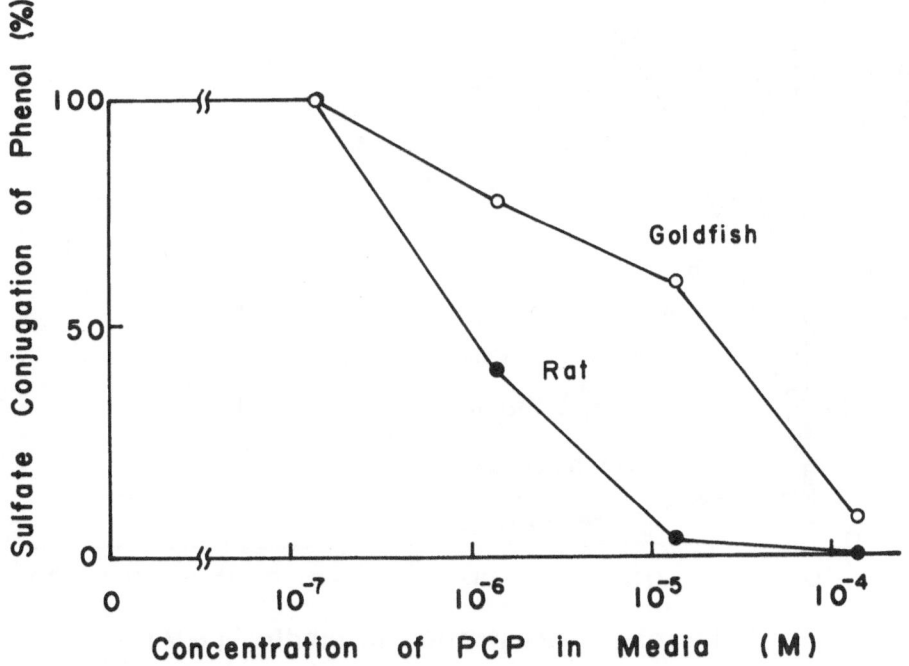

Figure 9. Effect of PCP on sulfate conjugation of phenol by liver slices of goldfish and albino rat (Kobayashi *et al.*, 1976a)

chlorophenols, showing activity for each (except *p*-chlorophenol) at low concentrations (0.07-0.13 mM in incubation media) and an abrupt decrease of the activity with increasing the substrate concentrations, as shown in Figure 10 (Kimura and Kobayashi, 1977).

An increase of the Cl-atom number in the chlorophenols caused an abrupt decrease of the sulfate conjugation activity of the soluble fraction, resulting in the lowest activity for PCP. On the other hand, an increase of the Cl-atom number in chlorophenols caused an increase in fish-toxicity (Kobayashi *et al.*, 1973).

From these results, it seems that a decrease of the sulfate conjugation activity in the liver promotes an accumulation of chlorophenols in fish and finally leads to the death of fish.

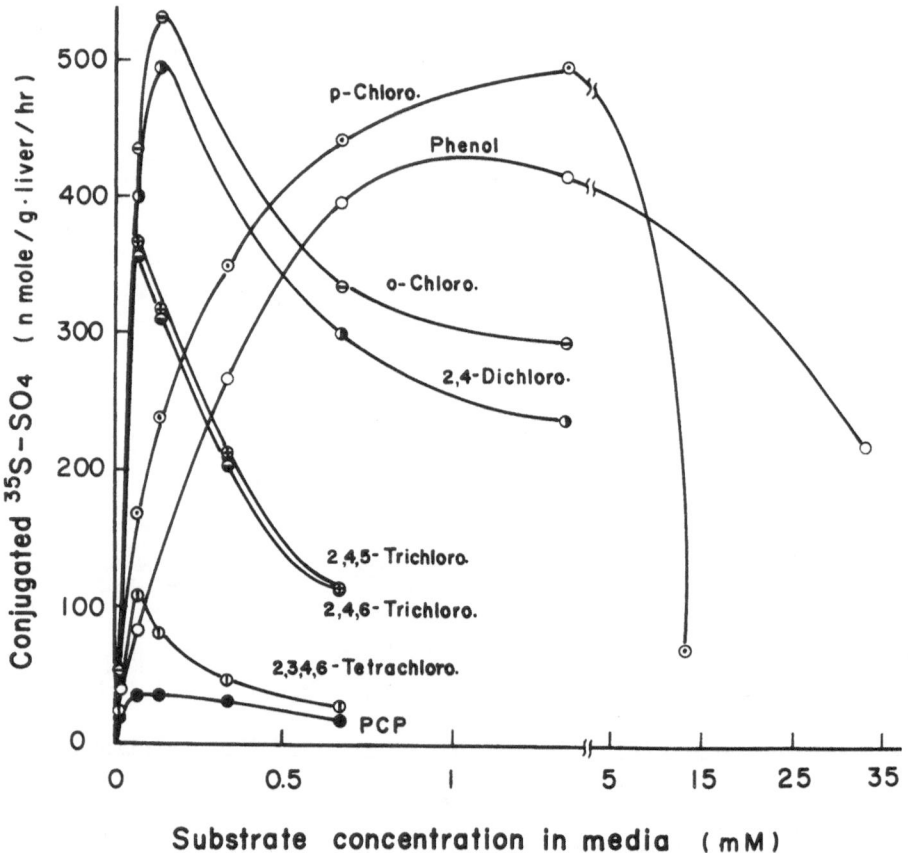

Figure 10. Sulfate conjugation of various chlorophenols by liver-soluble fraction of goldfish (Kimura and Kobayashi, 1977)

TABLE 2. Sulfate conjugation activities
with phenol by liver-soluble fractions of
several fishes

Species	Activity (nmole/g-liver/hr)
Red sea bream	603
Black sea bream	126
Parrot bass	146
Yellowtail	32
Goldfish	655
Carp	42
Rainbow trout	84
Albino rat	870

(Kimura and Kobayashi, 1977)

Table 2 shows the sulfate conjugation activities for phenol in the soluble fractions of livers from several fresh-water and marine fish. The data indicate that the sulfate conjugation is a common detoxification mechanism for phenols in fish.

2. *Glucuronic Acid Conjugation*

(1) Accumulation of a conjugated-PCP in bile of goldfish: As the presence of PCP-conjugates in bile had not been ascertained in the previous work (Fig. 6), a further study has been made regarding the confirmation of the accumulation of a conjugated-PCP in bile of goldfish exposed to PCP, and its isolation and identification (Kobayashi *et al.*, 1977).

During exposure of goldfish to 0.1 ppm PCP, the biliary concentration of the conjugated-PCP rapidly increased with time after 5 hours and reached a level of 4.46 μmole/g bile, corresponding to a concentration factor of 12000 at 48 hr, whereas the amount of free -PCP in the bile was negligible compared with that of the conjugate as shown in Figure 11 (Kobayashi *et al.*, 1977).

This indicates that the characteristic accumulation of PCP observed in the gall bladder of goldfish (Fig. 6) is due to the conjugated-PCP, and that free-PCP taken up by goldfish from media reaches the gall bladder within 1 hr even though at a much lower level, but it takes approximately 5 hr for PCP to be excreted into the bile after conjugation probably in liver, as was observed in the biliary excretion of phenol (Kobayashi *et al.*, 1976b).

(2) Isolation and identification of a conjugated-PCP from bile of goldfish: After 48 hr exposure to 0.1 ppm PCP, 30 goldfish (each approximately 110 g in body weight) were removed and their bile was collected by gall bladder puncture. A conjugated-PCP in the collected bile was isolated by treating the bile with activated charcoal columns, followed by elution with an acetone-ammonia mixture and finally by passing the concentrated eluate through a Sephadex G-10 column (Kobayashi *et al.*, 1977).

The isolated conjugate was identified as pentachlorophenyl-β-glucuronide by hydrolysis on incubation with β-glucuronidase, by thin-layer and gas-liquid chromatography, by determination of the molar ratio of PCP to glucuronic acid and by co-chromatography with PCP-Na, PCP-SO$_3$K and β-d-glucuronic acid (Kobayashi *et al.*, 1977). No other conjugates including the sulfate conjugate, which is directly excreted into surrounding water by goldfish as previously described, were detected in the bile.

According to Glickman *et al.* (1977) , the bile from rainbow trout exposed to PCP also contained high concentrations of PCP, mostly as the glucuronide

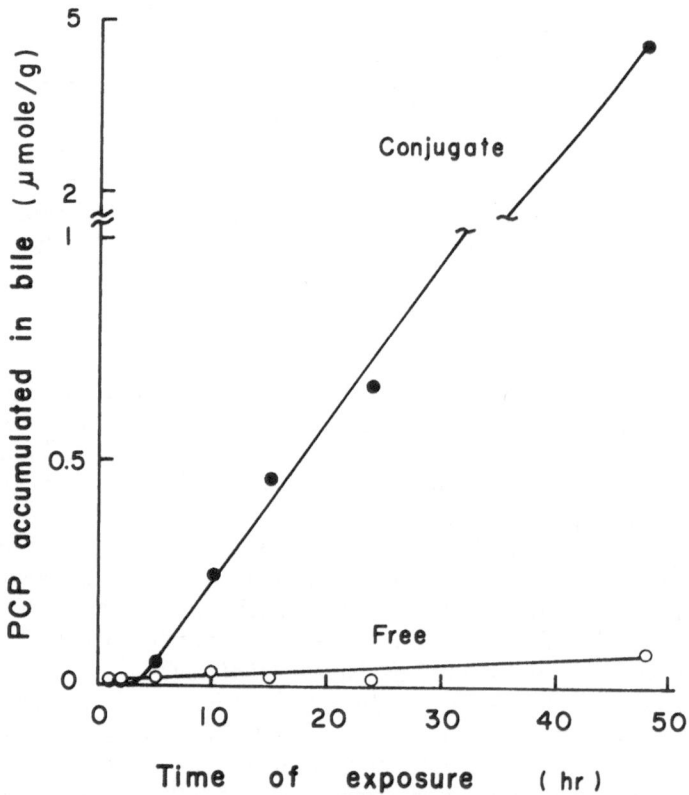

Figure 11. Accumulation of free- and conjugated-PCP in bile of goldfish exposed to PCP (Kobayashi *et al.*, 1977).

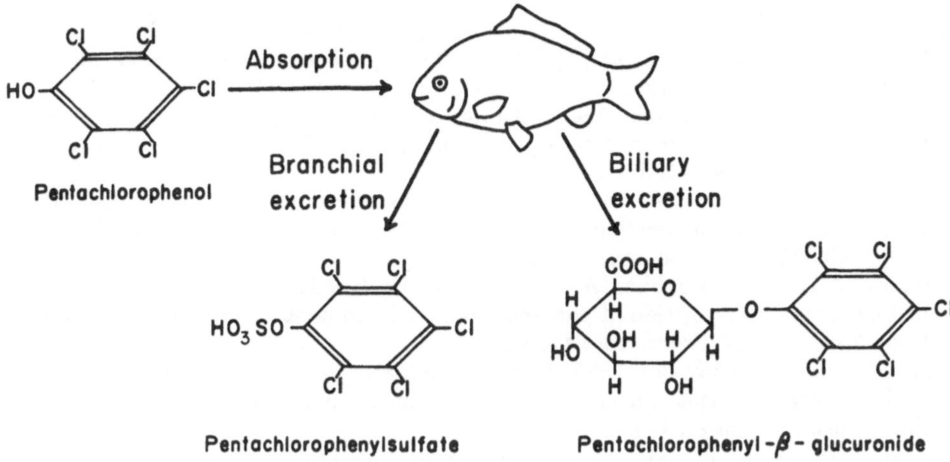

Figure 12. A schematic view of the major detoxification pathways for PCP in fish.

conjugate, but no other metabolites were detected. The PCP-glucuronide also has been found in the urine of rabbits after oral administration of PCP-Na (Tashiro *et al.*, 1970).

Not only PCP and phenol itself, but various phenolic compounds, such as *o*-aminophenol (Adamson, 1967), phenol red (Adamson and Guarino, 1972), *p*-nitrophenol (Adamson and Sieber, 1974), 3-trifluoromethyl-4-nitrophenol (Lech, 1973), and 2',5-dichloro-4'-nitrosalicylanilide (Statham and Lech, 1975) are also excreted as their glucuronides in bile of fishes exposed to these drugs. The biliary excretion after glucuronide conjugation must be one of the general detoxification mechanisms for phenolic compounds in fish.

Figure 12 shows a schematic view of the major detoxification pathways for PCP in fish. The renal excretion of the conjugates appears to be a minor route when compared with both the branchial and biliary excretion.

Our serial studies on the metabolism of PCP and phenol in fish revealed that fish, as well as mammals, possess both typical detoxification mechanisms for phenolic compounds by sulfate and glucuronide conjugations contrary to the conclusion of Brodie and Maickel (1962). However, the mechanism of the discriminative excretion between the sulfate and glucuronide conjugates in fish still remains to be investigated.

References

Adamson, R.H., 1967. Drug metabolism in marine vertebrates. *Federation Proc.*, **26:** 1047-1055.

Adamson, R.H., and A.M. Guarino, 1972. The effect of foreign compounds on elasmobranchs and the effect of elasmobranchs on foreign compounds. *Comp. Biochem. Physiol.*, **42 A:** 171-182.

Adamson, R.H., and S.M. Sieber, 1974. The disposition of xenobiotics by fishes. In, *Survival in Toxic Environments,* pp 203-211 (M.A.Q. Khan and J.P. Bederka, Jr., Ed.), Academic Press, New York.

Akitake, H., and K. Kobayashi, 1975. Studies on the metabolism of chlorophenols in fish—III. Isolation and identification of a conjugated PCP excreted by goldfish. *Bull. Japan. Soc. Sci. Fish.,* **41:** 321-327.

Brodie, B.B., and R.P. Maickel, 1962. Comparative biochemistry of drug metabolism. Proc. Ist. Intern. Pharmacol. Meeting, *Mode of Action of Drugs* (B.B. Brodie *et al.,* Ed.), Vol. **6,** pp. 299-324, Pergamon Press, Oxford.

Glickman, A.H., C.N. Statham, A. Wu, and J.J. Lech, 1977. Studies on the uptake, metabolism and disposition of pentachlorophenol and pentachloroanisole in rainbow trout. *Toxicol. Appl. Pharmacol.,* **41:** 649.

Kimura, S., and K. Kobayashi, 1977. Sulfate conjugation activity of soluble fraction of goldfish liver for various phenols. Presented at the Annual Meeting of Japan.Soc. Sci. Fish., April 2nd, Tokyo, Japan.

Kobayashi, K., and H. Akitake, 1975a. Studies on the metabolism of chlorophenols in fish—I. Absorption and excretion of PCP by goldfish. *Bull. Japan. Soc. Sci. Fish.,* **41:** 87-92.

Kobayashi, K., and H. Akitake, 1975b. Studies on the metabolism of chlorophenols in fish—II. Turnover of absorbed PCP in goldfish. *Bull. Japan. Soc. Sci. Fish.,* **41:** 93-99.

Kobayashi, K., and H. Akitake, 1975c. Studies on the metabolism of chlorophenols in fish—IV. Absorption and excretion of phenol by goldfish. *Bull. Japan. Soc. Sci. Fish.,* **41:** 1271-1276.

Kobayashi, K., H. Akitake, and K. Manabe, 1973. Accumulation and toxicity of chlorophenols in fish. Presented at the Annual Meeting of Japan.Soc. Sci. Fish., April 2nd, Tokyo, Japan.

Kobayashi, K., H. Akitake, C. Matsuda, and S. Kimura, 1975. Studies on the metabolism of chlorophenols in fish—V. Isolation and identification of a conjugated phenol excreted by goldfish. *Bull. Japan. Soc. Sci. Fish.,* **41:** 1277-1282.

Kobayashi, K., H. Akitake, and T. Tomiyama, 1969. Studies on the metabolism of pentachlorophenate, a herbicide, in aquatic organisms—I. Turnover of absorbed PCP in *Tapes philippinarum. Bull. Japan. Soc. Sci. Fish.,* **35:** 1179-1183.

Kobayashi, K., H. Akitake, and T. Tomiyama, 1970a. Studies on the metabolism of pentachlorophenate, a herbicide, in aquatic organisms—II. Biochemical change of PCP in sea water by detoxication mechanism of *Tapes philippinarum. Bull. Japan. Soc. Sci. Fish.,* **36:** 96-102.

Kobayashi, K., H. Akitake, and T. Tomiyama, 1970b. Studies on the metabolism of pentachlorophenate, a herbicide, in aquatic organisms—III. Isolation and identification of a conjugated PCP yielded by a shellfish, *Tapes philippinarum. Bull. Japan. Soc. Sci. Fish.,* **36:** 103-108.

Kobayashi, K., S. Kimura, and H. Akitake, 1976a. Studies on the metabolism of chlorophenols in fish—VII. Sulfate conjugation of phenol and PCP by fish livers. *Bull. Japan. Soc. Sci. Fish.,* **42:** 171-177.

Kobayashi, K., S. Kimura, and E. Shimizu, 1976b. Studies on the metabolism of chlorophenols in fish—VIII. Isolation and identification of phenyl-β-glucuronide accumulated in bile of goldfish. *Bull. Japan. Soc. Sci. Fish.,* **42:** 1365-1372.

Kobayashi, K., S. Kimura, and E. Shimizu, 1977. Studies on the metabolism of chlorophenols in fish—IX. Isolation and identification of pentachlorophenyl-β-glucuronide accumulated in bile of goldfish. *Bull. Japan. Soc. Sci. Fish.,* **43:** 601-607.

Lech, J.J., 1973. Isolation and identification of 3-trifluoromethyl-4-nitrophenyl glucuronide from bile of rainbow trout exposed to 3-trifluoromethyl-4-nitrophenol. *Toxicol. Appl. Pharmacol.,* **24:** 114-124.

Maickel, R.P., W.R. Jondorf, and B.B. Brodie, 1958. Conjugation and excretion of foreign phenols by fish and amphibia. *Federation Proc.,* **17:** 390.

Statham, C.N., and J.J. Lech, 1975. Metabolism of 2',5-dichloro-4'-nitrosalicylanilide (Bayer 73) in rainbow trout *(Salmo gairdneri). J. Fish. Res. Board Can.,* **32:** 515-522.

Tashiro, S., T. Sasamoto, T. Aikawa, S. Tokunaga, E. Taniguchi, and M. Eto, 1970. Metabolism of pentachlorophenol in mammals. *J. Agr. Chem. Soc. Japan,* **44:** 124-129.

Tomiyama, T., and K. Kawabe, 1962. The toxic effect of pentachlorophenate, a herbicide, on fishery organisms in coastal waters—I. The effect on certain fishes and a shrimp. *Bull. Japan. Soc. Sci. Fish.,* **28:** 379-382.

Tomiyama, T., K. Kobayashi, and K. Kawabe, 1962a. The toxic effect of pentachlorophenate, a herbicide, on fishery organisms in coastal waters—III. The effect on *Venerupis philippinarum. Bull. Japan. Soc. Sci. Fish.,* **28:** 417-421.

Tomiyama, T., K. Kobayashi, N. Uyeda, and K. Kawabe, 1962b. The toxic effect of pentachlorophenate, a herbicide, on fishery organisms in coastal waters—IV. The effect on *Venerupis philippinarum* of PCP which was constantly supplied or adsorbed on estuary mud. *Bull. Japan. Soc. Sci. Fish.,* **28:** 422-425.

Studies on the Uptake, Disposition and Metabolism of Pentachlorophenol and Pentachloroanisole in Rainbow Trout *(Salmo gairdneri)*

JOHN J. LECH*, ANDREW H. GLICKMAN and CHARLES N. STATHAM

Abstract—Pentachlorophenol (PCP) and pentachloroanisole (PCA) were rapidly taken up by rainbow trout when the concentrations of these compounds in water were 0.025 μg/ml. Following a 24 hr exposure of trout to ^{14}C-PCP the liver, blood, fat and muscle contained 16, 6.5, 6.0 and 1.0 μg/g PCP, respectively. Exposures of trout to ^{14}C-PCA resulted in tissue concentrations of a similar magnitude to PCP with the exception of fat where PCA concentrations attained levels of up to 80 μg/g. Elimination studies indicated that PCA was retained in most tissues for a considerably longer period of time than PCP. The half-lives for PCP residues in blood, liver, fat and muscle were 6.2, 9.8, 23 and 6.9 hr respectively, while the half-lives for PCA in the same tissues were 6.3, 6.9, 23 and 6.3 days. Thin-layer chromatographic and GC-MS analyses of the PCP-exposed trout failed to demonstrate methylated PCP (PCA) in any of the tissues studied. Bile from trout exposed to PCP showed high concentrations (250 μg/g) of PCP-glucuronide. The bile of trout exposed to PCA contained both PCA and PCP-glucuronide. The presence of PCP-glucuronide in bile indicated some demethylation of PCA *in vivo* by rainbow trout.

Introduction

Several recent studies have confirmed the presence of both pentachlorophenol (PCP) and pentachloroanisole (PCA) in fish raised in disinfected municipal effluents

JOHN J. LECH, ANDREW H. GLICKMAN and CHARLES N. STATHAM • Department of Pharmacology, Medical College of Wisconsin, Milwaukee, Wisconsin 53233, and Center for Great Lakes Studies, University of Wisconsin-Milwaukee, Milwaukee, Wisconsin. U.S.A. * NIEHS Research Career Development Awardee (ES-00002).

(Kopperman *et al.*, 1976) and also in several tissues of lake trout from Lake Michigan (Veith, 1977). Although PCP is known to be methylated in liquid cultures of *Trichoderma virgatum* (Cserjesi and Johnson, 1972) and by species of bacteria (Kopperman *et al.*, 1976), little is known of the methylation of phenols in fish. Investigations concerning the methylation of monohydric phenols in mammals *in vivo* are sparse and studies with the monophenols 3-trifluoromethyl-4-nitrophenol and 2′,5′-dichloro-4′-nitrosalicylamilide in rainbow trout did not detect methylated metabolites (Lech, 1973; Statham and Lech, 1975). PCP has been shown to be excreted in the urine of rats as PCP-glucuronide in addition to the oxidation

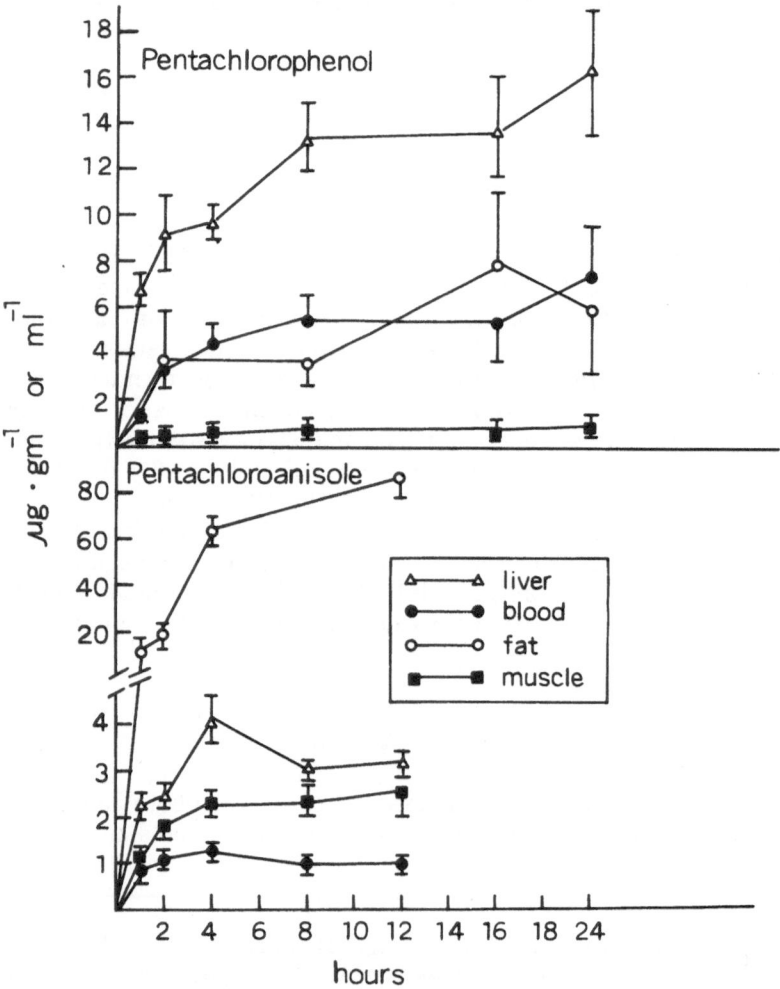

Figure 1. Time course of PCP and PCA in several tissues of rainbow trout. Data are calculated as micrograms of PCP or PCA per gram wet weight. Each point represents the Mean ± S.E. from at least 6 fish in two separate uptake studies (Glickman *et al.*, 1977).

product, tetrachlorohydroquinone (Braun *et al.*, 1976). Although several studies in goldfish demonstrated that PCP is excreted as a sulfate conjugate (Akitake and Kobayashi, 1975), little is known of the biological disposition of PCP in salmonids. In order to better understand the fate of PCP in rainbow trout, these studies were undertaken to examine the uptake, metabolism and disposition of PCP and PCA.

Uptake and Disposition of PCP and PCA

PCP and PCA were rapidly taken up by rainbow trout exposed to the ^{14}C-labeled compounds at 0.025 μg/ml in water. The time courses of the uptake of these

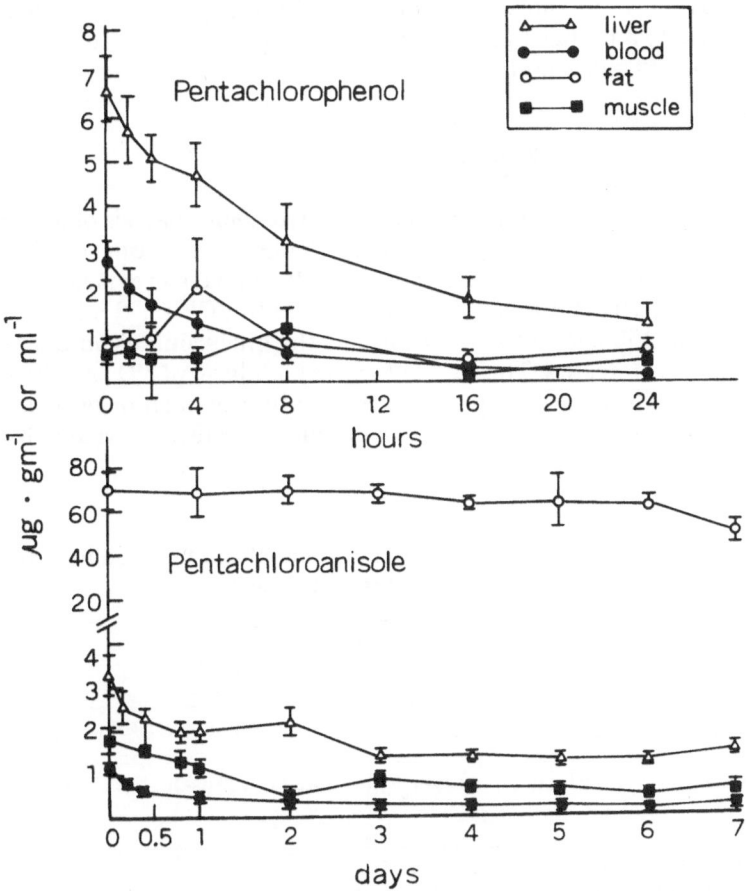

Figure 2. Elimination of ^{14}C from PCP and PCA exposed rainbow trout. After exposure to a medium containing 0.025 μg/ml for a period of 12 hours (PCA) or 24 hours (PCP) the fish were transferred to fresh running water and sampled at the indicated times. Note the difference between time scale in the upper and lower panels. Each data point represents the Mean ± S.E. from six fish in two separate experiments (Glickman *et al.*, 1977).

compounds are illustrated in Figure 1. It should be emphasized that these were static exposures and were intended to determine the time course of uptake prior to elimination studies, the plateau of the tissue concentrations therefore, should not be taken as steady state levels. The graphs do indicate however that PCA is rapidly accumulated in adipose tissue, reaching a concentration approximately ten times greater than PCP. On the other hand PCP appeared to reach higher concentrations than PCA in the liver.

Elimination studies of ^{14}C from tissues of PCA and PCP exposed trout were carried out following a loading period of 12 and 24 hrs, respectively, at a concentration of 0.025 μg/ml. It is obvious from the data shown in Figure 2 that PCA is retained in the body tissues, particularly in the fat, for a much longer period of time than is PCP. This is more clearly illustrated in the calculated half-lives for PCP and PCA shown in Table 1. The longer half-life of PCA when compared to PCP may be related to a higher lipid solubility of PCA and for the ability of PCP to be rapidly conjugated and excreted.

Metabolism of PCP and PCA

Thin layer chromatographic and mass spectroscopic analysis of acetone extracts of tissues from the PCP exposed trout failed to demonstrate methylated metabolites. The data are summarized in Table 2 and have previously been reported (Glickman *et al.*, 1977). Bile from trout exposed to PCP for 8 hrs contained high concentrations (250 μg/g) of PCP, mostly as the glucuronide conjugate. No other metabolites were detected. The reason for the high level of ^{14}C as PCP-glucuronide is most likely related to the fact that the conjugate is an organic anion and is capable of being concentrated by the biliary anionic transport system. Kobayashi *et*

TABLE 1. Half-Life (t ½) of Pentachlorophenol and Pentachloroanisole in Rainbow Trout Tissues[a]

| | t ½ | | | | | |
| | tissue | | | | | |
Chemical	Blood	Liver	Fat	Muscle	Gills	Heart
Pentachlorophenol	6.2 hr	9.8 hr	23.7 hr	6.9 hr	10.3 hr	6.9 hr
Pentachloroanisole	6.3 days	6.9 days	23.4 days	6.3 days	—	—

[a]Calculated from data on elimination studies shown in Figure 2. Glickman *et al.*, 1977.

TABLE 2. GC-MS and TLC Analysis of Tissue Extracts from Rainbow
Trout Exposed to ^{14}C-Pentachlorophenol[a]

Sample	Molecular ion (M+) observed	^{14}C-R$_f$
Pentachlorophenol (PCP) standard C$_6$HOCl$_5$	264,266,268,270	0.46
Pentachloroanisole (PCA) standard C$_7$H$_3$OCl$_5$	278,280,282,284	0.85
Liver extract	264,266,268,270	0.46
Methylated liver extract	278,280,282,284	0.85
Muscle extract	264,266,268,270	0.46
Methylated muscle extract	278,280,282,284	0.85
Bile extract[b]	264,266,268,270	0.46
Methylated bile extract	278,280,282,284	0.85

[a]See Glickman *et al.*, 1977 for experimental details.
[b]Hydrolysis by β-glucuronidase resulted in residue at an R$_f$ 0 to migrate to an R$_f$ of 0.46.

al.(1977) have recently reported that PCP can be conjugated with glucuronic acid
and sulfate in goldfish and that high concentrations of the glucuronide are excreted
in the bile. Previous investigations (Statham *et al.*, 1976; Lech *et al.*, 1973) have
shown that several phenols can be conjugated and concentrated in the bile and that
conjugation of the phenol and transport may be a rate limiting step in the
elimination of phenol.

Analysis of the tissues of PCA exposed trout showed only unchanged PCA with
the exception of liver where there was a trace amount of the free phenol. Bile, on
the other hand, contained a significant amount (10 µg/g) of PCP-glucuronide. GC-
MS data, shown in Table 3, substantiate this. The presence of PCP-glucuronide in
bile and what appears to be a minute amount of PCP in liver suggests the
demethylation of PCA *in vivo*. This finding is supported by the fact that several
investigators have confirmed the dealkylation of other foreign compounds in fish *in
vivo* (Hanson *et al.*, 1972; Olson *et al.*, 1977).

The 25-fold difference in magnitude of biliary excretion of ^{14}C derived from
PCA when compared to PCP may be the result of either or both of two factors.
First, demethylation of PCA may be a rate limiting step in the excretion of this
compound in bile since PCP is a phenol and is conjugated directly, while PCA must
first undergo demethylation before conjugation. The second possibility is that the
rate of biliary excretion of ^{14}C derived from PCA is determined by the rate of
transfer of PCA from tissue storage depots (i.e. adipose tissue) to sites of
metabolism and excretion (gills, liver, kidney). The observed long tissue half-lives of
^{14}C derived from PCA when compared to the half-lives of ^{14}C derived from PCP, in
addition to the relatively high concentration of PCA in the fat may tend to favor the
latter explanation.

TABLE 3. GC-MS Analysis of Chromatographically Separated ^{14}C from
β-Glucuronidase-Treated Bile from Rainbow Trout
Exposed to ^{14}C-Pentachloroanisole[a]

Sample	Molecular ion (M+) observed	(M+)-CH$_3$
Pentachlorophenol (PCP) standard	264,266,268,270	
Pentachloroanisole (PCA) standard	278,280,282,284	263,265,267,268
R_f = 0.46, untreated	264,266,268,270	
R_f = 0.46, methylated	278,280,282,284	263,265,267,269
R_f = 0.85, untreated	278,280,282,284	263,265,267,269
R_f = 0.85, methylated	278,280,282,284	263,265,267,269

[a]See Glickman *et al.,* 1977 for experimental details.

ACKNOWLEDGMENTS

This investigation was supported by Grant ES 01080 from the National Institute of Environmental Health Sciences and Grant R80-397-1010 from the United States Environmental Protection Agency.

References

Akitake, H., and K. Kobayashi, 1975. Studies on the metabolism of chlorophenols in fish. III. Isolation and identification of a conjugated PCP excreted by goldfish. *Bull. Japan. Soc. Sci. Fish.,* **41:** 1585-1588.

Braun, W.H., J.D. Young, M.W. Sauerhoff, G.E. Blau, and P.J. Gehring, 1976. The pharmacokinetics and metabolism of pentachlorophenol in rats and monkeys. *Toxicol. Appl. Pharmacol.,* **37:** 94.

Cserjesi, A.J., and E.L. Johnson, 1972. Methylation of pentachlorophenol by *Trichoderma virgatum. Can. J. Microbiol.,* **18:** 45-49.

Glickman, A.H., C.N. Statham, and J.J. Lech, 1977. Studies on the uptake, metabolism and disposition of pentachlorophenol and pentachloroanisole in rainbow trout. *Toxicol. Appl. Pharmacol.,* **41:** 649-658.

Hanson, L.G., I.P. Kapoor, and R.L. Metcalf, 1972. Biochemistry of selective toxicity and biodegradability: comparative 0-dealkylation by aquatic organisms. *Comp. Gen. Pharmacol.,* **3:** 339-344.

Kobayashi, K., S. Kimura, and E. Shimizu, 1977. Studies on the metabolism of chlorophenols in fish. IX. Isolation and identification of pentachlorophenyl-β-glucuronide accumulated in bile of goldfish. *Bull. Japan. Soc. Sci. Fish.,* **43:** 601-607.

Kopperman, H.L., D.W. Kuehl, and G.E. Glass, 1976. Impact of water chlorination. *Oak Ridge Nat. Lab. Conf.* 751096, p. 327.

Lech, J.J., 1973. Isolation and identification of 3-trifluoromethyl-4-nitrophenyl glucuronide from bile of rainbow trout exposed to 3-trifluoromethyl-4-nitrophenol. *Toxicol. Appl. Pharmacol.*, **24**: 114-124.

Lech, J.J., S. Pepple, and M. Anderson, 1973. Effect of novobiocin on the acute toxicity, metabolism and biliary excretion of 3-trifluoromethyl-4-nitrophenol in rainbow trout. *Toxicol. Appl. Pharmacol.*, **25**: 542-552.

Olson, L.E., J.L. Allen, and J.W. Hogan, 1977. Biotransformation and elimination of the herbicide, dinitramine in carp. *J. Agr. Food Chem.*, **25**: 554-555.

Statham, C.N., and J.J. Lech, 1975. Metabolism of 2′,5′-dichloro-4′-nitrosalicylamilide (Bayer 73) in rainbow trout. *J. Fish. Res. Bd. Can.*, **32**: 515-522.

Statham, C.N., M. Melancon, and J.J. Lech, 1976. Bioconcentration of xenobiotics in trout bile: A proposed monitoring aid for some water borne chemicals. *Science*, **193**: 680-681.

Veith, G., 1977. U.S. Environmental Protection Agency, Personal Communication.

Dechlorination of Pentachlorophenol *in Vivo* **and** *in Vitro*

ULF G. AHLBORG

Abstract—Pentachlorophenol was administered intraperitoneally to rats and the urinary excretory products were identified using thinlayer chromatography and mass spectrometry. Pentachlorophenol and tetrachloro-*p*-hydroquinone were found in the urine. Both of these compounds were excreted partly as conjugates with glucuronic acid. The level of conjugated pentachlorophenol was 9 to 16%. A greater part of tetrachloro-*p*-hydroquinone occurred as a conjugate. Bacterial *β*-glucuronidase failed to split the conjugates because, at the concentrations found in urine, tetrachloro-*p*-hydroquinone inhibited this enzyme.

Further *in vivo* and *in vitro* studies indicated that rapid dechlorination of pentachlorophenol occurred in rats. The dechlorination was found to be mediated by liver microsomal enzymes and their activity can be enhanced by pretreatment with inducing agents such as phenobarbital, 3-methylcholanthrene and 2,3,7,8-tetrachlorodibenzo-*p*-dioxin (TCDD). The dechlorination products formed were tetrachloro-*p*-hydroquinone and trichloro-*p*-hydroquinone. The latter was not due to any contaminants of the PCP preparation used. Although previous reports by other investigators indicated the formation of dechlorination products such as tetrachlorophenol and trichlorophenol in mammals treated with pentachlorophenol, these results could not be verified by this investigation.

ULF G. AHLBORG • Department of Toxicology, Swedish Medical Research Council, Karolinska Institute, S-104 01 Stockholm, Sweden; Laboratory of Toxicology, National Food Administration, S-751 26 Uppsala, Sweden.

Introduction

Toxicological and analytical data concerning pentachlorophenol up to 1967 have been reviewed by Bevenue and Beckman (1967). Human exposure occurs frequently and there are several reports of intoxications in the literature ranging from accidental contamination of drinking water to an epidemic with two deaths that occurred in a nursery in St. Louis after the use of pentachlorophenol as a mildew preventative in a laundry detergent (Barthel *et al.*, 1969).

Commercial pentachlorophenol has been shown to contain different impurities such as tetrachlorophenols and polychlorinated dioxins (Johnson *et al.*, 1973). Chlorinated 2-phenoxyphenols (predioxins) have also been reported as contaminants (Rappe and Nilsson, 1972). Production and use of highly contaminated pentachlorophenol have been reported to cause chloracne in humans but pentachlorophenols with a low content of chlorinated dioxins have been proven not to be chloracnegenic (Johnson *et al.*, 1973).

Acute toxicity studies have indicated single oral LD_{50}'s in the rat ranging from 27 to 210 mg/kg depending on the solvent used, water solution of pentachlorophenolate being the least toxic preparation.

Metabolism of Pentachlorophenol

Early studies on the metabolism of pentachlorophenol in the rabbit and the rat (Deichman *et al.*, 1942; Betts *et al.*, 1955) indicated that pentachlorophenol was excreted unchanged and unconjugated from the body. Our own interest in the metabolism of pentachlorophenol started when, in preliminary experiments, we found that when ^{14}C-pentachlorophenol was administered intraperitoneally to rats, only about 40% of the radioactivity could be recovered after adjustment of the urine to pH 2 and extraction with diethyl ether. Several other methods for extraction of pentachlorophenol (Bevenue *et al.*, 1966; Rudling, 1970; Rivers, 1972) were tested but gave approximately the same recovery. By contrast, when pentachlorophenol was added to urine from unexposed rats the recoveries after these extractions were almost complete.

At that time, Jacobson and Yllner (1971) reported that after the administration of ^{14}C-pentachlorophenol to mice, the compound was excreted both unchanged and conjugated, and unconjugated tetrachloro-*p*-hydroquinone was identified as one excretion product.

In our early efforts to identify urinary excretion products thinlayer radiochromatography was used. In a solvent system with methyl acetate-isopropanol-concentrated ammonia (9 : 7 : 4) three distinct peaks occurred (Fig. 1, lower panel). The first peak was consistent with tetrachloro-*p*-hydroquinone and the last one with pentachlorophenol. The second peak was believed to represent a conjugate but treatment of the urine with bacterial β-glucuronidase, aryl sulfatase or concentrated hydrochloric acid at room temperature failed to change the distribution of radioactivity. However, the second peak completely disappeared

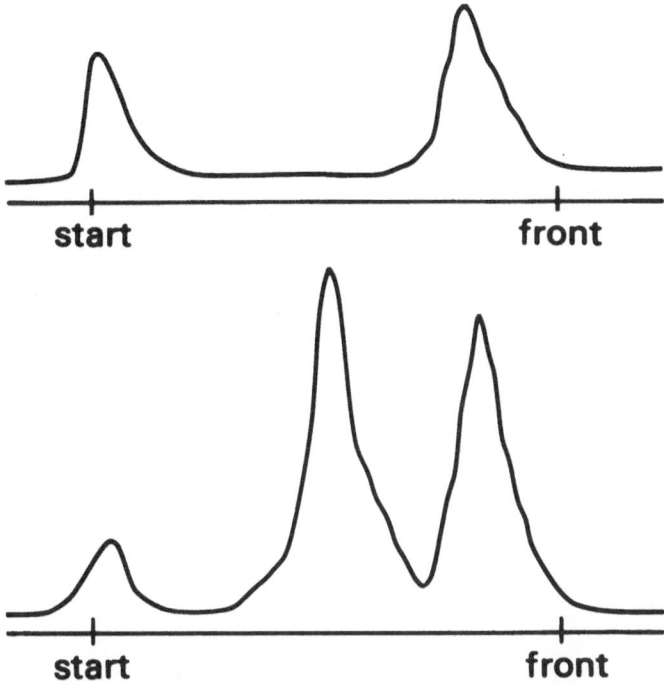

Figure 1. Thin layer radiochromatograms of urine from rats given ¹⁴C-pentachlorophenol (10 mg/kg) intraperitoneally. Solvent system: Methyl acetate-isopropanol-concentrated ammonia. (9 : 7 : 4).

       Lower panel: Untreated urine.
       Upper panel: Urine boiled with an aliquot of concentrated hydrochloric acid.

after treatment of the urine with concentrated hydrochloric acid at 100 °C (Fig. 1, upper panel). Quantitatively the result is an increase of the activity of the other two peaks, the increase being more pronounced for the first peak. The findings from the thin layer chromatographic studies were subsequently verified by gas chromatography — mass spectrometry (Ahlborg *et al.,* 1974).

    There seemed to be good reason to believe that both pentachlorophenol and tetrachloro-*p*-hydroquinone were excreted partly as conjugates with glucuronic acid and the failure to split the conjugates by means of treatment with bacterial β-glucuronidase was rather puzzling. However, in experiments with bacterial β-glucuronidase using phenolpthalein-β-glucuronide or p-nitrophenyl-β-D-glucuronide as substrate we could show that tetrachloro-*p*-hydroquinone was a potent inhibitor of bacterial β-glucuronidase activity.

    In a subsequent study we could show that several chlorinated hydroquinones and benzoquinones were potent inhibitors of β-glucuronidase of bacterial origin with I_{50} values in the range $10^{-6} - 10^{-7}$ M (Ahlborg *et al.,* 1977a). These concentrations are far below those occurring in the urine from the pentachlorophenol-treated rats and the failure of bacterial β-glucuronidase to split possible glucuronides is thus explained. The inhibition was found to be competitive

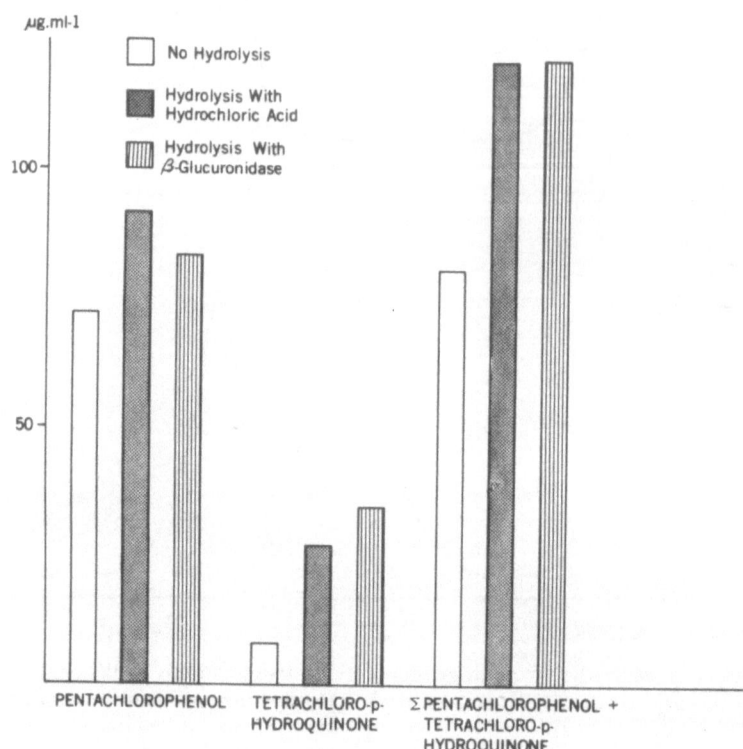

Figure 2. Levels of pentachlorophenol and tetrachloro-*p*-hydroquinone in urine from rats administered pentachlorophenol intraperitoneally (10 mg/kg) after different treatment of the urine.

in nature. No inhibitory effect of the benzo- and hydroquinones studied *in vitro* or *in vivo* could be demonstrated on β-glucuronidase from liver.

In order to solve the question of conjugation we have treated urine from pentachlorophenol dosed rats in three different ways:

a)   Acidification with concentrated hydrochloric acid to pH 2.
b)   Boiling with concentrated hydrochloric acid.
c)   Incubation with β-glucuronidase from bovine liver followed by acidification to pH 2.

The results of extraction and analysis by means of mass fragmentography are shown in Figure 2. Both types of hydrolytic treatment increased the recovery of free pentachlorophenol and tetrachloro-*p*-hydroquinone and the total increase was about 50%. The major part of this increase was due to a three to four-fold increase in the recovery of tetrachloro-*p*-hydroquinone. This is indirect evidence for the presence of both compounds as conjugates with glucuronic acid.

Recently the presence of pentachlorophenol-glucuronide in urine from exposed rats has been reported by Braun and Sauerhoff (1976). The level of conjugated

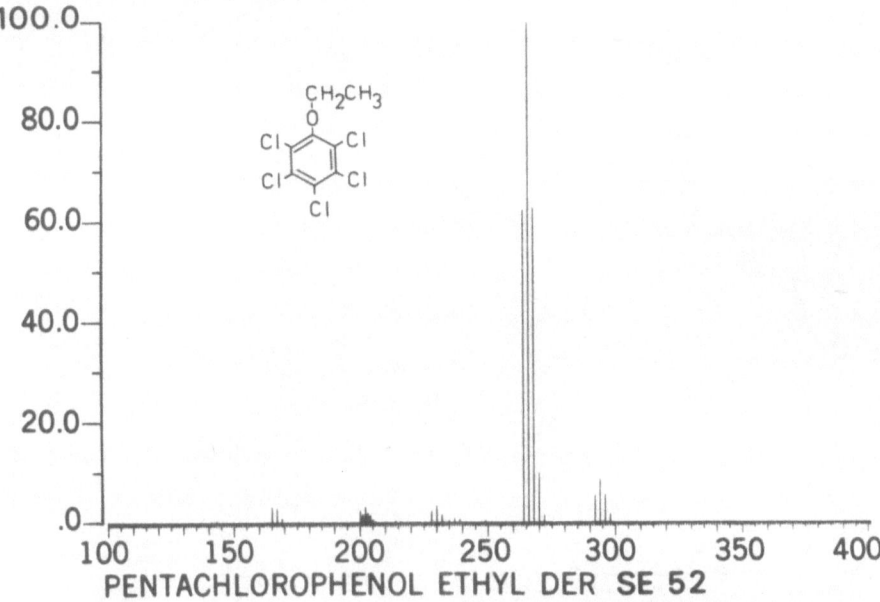

Figure 3. Mass spectrum of pentachlorophenol ethyl ether.

pentachlorophenol they found (9.4%) is in the same range as found by us (9-16%). However, tetrachloro-*p*-hydroquinone was only found unconjugated which is in clear contrast to our findings where the greater part of tetrachloro-*p*-hydroquinone occurs as a conjugate.

In subsequent studies gas chromatography — mass spectrometry was utilized and a computerized system for mass fragmentography was developed (Elkin *et al.*, 1973). The system consists of a LKB 9000 combination instrument gas chromatograph — mass spectrometer connected to a Digital Equipment PDP-12 computer for the control of the mass spectrometer and the recording and handling of the data. The system has also been developed and adopted for use on the LKB 2091 combination instrument interfaced to Digital Equipment GT40 and PDP 11/45.

All the data reported below are based on quantitative analysis by means of mass fragmentography. All urinary specimens have been boiled with concentrated hydrochloric acid in order to split conjugates present. Extractions have been done with diethyl ether and the extracts have been derivatized with N,O-bis(trimethylsilyl)acetamide to obtain trimethylsilylethers or derivatized with diazoethane to obtain ethyl ethers. The latter derivatives were prepared in hexane and subjected to cleaning with methanol/water as described by Cranmer and Freal (1970). Standard curves have been prepared for the determinations and tetrachloropyrocatechol or 2,3,4,5-tetrachlorophenol has been used as an internal standard. Gas chromatography has been performed on all glass columns packed with 5% SE-52 on Chromosorb W (60-80 mesh) or on Gas Chrom Q (100-120 mesh).

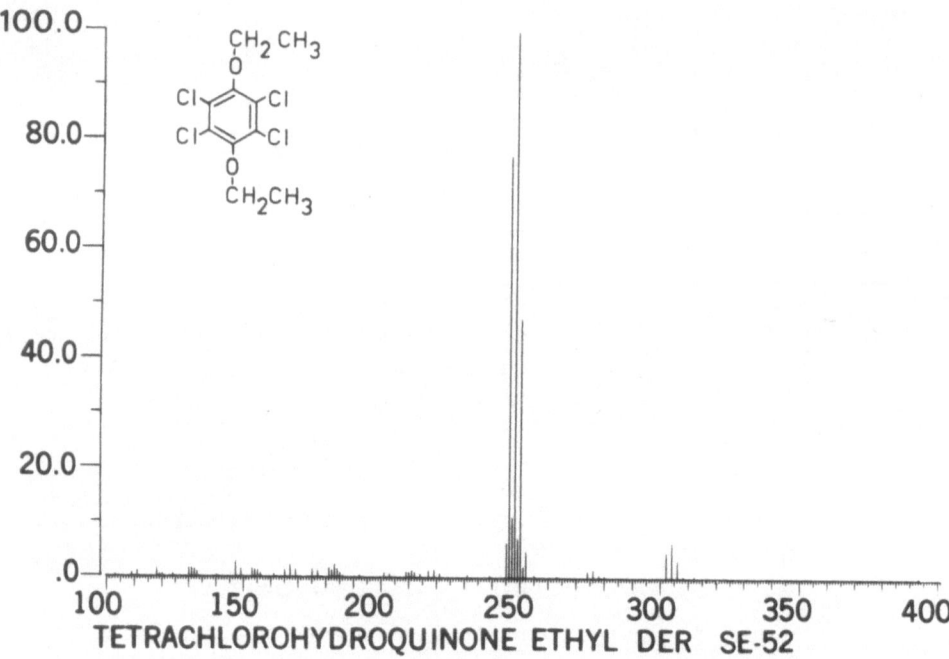

Figure 4. Mass spectrum of tetrachloro-*p*-hydroquinone ethyl ether.

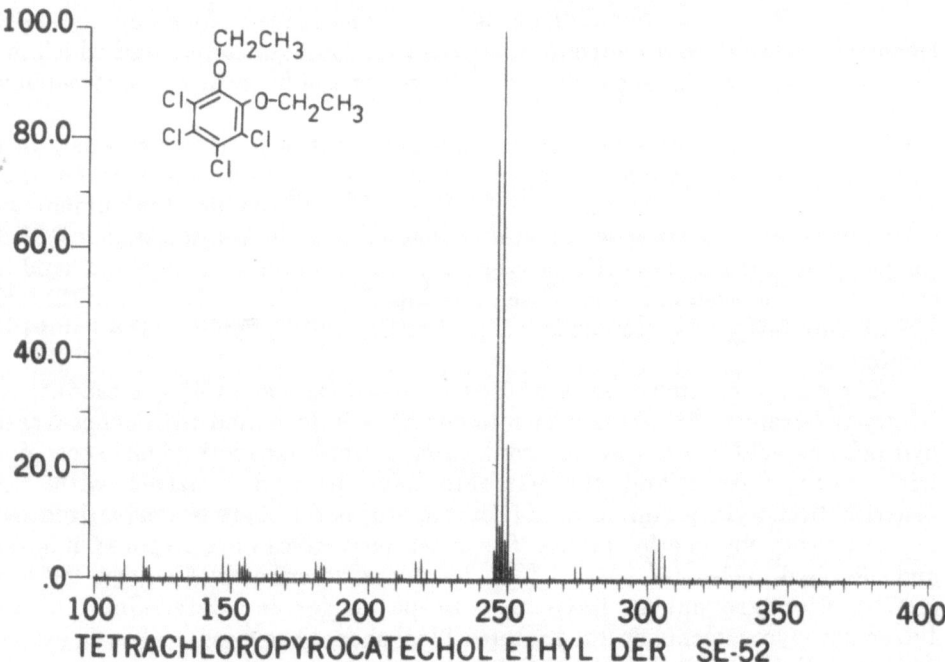

Figure 5. Mass spectrum of tetrachloropyrocatechol ethyl ether.

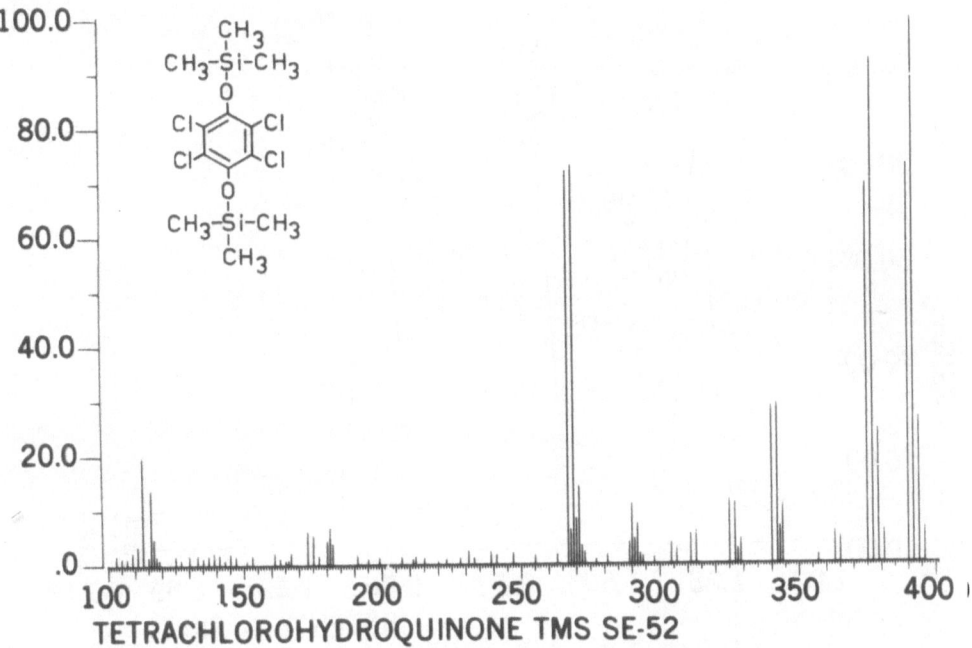

Figure 6. Mass spectrum of tetrachloro-*p*-hydroquinone trimethylsilylether.

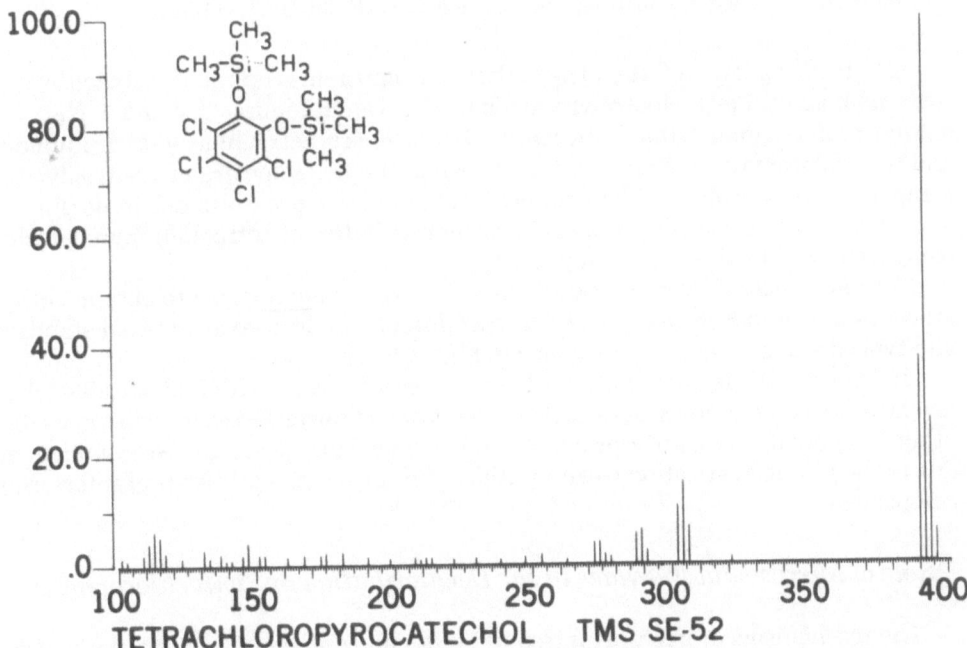

Figure 7. Mass spectrum of tetrachloropyrocatechol trimethylsilylether.

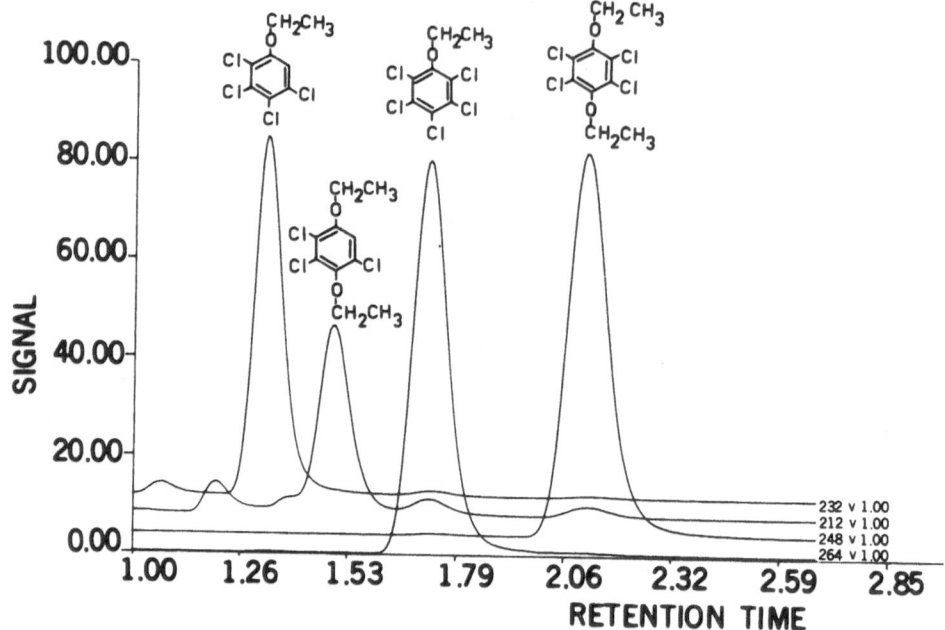

Figure 8. Mass fragmentogram of urine from rats given pentachlorophenol (10 mg/kg) intraperitoneally. Urine boiled with concentrated hydrochloric acid and extracted and derivatized to obtain ethyl ether derivatives. 2,3,4,5-Tetrachlorophenol was used as an internal standard. Mass fragmentogram recorded on LKB 2091/PDP 11/45.

The mass spectra of the ethyl ethers of pentachlorophenol, tetrachloro-*p*-hydroquinone and tetrachloropyrocatechol are given in Figures 3, 4 and 5. Note the similarity in fragmentation between the two isomers, tetrachloro-*p*-hydroquinone and tetrachloropyrocatechol. In contrast, when the mass spectra of the trisilylated compounds are compared, the fragmentation patterns are quite different (Figs. 6 and 7), this has been utilized to exclude the possibility of tetrachloropyrocatechol being a metabolite of pentachlorophenol.

A typical mass fragmentogram of a urine extract derivatized to obtain   ethyl ethers is shown in Figure 8. 2,3,4,5-Tetrachlorophenol is used as internal standard and typical standard curves are shown in Figure 9.

In general, trimethylsilylderivatives are easy to work with and suitable when levels of the compounds studied are not too low. At lower levels interference from other compounds presents a problem and ethylderivatives are to be recommended due to the postderivatization clean-up which diminishes the influence of interfering compounds.

Role of Microsomal Enzymes in the Dechlorination of Pentachlorophenol

The mechanisms of dechlorination of pentachlorophenol seems not to be very well known and we have performed some experiments to achieve a better

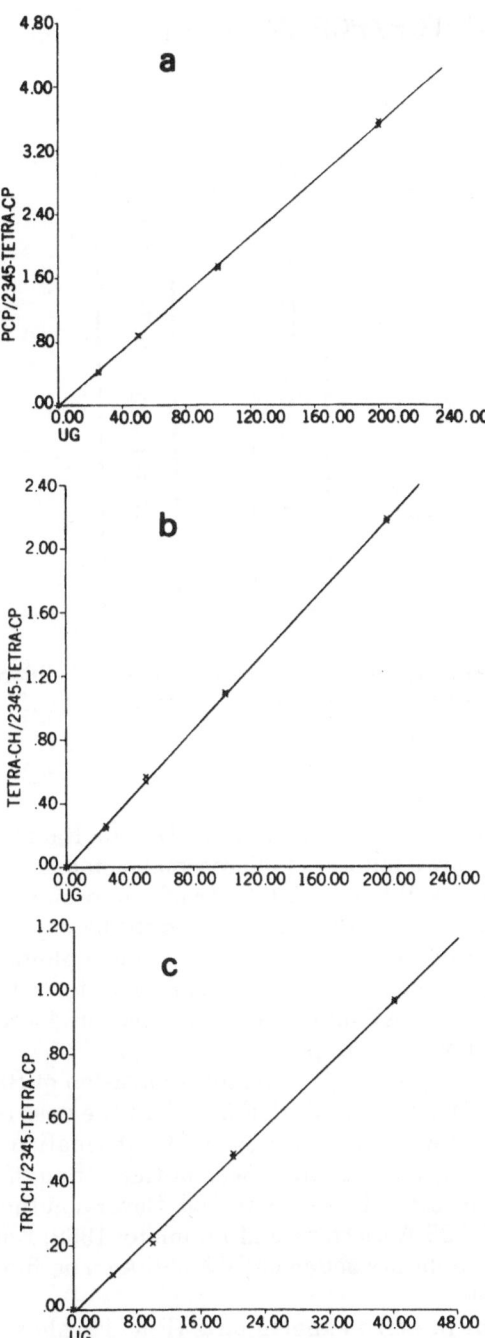

Figure 9. Computer-generated standard curves for a) pentachlorophenol (0-200 μg/ml), and b) tetrachloro-*p*-hydroquinone (0-200 μg/ml), and c) trichloro-*p*-hydroquinone (0-40 μg/ml) when using 2,3,4,5-tetrachlorophenol as an internal standard.

RATIO TCH/PCP IN URINE

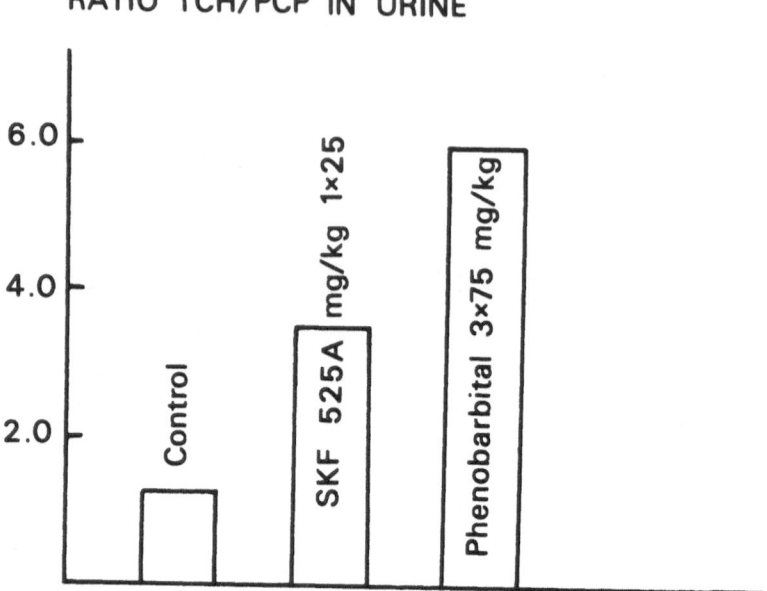

Figure 10. Ratio of tetrachloro-*p*-hydroquinone versus pentachlorophenol in urine from rats given pentachlorophenol (10 mg/kg) intraperitoneally after pretreatment with phenobarbital or SKF 525 A.

understanding of this process (Ahlborg *et al.*, 1977b). Rats were pretreated with phenobarbital, a well known inducer of microsomal metabolizing enzymes, and with SKF 525 A, a similarly well known inhibitor of microsomal enzymes. The result came somewhat as a surprise (Fig. 10). As could be expected, phenobarbital increased the metabolic conversion of pentachlorophenol to tetrachloro-*p*-hydroquinone as demonstrated by an increased ratio of tetrachloro-*p*-hydroquinone versus pentachlorophenol. SKF 525 A, however, also produced a striking increase in the dechlorination of pentachlorophenol.

Further experiments with repeated administration of SKF 525 A (Fig. 11), revealed that the administration of SKF 525 A in the dose of 50 mg/kg every 6 hours was necessary to supress the increase in dechlorination. On the second day, however, a decrease in the metabolism was noticed though no SKF 525 A was administered during the second 24-hour period. Several authors have reported a biphasic effect of SKF 525 A (Serrone and Fujimoto, 1962; Kato *et al.*, 1964). Our data indicate that the inhibitory action of SKF 525 A on dechlorination seems to be of rather short duration.

The data from our *in vitro* experiments (Fig. 12) show an increase in the metabolic conversion of pentachlorophenol to tetrachloro-*p*-hydroquinone when using liver microsomes from rats pretreated with phenobarbital, and a pronounced inhibitory action of SKF 525 A both on the activity of control microsomes and

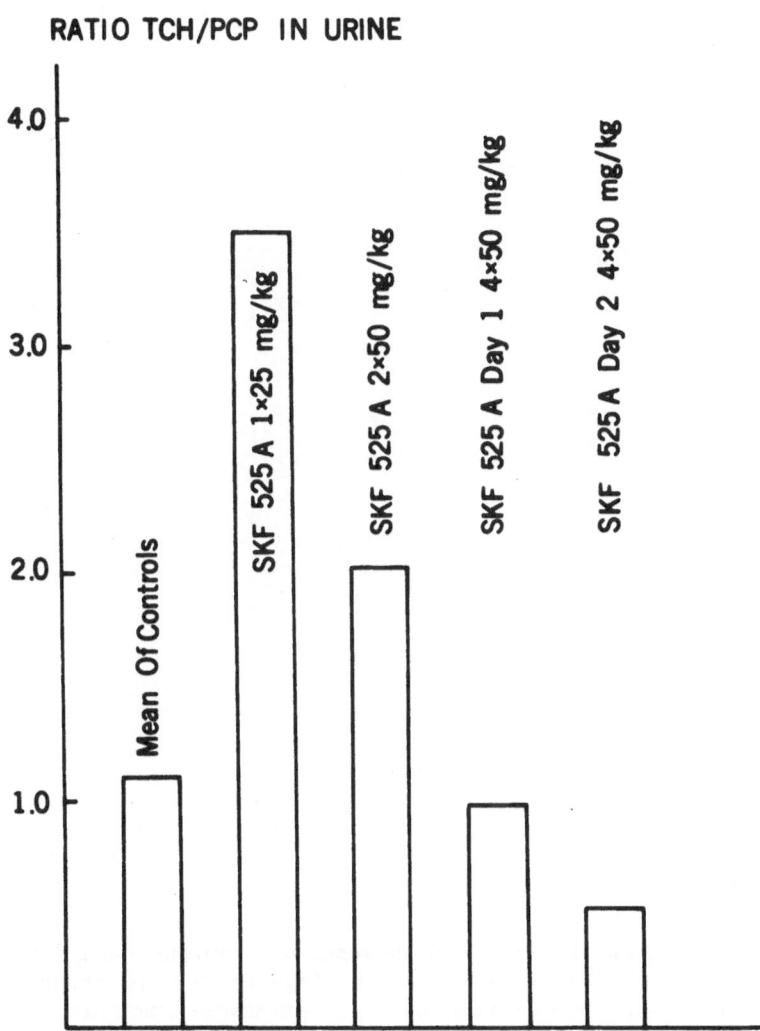

Figure 11. Ratio of tetrachloro-*p*-hydroquinone versus pentachlorophenol in urine from rats given pentachlorophenol (10 mg/kg) intraperitoneally in various combinations with SKF 525 A.

induced microsomes. The lack of inhibition *in vivo* thus seems to be due to the induction being more significant than the inhibition occurring during the time interval studied.

Poland and Glover (1973, 1974) have shown that 2,3,7,8-tetrachlorodibenzo-*p*-dioxin (TCDD) is a potent inducer of e.g. aryl hydrocarbon hydroxylase and that

PER CENT TETRA-CH PRODUCED IN VITRO

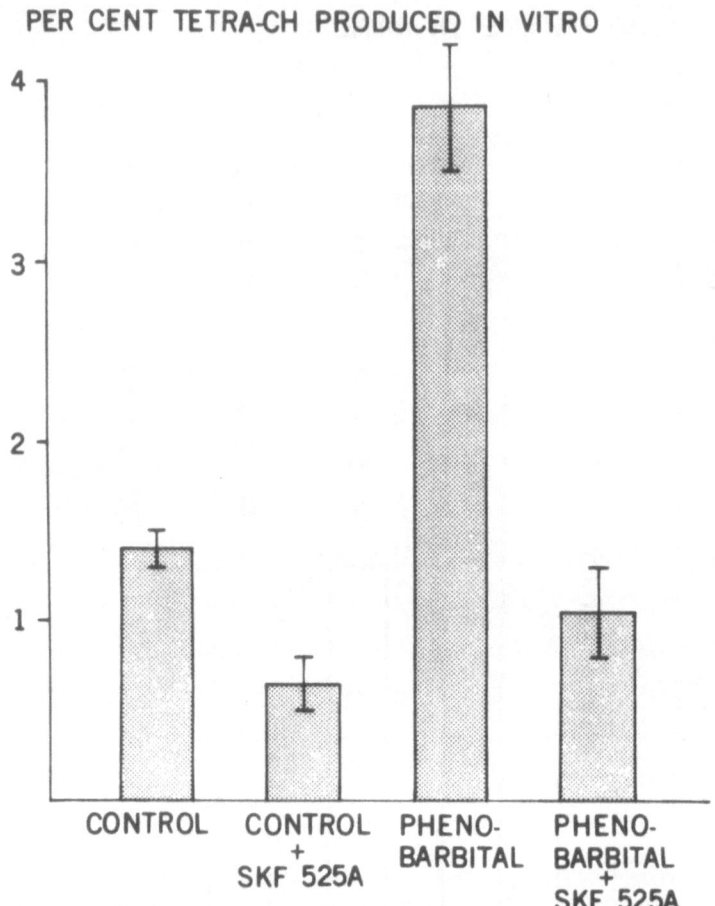

Figure 12. Per cent metabolism of pentachlorophenol to tetrachloro-*p*-hydroquinone in *in vitro* incubations with liver microsomes from rats pretreated with phenobarbital. The inhibition caused by the addition of SKF 525 A to the incubates is indicated in the diagram. (Ahlborg *et al.,* 1977b; Arch. Toxicol., in press).

TCDD is more potent than any other inducer on a dose per unit body weight basis. In a comparison between TCDD and 3-methylcholanthrene (3-MC) they found that the maximal induction that can be achieved was the same for the two compounds. The effect of TCDD was, however, considerably more longlasting.

Using the same dosage of TCDD and 3-MC as did Poland and Glover (1974) we have compared the dechlorination activity *in vivo* and *in vitro* after pretreatment with phenobarbital, TCDD and 3-MC. In the *in vivo*   experiments with rats induced with TCDD we noticed the occurrence of trichloro-*p*-hydroquinone   as an excretion product in the urine from animals dosed with pentachlorophenol. The compound had earlier been tentatively identified in urine (Ahlborg *et al.,* 1974) but was at that

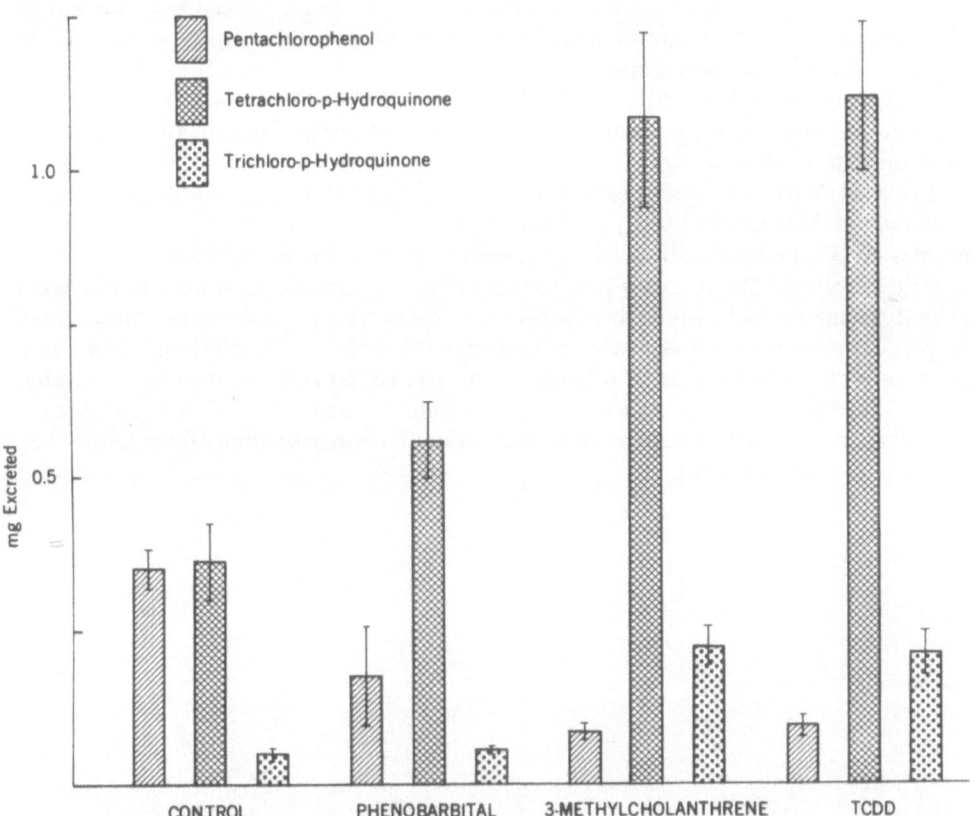

Figure 13. Total amounts excreted (mg) of pentachlorophenol, tetrachloro-*p*-hydroquinone and trichloro-*p*-hydroquinone in urine from rats given pentachlorophenol (10 mg/kg) intraperitoneally after pretreatment with phenobarbital, 3-methylcholanthrene or 2,3,7,8-tetrachlorodibenzo-*p*-dioxin. (Ahlborg and Thunberg, 1977; Arch. Toxicol., in press).

time thought to be due to the presence of tetrachlorophenols occurring as impurities in the pentachlorophenol used. That minor amounts of trichloro-*p*-hydroquinone appear as a metabolite 2,3,4,5- and 2,3,4,6-tetrachlorophenol has also later been verified (Ahlborg and Larsson, 1977).

Trichloro-*p*-hydroquinone could now be demonstrated to occur as a minor metabolite of pentachlorophenol. The pentachlorophenol used in this study has a purity of 99.9% and the fact that the amounts of trichloro-*p*-hydroquinone formed are related to the different inducing pretreatments can only be explained as the result of a continued dechlorination of pentachlorophenol (Ahlborg and Thunberg, 1977).

The total amounts of pentachlorophenol, tetra- and trichloro -*p*- hydroquinone (in mg) excreted in the urine during the first 24 hours after the administration of pentachlorophenol intraperitoneally under the various pretreatment conditions are given in Figure 13. Pretreatment with phenobarbital increased the amount of tetrachloro-*p*-hydroquinone excreted but had no influence on the amount of trichloro-*p*-hydroquinone formed.

Pretreatment with 3-MC or TCDD increased the amount of tetrachloro-*p*-hydroquinone excreted even more, but also increased the amount of trichloro-*p*-hydroquinone.

Pretreatment with phenobarbital did not affect the summarized excretion as compared to the control but pretreatment with 3-MC or TCDD increased the summarized excretion by about 60%, essentially during the first 24 hours.

The results of the *in vitro* incubations (Fig. 14) are in good agreement with those demonstrated *in vivo;* the dechlorination activity being more pronounced when using microsomes from rats pretreated with 3-MC or TCDD than when using microsomes from phenobarbital-pretreated rats. In the *in vitro* experiments SKF 525 A was added to the incubations in a concentration of 10^{-4} M. The inhibitory effect was most pronounced for the incubations with microsomes from control or phenobarbital-treated rats.

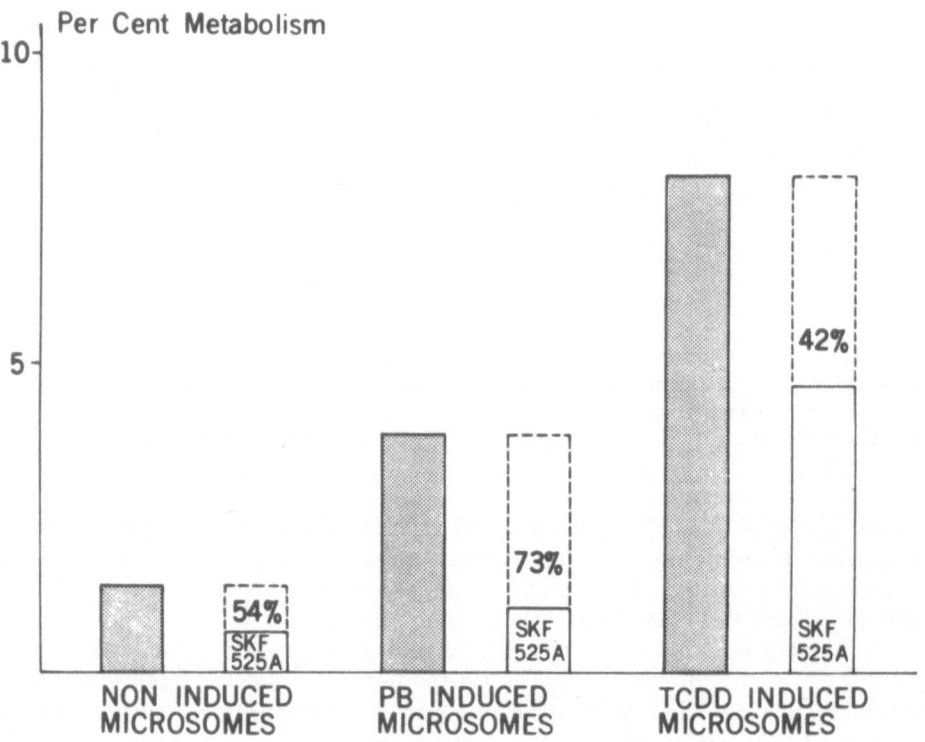

Figure 14. Per cent metabolism of pentachlorophenol to tetrachloro-*p*-hydroquinone in *in vitro* incubations with liver microsomes from rats pretreated with phenobarbital or 2,3,7,8-tetrachlorodibenzo-*p*-dioxin. The inhibition caused by the addition of SKF 525 A to the incubates is indicated in the diagram.(Ahlborg and Thunberg, 1977; Arch. Toxicol., in press).

Conclusions

In our studies we have shown that rapid dechlorination of pentachlorophenol occurs in rats. The dechlorination is mediated by liver microsomal enzymes and their activity can be enhanced by pretreatment with inducing agents such as phenobarbital, 3-methylcholanthrene and TCDD. The dechlorination products formed are tetrachloro-*p*-hydroquinone and trichloro-*p*-hydroquinone. Other reports have indicated the presence of dechlorination products such as tetrachloro- and trichlorophenols (Engst *et al.*, 1976) and tetrachloro-*p*-benzoquinone (Tashiro *et al.*, 1970). We have not been able to verify these findings.

The pharmacokinetics of pentachlorophenol have been studied in rats and monkeys (Braun and Sauerhoff, 1976). From their studies it is evident that rats and mice excrete pentachlorophenol more rapidly than monkeys. In addition, no dechlorination seemed to occur in the monkey. However, there is no evidence available supporting that pentachlorophenol will cause cumulative toxicity in mammals with repeated exposure to moderate doses.

A recent study in Sweden (Ahling *et al.*, 1977) has surveyed the content of chlorinated dioxins, dibenzofurans and chlorinated 2-phenoxyphenols (predioxins) in commercial chlorophenol preparations on the Swedish market. The preparations have also been studied in pyrolysis experiments, and these experiments indicate that burning of chlorophenol-preserved wood may produce considerable amounts of chlorinated dioxins. This in combination with the knowledge that the commercial preparations have been handled carelessly with regard to work hygiene, has recently led to the withdrawal of the licensing for wood preservatives based on chlorophenols in Sweden. The Swedish use of chlorophenol formulations, however, is restricted to wood preservation and the chlorophenols have been judged to be replaceable with other fungicides such as arsenics, fluorides and benzimidazoles.

ACKNOWLEDGMENTS

This work was supported by grants from the Swedish Environment Protection Board (7-1/76 and PK-345/76).

References

Ahlborg, U.G., and K. Larsson, 1977. Metabolism of tetrachlorophenols in the rat. *Arch. Toxicol.*, in press.

Ahlborg, U.G., K. Larsson, and T. Thunberg, 1977b. Metabolism of pentachlorophenol *in vivo* and *in vitro*. *Arch. Toxicol.*, in press.

Ahlborg, U.G., J.E. Lindgren, and M. Mercier, 1974. Metabolism of pentachlorophenol. *Arch. Toxicol.*, **32**: 271-281.

Ahlborg, U.G., E. Manzoor, and T. Thunberg, 1977a. Inhibition of β-glucoronidase by chlorinated hydroquinones and benzoquinones. *Arch. Toxicol.*, **37**: 81-87.

Ahlborg, U.G., and T. Thunberg, 1977. Effect of 2,3,7,8-tetrachlorodibenzo-*p*-dioxin on the *in vivo* and *in vitro* dechlorination of pentachlorophenol. *Arch. Toxicol.*, in press.

Ahling, B., B. Jansson, and G. Sundstrom, 1977. Preliminary report to the Swedish Environmental Protection Board.

Barthel, W.F., A. Curley, C.L. Trasher, V.A. Sedlak, and R. Armstrong, 1969. Determination of pentachlorophenol in blood, urine, tissue and clothing. *J. Ass. Offic. Anal. Chem.*, **52**: 294-298.

Betts, J.J., S.P. James, W.V. Thorpe, 1955. The metabolism of pentachloronitrobenzene and 2,3,4,6-tetrachloronitrobenzene and the formation of mercapturic acids in the rabbit. *Biochem. J.*, **61**: 611-617.

Bevenue, A., and H. Beckman, 1967. Pentachlorophenol: A discussion of its properties and its occurrence as a residue in human and animal tissues. *Residue Rev.*, **19**: 83-145.

Bevenue, A., J.R. Wilson, E.F. Potter, and K. Moon, 1966. A method for the determination of pentachlorophenol in human urine in picogram quantities. *Bull. Environ. Contam. Toxicol.*, **1**: 257-266.

Braun, W.H., and M.W. Sauerhoff, 1976. The pharmacokinetic profile of pentachlorophenol in monkeys. *Toxicol. Appl. Pharmacol.*, **38**: 525-533.

Cranmer, M., and J. Freal, 1970. Gas chromatographic analysis of pentachlorophenol in human urine by formation of alkyl ethers. *Life Sci.*, **9**: 121-128.

Deichmann, W., W. Machle, K.V. Kitzmiller, and G. Thomas, 1942. Acute and chronic effects of pentachlorophenol and sodium pentachlorophenate upon experimental animals. *J. Pharmacol. Exp. Therap.*, **76**: 104-117.

Elkin, K., L. Pierrou, U.G. Ahlborg, B. Holmstedt, and J.E. Lindgren, 1973. Computer-controlled mass fragmentography with digital signal processing. *J. Chromatogr.*, **81**: 47-55.

Engst, R., R.M. Macholz, M. Kujawa, H.J. Lewerenz, and R. Plass, 1976. The metabolism of lindane and its metabolites gamma-2,3,4,5,6-pentachlorocyclohexene, pentachloro-benzene and pentachlorophenol in rats and the pathways of lindane metabolism. *J. Environ. Sci. Hlth.*, **11B**: 95-117.

Jacobson, I., and S. Yllner, 1971. Metabolism of ^{14}C-pentachlorophenol in the mouse. *Acta Pharmacol. Toxicol.*, **29**: 513-524.

Johnson, R.L., P.J. Gehring, R.J. Kociba, and B.A. Schwetz, 1973. Chlorinated dibenzodioxins and pentachlorophenol. *Environ. Hlth. Persp.*, **5**: 171-176.

Kato, R., E. Chiesara, and P. Vassahelli, 1964. Further studies on the inhibition and stimulation of microsomal drug-metabolizing enzymes of rat liver by various compounds. *Biochem. Pharmacol.*, **13**: 69-83.

Poland, A., and E. Glover, 1973. Chlorinated dibenzo-*p*-dioxins: Potent inducers of delta-aminolevulinic acid synthetase and aryl hydrocarbon hydroxylase. *Mol. Pharmacol.*, **9**: 736-747.

Poland, A., and E. Glover, 1974. Comparison of 2,3,7,8-tetrachlorodibenzo-*p*-dioxin, a potent inducer of aryl hydrocarbon hydroxylase, with 3-methylcholanthrene. *Mol. Pharmacol.*, **10**: 349-359.

Rappe, C., and C.A. Nilsson, 1972. An artifact in the gas chromatographic determination of impurities in pentachlorophenol. *J. Chromatogr.*, **67**: 247-253.

Rivers, J., 1972. Gas chromatographic determination of pentachlorophenol in human blood and urine. *Bull. Environ. Contam. Toxicol.*, **8**: 294-296.

Rudling, L., 1970. Determination of pentachlorophenol in organic tissues and water. *Water Res.*, **4**: 533-537.

Serrone, D.M., and J.M. Fujimoto, 1962. The effect of certain inhibitors in producing shortening of hexobarbital action. *Biochem. Pharmacol.*, **11**: 609-615.

Tashiro, S., T. Sasmoto, T. Aikawa, S. Tokunaga, E. Taniguchi, and E. Morifusa, 1970. Metabolism of pentachlorophenols in mammals. *J. Agric. Chem. Soc. Japan*, **44**: 124-129. (in Japanese)

Pentachlorophenol in Different Species of Vertebrates after Administration of Hexachlorobenzene and Pentachlorobenzene

G. KOSS and W. KORANSKY

Abstract—Different animal species (rat, mouse, guinea pig, laying hen, rainbow trout) were found to convert hexa- and pentachlorobenzene to pentachlorophenol. Other major metabolites were shown to be tetrachlorohydroquinone and sulphur containing compounds.

Pentachlorophenol as well as the other metabolites were detected not only in the excreta but also in the tissues of animals treated with hexa- and pentachlorobenzene.

The possible role of both chlorobenzenes for the contamination of man with pentachlorophenol and the toxicologic relevance of its formation are discussed.

Introduction

There is increasing concern today about environmental contamination with pentachlorophenol. One of the major questions rising out of the debate on this chemical is focused on its origin. The utilization of pentachlorophenol in the wood industry and in plant protection (Melnikov, 1971) is generally recognized as a main source of environmental contamination. Another source, we believe, is the biotransformation of hexachlorobenzene and pentachlorobenzene.

G. KOSS and W. KORANSKY • Institut für Toxikologie und Pharmakologie, Philipps-Universität, Pilgrimstein 2, D-3550 Marburg, Federal Republic of Germany.

Metabolism of Hexachlorobenzene

Studies on the metabolism of hexachlorobenzene in rats (Koss *et al.*, 1976; Mehendale *et al.*, 1975; Lui and Sweeney, 1975; Engst *et al.*, 1976a) and in monkeys (Rozman *et al.*, 1975) showed pentachlorophenol to be a metabolite of this chlorobenzene.

In our experiments with labeled hexachlorobenzene we obtained evidence for an unexpectedly high extent of its conversion to metabolites. Four weeks after administration, 7% of the radioactivity was excreted via the kidneys and 27% with the feces. As shown in Figure 1, nearly the total of label in the urine was contained in metabolites of hexachlorobenzene, while 69% of the radioactivity in the feces was represented by the unchanged drug. From the data in Figure 1 it was calculated that the rat eliminates almost half of the amount of hexachlorobenzene in the form

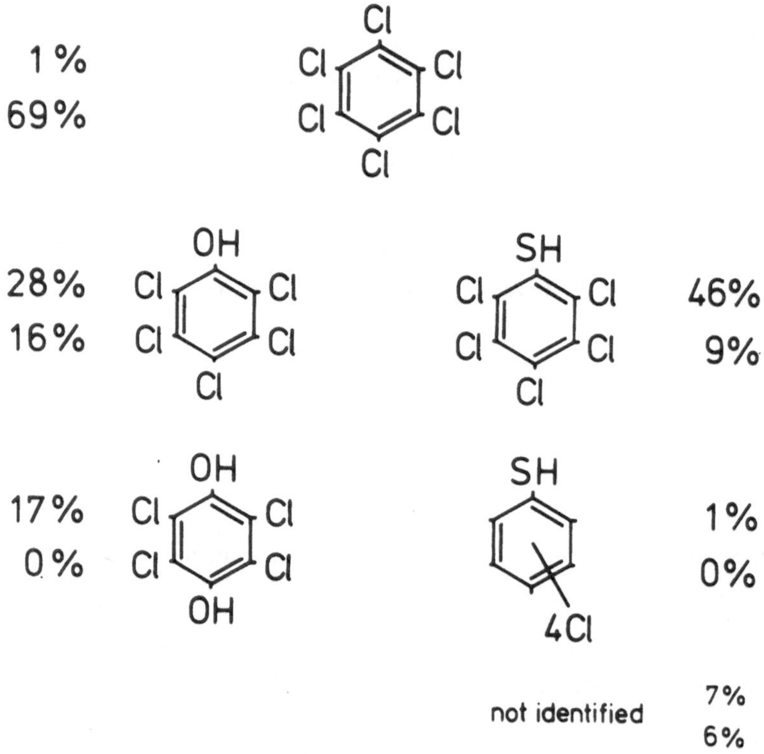

Figure 1. Relative content of hexachlorobenzene and its metabolites in the urine and feces of female rats dosed intraperitoneally with 1.42 mMol ^{14}C-hexachlorobenzene/Kg. The excreta were collected for 4 weeks after administration. The amount of label contained in hexachlorobenzene and its metabolites was expressed as a percentage of the radioactivity excreted with the urine (upper figures) and feces (lower figures) over a period of 4 weeks after administration. Based on data from Koss *et al.*, 1976.

cf metabolites including pentachlorophenol which accounts for about 1/5 of the total excreted. The other major metabolites identified are tetrachlorohydroquinone and pentachlorothiophenol. Both substances represent 3% and 16%, respectively, of the label excreted.

Other animal species treated with hexachlorobenzene were also found to convert this substance to hydrophilic metabolites (Koss *et al.*, 1977). In the excreta of guinea pigs and laying hens pentachlorophenol, tetrachlorohydroquinone and pentachlorothiophenol were detected. Mouse and rainbow trout excreted mainly pentachlorophenol. In the feces of the Japanese quail, however, only pentachlorothiophenol was found in measurable amounts.

Aside from the excreta, metabolites of hexachlorobenzene were also detected in the tissues of the experimental animal. By means of combined gas chromatography — mass spectrometry, pentachlorophenol was identified in the livers of rats treated with hexachlorobenzene (Fig. 2). Evidence for the occurrence of tetrachlorohydroquinone and pentachlorothiophenol in this organ was also obtained.

Continuous administration even of very low doses of hexachlorobenzene resulted in measurable amounts of pentachlorophenol. Its content in the blood was measured to be tenfold less than the concentration of hexachlorobenzene (Fig. 3).

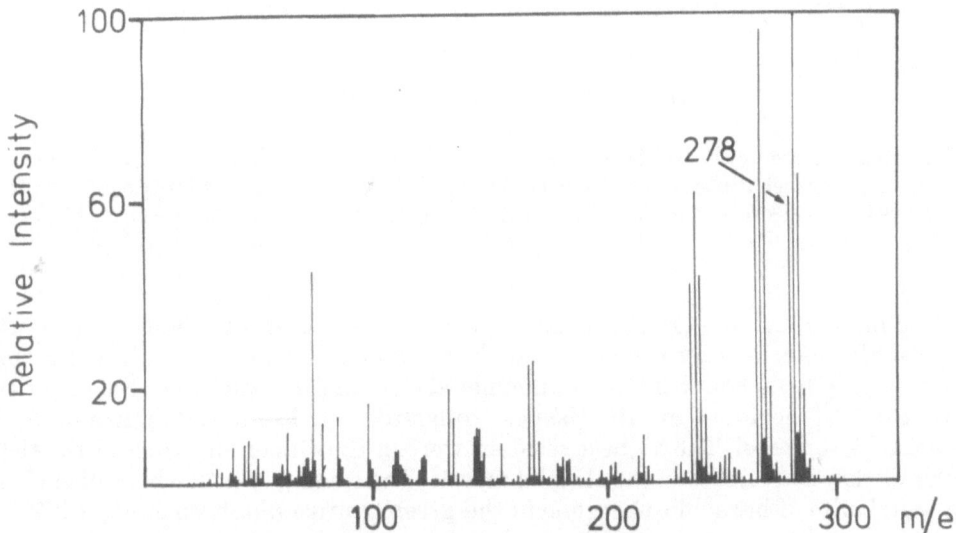

Figure 2. Mass spectrum of pentachlorophenol (as methylether) detected in the liver of female rats dosed with 114 μM hexachlorobenzene/Kg. This mass spectrum is identical with that of authentic pentachlorophenyl methyl ether. The drug dissolved in olive oil was administered by gastric tube twice a week over a period of 16 weeks.

Instrument: LKB 2091 Gas chromatograph — Mass spectrometer equipped with a PDP 11 computer.

GC conditions: glass column (270 × 2.2 mm i.d.) with 3% QF-1 on 100-120 mesh Gas chrom Q, oven temperature 150 °C, injection port temperature 220 °C, carrier gas helium.

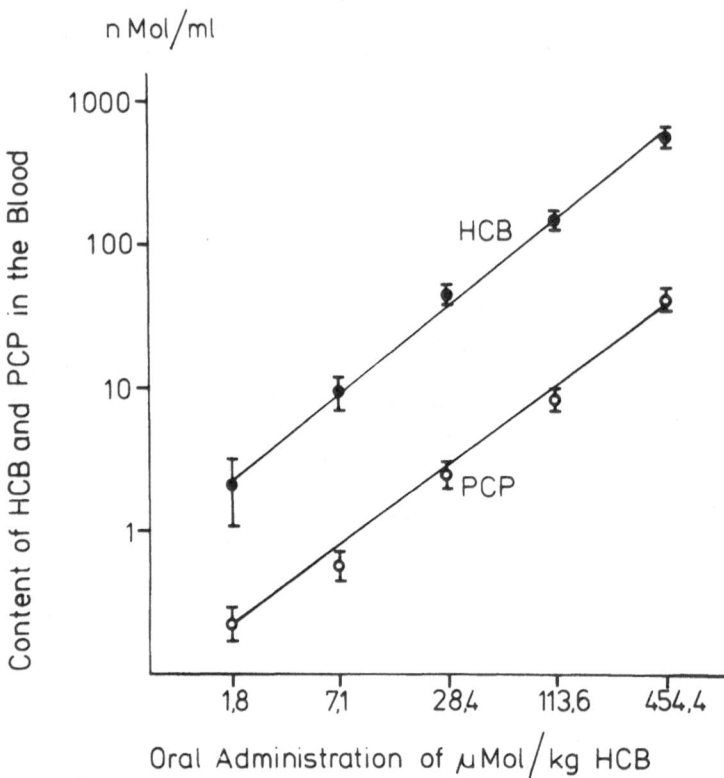

Figure 3. Content of hexachlorobenzene and pentachlorophenol in the blood of female rats dosed with hexachlorobenzene. The drug dissolved in olive oil was administered by gastric tube twice a week over a period of 16 weeks. Each point represents the mean ± S.D. of at least 3 animals.

Sanborn *et al.* (1977) studied the uptake and elimination of ^{14}C-hexachlorobenzene by the green sunfish, *Lepomis cyanellus*. Following the ingestion of hexachlorobenzene contaminated (1,10 and 10 ppm) food, the majority of the ^{14}C residue in all tissues consisted of hexachlorobenzene and pentachlorophenol. The highest residues were in the alimentary tract of the fish and the lowest residues in the skeletal muscle. The finding of pentachlorophenol as a metabolite of hexachlorobenzene in the green sunfish (Sanborn *et al.*, 1977) is consistent with our findings in the rainbow trout.

Metabolism of Pentachlorobenzene

To answer the question whether or not pentachlorobenzene contributes to environmental contamination with pentachlorophenol, it was necessary to find out whether this chlorobenzene is converted to pentachlorophenol to a significant extent. Parke and Williams (1960) and Engst *et al.* (1976b) detected pentachlorophenol in the excreta of the animals treated with pentachlorobenzene. Quantitative data, however, were not reported.

We examined the urine, feces and tissues of rats after administration of pentachlorobenzene for their content of this drug and its metabolites (Koss and Koransky, manuscript in preparation). The data obtained in this study documented that pentachlorobenzene undergoes an almost complete biodegradation. Only 3% of the compound administered to the rats was excreted in its unchanged form, while by far the major portion was eliminated as hydrophilic metabolites, including pentachlorophenol, which accounted for about 9% (Fig. 4). Other metabolites identified were 2,3,4,5-tetrachlorophenol, tetrachlorohydroquinone, and a hydroxylated chlorothiocompound. In addition to these substances traces of another isomer of tetrachlorophenol were detected in the urine and feces. Examination of

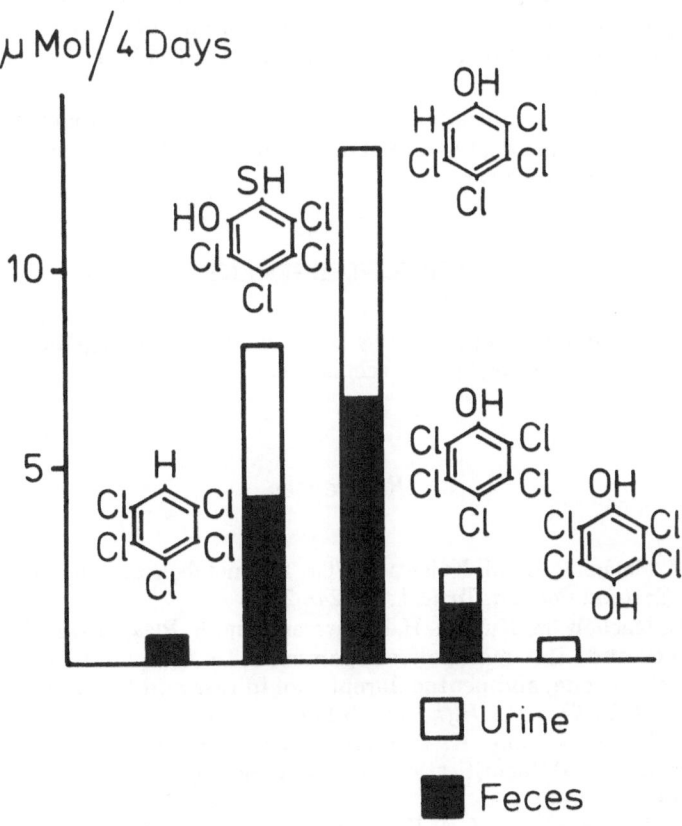

Figure 4. Content of pentachlorobenzene and its metabolites in the urine and feces of female rats dosed with pentachlorobenzene. A single dose of 403 μM pentachlorobenzene/Kg dissolved in olive oil was administered intraperitoneally and the excreta were collected for a period of 4 days. (n = 3)

the animal for the metabolites which were detected in the excreta showed pentachlorophenol, tetrachlorohydroquinone, tetrachlorophenol, and the hydroxylated chlorothiocompound to be present in its tissues.

Conclusions

Our findings confirm that there are indirect sources of environmental contamination with pentachlorophenol. Among them are hexa- and pentachlorobenzene as important parent compounds for the formation of pentachlorophenol. Its amount in the organism deriving from the biotransformation of both chlorobenzenes is probably smaller than the amount of pentachlorophenol originating from its utilization in the wood industry and plant protection. It is important, however, to consider that the metabolite is formed where the possible toxic effects occur. In this connection we would like to point to the high residues of hexachlorobenzene in the human body and in human milk, which may lead to a formation of pentachlorophenol not only in the adult but also in the newborn child.

In addition to pentachlorophenol, it also appears to be important to take into consideration the formation and spread of the toxic sulphur containing metabolites of hexa- and pentachlorobenzene.

ACKNOWLEDGEMENT

This investigation was supported by Deutsche Forschungsgemeinschaft and by Bundesministerium für Forschung und Technologie.

References

Engst, R., R.M. Macholz, and M. Kujawa, 1976a. The metabolism of hexachlorobenzene in rats. *Bull. Environ. Contam. Toxicol.,* **16:** 248-252.

Engst. R., R.M. Macholz, M. Kujawa, H.J. Lewerenz, and R. Plass, 1976b. The metabolism of lindane and its metabolites gamma-2,3,4,5,6-pentachlorocyclohexene, pentachlorobenzene, and pentachlorophenol in rats and the pathways of lindane metabolism. *J. Environ. Sci. Hlth.,* **11:** 95-117.

Koss, G., W. Koransky and K. Steinbach, 1976. Studies on the toxicology of hexachlorobenzene. II. Identification and determination of metabolites. *Arch. Toxicol.,* **35:** 107-114.

Koss, G., J.J.T.W.A. Strik, and C.A. Kan, 1977. Metabolites of hexachlorobenzene in the excreta of different animal species. Abstract presented at the International Conference on *in vivo* aspects of Biotransformation and Toxicity of Industrial and Environmental Xenobiotics. Prague, Sept. 13-15, 1977.

Lui, H., and G.D. Sweeney, 1975. Hepatic metabolism of hexachlorobenzene in rats. *FEBS Letters,* **51:** 225-226.

Mehendale, H.W., M. Fields, H.B. Matthews, 1975. Metabolism and effects of hexachlorobenzene on hepatic microsomal enzymes in the rat. *J. Agric. Food Chem.*, **23:** 261-264.

Melnikov, N.N., 1971. Chemistry of pesticides. *Residue Rev.*, **36:** 89-110.

Parke, D.V., and R.T. Williams, 1960. Studies in detoxication. 81. The metabolism of halogenobenzenes. Penta- and hexachlorobenzenes. Further observations on 1 : 3 : 5 - trichlorobenzene. *Biochem. J.*, **74:** 5 - 9.

Rozman, K., W. Mueller, M. Iatropoulos, F. Coulston, F. Korte, 1975. Ausscheidung, Koerperverteilung und Metabolisierung von Hexachlorbenzol nach oraler Einzeldosis in Ratten und Rhesusaffen. *Chemosphere,* **5:** 289-298.

Sanborn, J.R., W.F. Childers, and L.G. Hansen, 1977. Uptake and elimination of [^{14}C] Hexachlorobenzene (HCB) by the green sunfish, *Lepomis cyanellus* Raf., after feeding contaminated food. *J. Agr. Food Chem.*, **25:** 551-553.

II

Pharmacology and Toxicology of Pentachlorophenol

B. Comparative Toxicology and Pharmacology

Toxicity of Pentachlorophenol and Related Compounds to Early Life Stages of Selected Estuarine Animals

PATRICK W. BORTHWICK and STEVEN C. SCHIMMEL

Abstract—Newly hatched individuals of four estuarine species were exposed to pentachlorophenol (PCP), sodium pentachlorophenate (Na-PCP), or Dowicide® G (79% Na-PCP), in static toxicity tests.

The 96-hour LC_{50} values for sheepshead minnow *(Cyprinodon variegatus)* fry exposed to PCP at ages 1-day, 2-wk, 4-wk, and 6-wk were 329, 392, 240, and 223 $\mu g/l$, respectively. The 96-hr LC_{50} value for 2-wk-old fry exposed to Dowicide® G was 516 $\mu g/l$. The larvae (48-hr post hatch) of pinfish, *Lagodon rhomboides,* were particularly sensitive to Na-PCP (96-hr LC_{50}: 38 $\mu g/l$) and Dowicide® G (96-hr LC_{50}: 66 $\mu g/l$). For 24-hr-old grass shrimp *(Palaemonetes pugio)* larvae exposed to Na-PCP the 96-hr LC_{50} was 649 $\mu g/l$. Na-PCP caused abnormal development of eastern oyster *(Crassostrea virginica)* embryos, the 48-hr EC_{50} being 40 $\mu g/l$.

Introduction

Pentachlorophenol (PCP) and its water soluble sodium salt, sodium pentachlorophenate (Na-PCP), are broad spectrum biocides used in an array of pesticide formulations such as Dowicide® G (79% Na-PCP). Although there have been numerous reports on the toxicity of these compounds to freshwater organisms,

Contribution No. 343 from the Environmental Research Laboratory, Gulf Breeze.

Mention of registered tradenames does not constitute endorsement by the United States Environmental Protection Agency.

PATRICK W. BORTHWICK and STEVEN C. SCHIMMEL • United States Environmental Protection Agency, Environmental Research Laboratory, Sabine Island, Gulf Breeze, Florida 32561, U.S.A.

little is known about their effects on estuarine and marine organisms. The latter is of particular interest because of the use of PCP formulations in oil and gas exploration. Pentachlorophenol and its sodium salt are used in drilling muds and well completion fluids to control bacterial growth (Robichaux, 1975), and Dowicide® G is used as a preservative in drilling fluids (Shaw, 1975). The impact of such biocides, used in land-based, coastal, and offshore oil and gas well-drilling operations, on estuarine and marine animals is not yet established. Further, we lack sufficient monitoring information, and data on the quantities used, point source disposal, and dispersal patterns of PCP formulations used in the coastal and marine environment. Land (1974) noted that microorganisms in drilling fluids are less sensitive than fish to Dowicide® G and proposed the use of formaldehyde as an alternative. Zitko (1975) suggested that PCP be eliminated as a bactericide in well-drilling operations due to the presence of highly toxic impurities (e.g. dioxins) and their acute toxicity to aquatic animals.

Because of its consistent toxic effect, sodium pentachlorophenate was selected by Davis and Hoos (1975) as an ideal reference toxicant for static freshwater salmonid bioassays: 96-hr LC_{50} values reported for juvenile rainbow trout *(Salmo gairdneri)*, coho salmon *(Oncorhynchus kisutch)*, and sockeye salmon *(Oncorhynchus nerka)* ranged from 37 to 130 μg/l (37 to 130 ppb).

The purpose of our study was to determine the toxicity of PCP and related compounds to sheepshead minnow *(Cyprinodon variegatus)* fry, prolarval pinfish *(Lagodon rhomboides)*, grass shrimp *(Palaemonetes pugio)* larvae, and developing embryos of the eastern oyster *(Crassostrea virginica)*.

Developmental stages are selected for testing because they are often more sensitive than adult stages to toxic substances. Planktonic larvae and demersal fish embryos and fry usually have inadequate mobility to avoid polluted habitats. Further, small individuals are optimal for static toxicity tests.

Experimental Animals and Bioassay Procedures

Acute 96-hr static toxicity tests were conducted following standard methods (American Public Health Association, 1976). Toxicity tests were performed using pentachlorophenol (analytical grade, Baker), sodium pentachlorophenate (analytical grade, City Chemical Company), and Dowicide® G (Dow Chemical Company). Stock solutions of PCP were made in triethylene glycol and diluted to the desired concentrations with filtered (5 μm) seawater. Solutions of Na-PCP and Dowicide® G were made by dissolving them in filtered seawater. Jars containing the exposure media and test animals were placed randomly in a temperature controlled recirculating water bath under 12 hr light: 12 hr dark conditions. The salinity and temperature at which the tests were performed varied with the species and this information is summarized in Table 1.

Sheepshead minnows were seined from tidal ditches and ponds on Santa Rosa Island, Florida, and adapted to exposure conditions for two weeks in flowing seawater which was diluted to 10°/₀₀ salinity with deionized water and heated to

30 °C. Brood stock were fed a diet of frozen adult *Artemia* and Biorell® Tropical Flake Food daily.

Sheepshead minnow fry were obtained by methods outlined by Schimmel *et al.* (1974). Adult females adapted to laboratory conditions were injected intraperitoneally with 50 I.U. of human chorionic gonadotropin to insure egg maturation. Manually stripped ova were incubated for 30 minutes in seawater with macerated testes taken from adult male fish to produce embryos, and they were tested at specified post-hatch ages (1 day, 2 wk, 4 wk, and 6 wk). Sheepshead minnow fry were fed live *Artemia* nauplii daily until their use in toxicity tests. Each bioassay consisted of five logarithmic toxicant concentrations (100 to 1000 μg/l) and appropriate controls. Sheepshead minnow fry were transferred to clean jars containing freshly prepared media, and observed every 24 hours to determine mortality or aberrant behavior.

Prolarval pinfish (48-hr post-hatch) used in the toxicity tests were obtained by using the method described by Schimmel (1977). The concentrations of Na-PCP used were: 5.6, 10.0, 18.0, 32.0, 56.0 and 100 μg/l; for Dowicide® G: 18.0, 32.0, 56.0, 100 and 180 μg/l. Pinfish larvae were not fed during the 96-hr test.

Grass shrimp larvae were obtained following the method described by Tyler-Schroeder (1976). Grass shrimp larvae (< 24 hr) were exposed to Na-PCP for 96 hours under static conditions. The larvae were fed brine shrimp daily, approximately 10 brine shrimp/larva.

Spawning was induced in the eastern oysters under laboratory conditions by an elevation of temperature to 32 °C and by the addition of 'stripped' spermatozoa to the medium. The methods of Woelke (1972) were used for the 48-hr exposure of oyster larvae to Na-PCP. The concentrations of Na-PCP tested were: 3.2, 10.0, 32.0 and 100 μg/l. After the 48-hr exposure the embryos were examined for abnormalities in development to the straight-hinged stage. Larvae were considered normal if their shells were fully formed.

Probit analysis (Finney, 1971) was used to determine the 96-hr LC_{50} value, using death as the quantal response for sheepshead minnows, pinfish and grass shrimp. Similarly, the 48-hr EC_{50} value was determined for eastern oyster embryos using abnormal development of experimental embryos as the quantal response.

Toxicity of PCP and Related Compounds

The acute toxicity of PCP and related compounds to laboratory-reared sheepshead minnow fry at various ages is shown in Table 1. The sheepshead minnow fry of 4- and 6-weeks age appeared to be more sensitive to PCP than did the day-old and 2-week-old individuals. Death of sheepshead minnow fry generally occurred during the first 24 hours of exposure. Death was characterized by laterally extended opercula and reddened areas around the pectoral fins. Aberrant behavior included hyperventilation and sluggish response to mechanical stimuli.

For 2-wk-old fry exposed to Dowicide® G, the 96-hr LC_{50} value was 516 (374-887) μg/l, while a 96-hr LC_{50} of 392 (307-489) μg/l was obtained with PCP. These

TABLE 1. Toxicity of PCP, Na-PCP and Dowicide® G to Early Life Stages of Estuarine Animals

Species	Age and stage	Chemical	96-hr LC$_{50}$ µg/l (ppb)	95% C.I.	Temp. °C	Salinity ‰
Cyprinodon variegatus (sheepshead minnow)	1-day fry	PCP	329[a]	(305-360)	30	10
	2-wk fry	PCP	392	(307-489)	30	10
		Dowicide® G	516	(374-887)	30	24
	4-wk fry	PCP	240	(192-284)	30	10
	6-wk fry	PCP	223	(163-291)	30	10
Lagodon rhomboides (pinfish)	48-hr prolarvae	Na-PCP	38	(26-57)	20	26
		Dowicide® G	66	(48-94)	20	26
Palaemonetes pugio (grass shrimp)	24-hr larvae	Na-PCP	649	(494-915)	25	24
Crassostrea virginica (eastern oyster)	developing embryos	Na-PCP	40[b]	(36-44)	25	17

[a] The values given are based on nominal concentrations.
[b] 48-hr EC$_{50}$ value for abnormal embryonic development.

differences may have been due to the fact that only 79% of the former formulation is Na-PCP while the latter is an analytical grade PCP with less contamination.

Parrish *et al.* (1977) reported the 96-hr LC_{50} and 95% confidence limits for juvenile (10-20 mm) sheepshead minnows exposed to PCP in flowing seawater to be 442 (308-635) $\mu g/l$. These values are somewhat close to the LC_{50} values obtained in the present investigation for 1-day-old and 2-week-old sheepshead minnow fry exposed to PCP for 96 hours.

Pinfish prolarvae (48-hr-old) were particularly sensitive to Na-PCP and Dowicide® G, the 96-hr LC_{50} values being 38 (26-57) and 66 (48-94) $\mu g/l$, respectively.

The 96-hr LC_{50} for laboratory-reared grass shrimp larvae exposed to Na-PCP was 649 (494-915) $\mu g/l$. Eastern oyster embryos exposed for 48 hours to Na-PCP exhibited abnormal development, the 48-hr EC_{50} being 40 (36-44) $\mu g/l$. Woelke (1972) exposed Pacific oyster *(Crassostrea gigas)* larvae to Na-PCP. Based on his tabular data on abnormal embryonic development the 48-hr EC_{50} appears to be 48 $\mu g/l$.

In conclusion, our study indicates that PCP and related compounds are toxic to developmental stages of several estuarine animals. Among the individuals tested the prolarval pinfish are extremely sensitive to Na-PCP (LC_{50}: 38 $\mu g/l$). At a similar concentration, Na-PCP caused abnormal development in eastern oysters. Na-PCP and PCP appear to be more toxic than Dowicide® G to prolarval pinfish and sheepshead minnow fry.

ACKNOWLEDGMENTS

The authors wish to thank Steven S. Foss, James M. Patrick, Jr., and Monte R. Tredway for assistance in conducting toxicity tests and manuscript preparation.

References

American Public Health Association, 1976. *Standard Methods for the Examination of Water and Wastewater.* 14th ed., pp. 685-740., American Public Health Association, Washington, D.C.

Davis, C., and A.W. Hoos, 1975. Use of sodium pentachlorophenate and dehydroabietic acid as reference toxicants for salmonid bioassays. *J. Fish. Res. Bd. Can.,* **32:** 411-416.

Finney, D.J., 1971. *Probit Analysis,* 3rd ed., Cambridge Univ. Press, London and New York.

Land, B., 1974. The toxicity of drilling fluid components to aquatic biological systems. A literature review. *Fish. Mar. Ser., Res. Dev. Tech. Rep.,* **487:** 1-33, Environment Canada, Winnipeg.

Parrish, P.R., E.E. Dyar, J.M. Enos, and W.G. Wilson, 1977. Chronic toxicity of chlordane, trifluralin, and pentachlorophenol to sheepshead minnows *(Cyprinodon variegatus). EPA Ecological Research Series,* EPA 600/11-77-062, In press.

Robichaux, T.J., 1975. Bactericides used in drilling and completion operations. *Conference on Environmental Aspects of Chemical Use in Well-Drilling Operations.* Houston, Texas, May 21-23, 1975. EPA-560-1-75-004: pp. 183-198. Office of Toxic Substances, United States Environmental Protection Agency, Washington, D.C.

Schimmel, S.C., D.J. Hansen, and J. Forester, 1974. Effects of Aroclor® 1254 on laboratory-reared embryos and fry of sheepshead minnows *(Cyprinodon variegatus)*. *Trans. Am. Fish. Soc.,* **103:** 582-586.

Schimmel, S.C., 1977. Notes on the embryonic period of the pinfish *Lagodon rhomboides* (Linnaeus). *Florida Sci.,* **40:** 3-6.

Shaw, D.R., 1975. The toxicity of drilling fluids, their testing and disposal. *Conference on Environmental Aspects of Chemical Use in Well-Drilling Operations.* Houston, Texas, May 21-23, 1975. EPA-560-1-75-004: pp. 463-471. Office of Toxic Substances, United States Environmental Protection Agency, Washington, D.C.

Tyler-Schroeder, D.B., 1976. Effects of two polychlorinated biphenyls, Aroclor® 1016 and 1242, on the grass shrimp, *Palaemonetes pugio:* Masters Thesis, The University of West Florida, Pensacola.

Woelke, C.E., 1972. Development of receiving water quality bioassay criterion based on the 48-hour Pacific oyster *(Crassostrea gigas)* embryo. *Washington Dept. Fish. Tech. Rep.,* **9:** 1-93, Olympia, Washington.

Zitko, V., 1975. Toxicity and environmental properties of chemicals used in well-drilling operations. *Conference on Environmental Aspects of Chemical Use in Well-Drilling Operations.* Houston, Texas, May 21-23, 1975. EPA-560-1-75-004: pp. 311-332. Office of Toxic Substances, United States Environmental Protection Agency, Washington, D.C.

Effects of Sodium Pentachlorophenate on Several Estuarine Animals: Toxicity, Uptake, and Depuration

STEVEN C. SCHIMMEL, JAMES M. PATRICK, JR.
and LINDA F. FAAS

Abstract—Several estuarine animals were exposed to sodium pentachlorophenate (Na-PCP), in flow-through toxicity tests. The following are test animals and their 96-hr LC_{50} values: grass shrimp *(Palaemonetes pugio)*, > 515 µg/l; brown shrimp *(Penaeus aztecus)*, > 195 µg/l; longnose killifish *(Fundulus similis)*, > 306 µg/l; pinfish *(Lagodon rhomboides)*, 53.2 µg/l; and striped mullet *(Mugil cephalus)*, 112 µg/l. The 192-hr EC_{50} (effect measured was shell deposition) for the eastern oyster *(Crassostrea virginica)* was 76.5 µg/l.

Eastern oysters exposed to Na-PCP concentrations of 25.0 and 2.5 µg/l accumulated the chemical in their tissues an average of 41 and 78 times, respectively. After Na-PCP delivery was discontinued, however, the oysters purged themselves of the pesticide within four days.

Introduction

Pentachlorophenol (PCP) and sodium pentachlorophenate (Na-PCP) have been used as defoliants, herbicides, insecticides, molluscicides, wood preservatives and other applications for many years (Bevenue and Beckman, 1967). Over 23 million kg PCP were produced in the United States in 1971: approximately 18 million kg

---

Contribution No. 336 of the Environmental Research Laboratory, Gulf Breeze.

Mention of a commercial product does not constitute endorsement by the United States Environmental Protection Agency.

STEVEN C. SCHIMMEL, JAMES M. PATRICK, Jr., and LINDA F. FAAS • United States Environmental Protection Agency, Environmental Research Laboratory, Sabine Island, Gulf Breeze, Florida 32561, U.S.A.

(78%) was used in wood preservation; and nearly 17% (3.9 million kg) was manufactured as Na-PCP (von Rumker *et al.*, 1975).

PCP has been reported in numerous monitoring studies. The compound was found in drinking water (Abrams *et al.*, 1975) and sewage effluents (Abrams *et al.*, 1975; Benvenue *et al.*, 1972; Buhler *et al.*, 1973), as well as in aquatic biota (Zitko *et al.*, 1974). The occurrence of PCP in blood and urine of humans has been documented (Rivers, 1972).

PCP was implicated in severe fish kills in recent years in Europe (Holmberg *et al.*, 1972) and North America (Personal Communication, Dr. O.E. Langer, Environmental Protection Service, Dept. of Fisheries and the Environment, West Vancouver, B.C., Canada). To our knowledge, all reported cases were associated with the pulpwood industry.

Few data have been published on the toxic effects of PCP or Na-PCP on estuarine organisms. Testing PCP on the yellow eel *(Anguilla anguilla)* under static conditions, Boström and Johansson (1972) described its biochemical effects and Holmberg *et al.* (1972) described uptake and depuration.

In this paper, we report the results of: (1) the evaluation of toxicity of Na-PCP to brown shrimp *(Penaeus aztecus)*, grass shrimp *(Palaemonetes pugio)*, longnose killifish *(Fundulus similis)*, pinfish *(Lagodon rhomboides)*, and striped mullet *(Mugil cephalus)*; (2) the effect of Na-PCP on shell deposition of the eastern oyster *(Crassostrea virginica)*; (3) uptake of Na-PCP by animals that survived the 96-hr exposure; and (4) uptake and depuration of Na-PCP by eastern oysters in a 56-day study.

Test Animals

All test animals were collected near the Environmental Research Laboratory, Gulf Breeze, Florida. The shrimp and oysters were acclimated to laboratory conditions for at least 10 days prior to testing; fishes, at least 14 days. If mortality in a specific lot of animals exceeded 1% in the 48 hours immediately preceeding the test or if abnormal behavior was observed during acclimation, the entire lot was discarded. The shrimp and fishes were not fed during the 96-hour tests. The oysters used in the shell deposition study and in the 56-day uptake/depuration study obtained food (plankton) from the unfiltered seawater in which they were held.

96-hr Toxicity Tests

Test methods for shrimp and fishes were similar to those described by Lowe *et al.* (1972), except that seawater was filtered through swimming pool and sand filters and heated to at least 24 °C for all tests. Maximum temperature was 25.1 °C. Salinity was allowed to vary with that of Santa Rosa Sound (18 to 31 ⁰/₀₀). Test methods were also compatible with those of American Public Health Association (1975). Acute toxicity of sodium pentachlorophenate was determined by exposing 20 animals per aquarium (70-liter capacity) to each concentration for 96 hours.

Analytical grade Na-PCP was dissolved in deionized water and metered by pumps at 30 ml/hr into filtered seawater that entered each aquarium by siphons calibrated to deliver 31 l/hr. One control aquarium received the same quantity of seawater and no Na-PCP. Mortality was recorded daily and dead animals were removed when discovered. At termination, surviving brown shrimp, grass shrimp, longnose killifish, and mullet were analyzed chemically for PCP residue content.

The eastern oyster *(Crassostrea virginica)* shell deposition study was conducted in unfiltered seawater at ambient temperature using the procedures of Butler (1965), except that the test was extended an additional 96 hours (192 hours total exposure) to allow for reduced shell deposition caused by exceptionally low seawater temperatures (temperature of seawater ranged from 7.3 to 8.3 °C; salinity, from 18.6 to 23.2 ‰). Shell deposition was measured to the nearest mm at the end of the exposure.

Bioconcentration of Na-PCP was studied by exposing 75 oysters (52 mm to 108 mm, umbo to distal valve height, $\bar{x} = 73$ mm) to Na-PCP concentrations of 5.0 and 50.0 μg/l (average measured PCP concentrations: 2.5 and 25 μg/l, respectively) in flowing seawater. Stock solutions of Na-PCP, made by mixing the compound with appropriate volumes of deionized water, were delivered to each aquarium by syringe pumps. Unfiltered ambient seawater (salinity 23 to 30.5 ‰, $\bar{x} = 26.6$ ‰; temperature 23.2 to 25.7 °C, $\bar{x} = 24.2$ °C) was delivered to the two experimental and one control aquaria (165 l) at the rate of 440 l/hr. Three oysters from each aquarium were sampled twice each week to determine PCP in edible oyster tissues. The 28-day exposure to Na-PCP was followed by a depuration period in which the oysters were held in PCP-free seawater. The depuration period lasted until no PCP (< 0.01 μg/g) was detected in oyster samples from all aquaria in two successive sampling periods.

Chemical Analysis of PCP

One to eight grams of tissues were weighed into 150 mm (O.D.) screw-top test tubes and extracted four times with 5 ml portions of acetonitrile for 30 seconds using a model PT 10-ST Willems Polytron (Brinkman Instruments, Westbury, New York). The test tube was centrifuged after each extraction and the acetonitrile transferred to a 100-ml graduated cylinder with ground glass stopper. To the combined extracts were added 70 ml of 2.0% aqueous sodium sulfate; 4 drops of 6N hydrochloric acid were added to convert the Na-PCP to PCP for measurement. (The aqueous sodium sulfate solution was prepared with distilled water that had been acidified and extracted with hexane to remove interfering impurities). This solution was then extracted with one 10-ml portion of hexane, followed by two 5-ml portions. The graduated cylinder was stoppered and shaken for one minute, the solvent phases allowed to separate, and the upper hexane layer pipetted into a 25-ml Kuderna-Danish concentrator tube. The combined extracts were then concentrated to 0.5 ml by evaporation with a gentle stream of nitrogen in a water bath maintained at a temperature less than 50 °C. The PCP-ethyl ether derivative was formed by addition of 10 to 15 drops of diazoethane that had been previously generated and stored at -20°C (Stanley, 1966). Gloves and a high-draft hood were used in handling diazoethane, a toxic, potentially explosive carcinogen. This mixture was allowed to stand at room temperature for 20 minutes, and the excess diazoethane was evaporated with a gentle stream of nitrogen. The concentrate was quantitatively transferred with three 0.5 ml portions of hexane to 9-mm Chromoflex column (Kontes Glass Company, Vineland, N.J.) containing 3.0 g Florisil topped with 2.0 g anhydrous sodium sulfate that had been washed previously with 10 ml hexane. PCP-ethyl ether was eluted from the column with 20 ml of 5% diethyl ether/hexane solution and the volume adjusted to the appropriate level for analysis by electron-capture gas chromatography.

Water samples were analyzed in 100-ml aliquots, adjusted to a pH < 2 with 6N HCl. Each sample was extracted twice with 50 ml 1:1 diethyl ether/petroleum ether with a 250 ml separatory funnel. The extract, collected in a 25-ml Kuderna-Danish concentrator tube with 100-ml Kuderna-Danish evaporator, was concentrated to 5 ml on a steam table. The concentrator tube was transferred to a nitrogen evaporator, concentrated to 0.5 ml, derivatized with diazoethane and analyzed as for tissue samples, without the Florisil cleanup.

Determinations were performed on Hewlett - Packard Model 5730A electron capture gas chromatographs equipped with 182 cm x 2 mm (I.D.) glass columns packed with 2% SP2100 on 100/120 mesh Supelcoport and 5% QF-1 on 80/100 mesh Gas-Chrom Q. The operating parameters were: oven temperature, 200°C; injector temperature, 200°C; detector temperature, 300°C; carrier gas 10% methane/argon flowing at 25 ml/minute.

Residue concentrations are reported as pentachlorophenol, calculated on a wet-weight basis without a correction factor for percentage recovery. To evaluate the integrity of the results, all samples were fortified with an internal standard (O, P'DDE) prior to analysis. Average recovery rates of PCP and O, P'DDE from fortified tissue and water samples were greater than 90%. A new PCP standard was prepared and a glassware blank run with each set of samples. Also, all glassware was previously washed with 1N KOH solution and rinsed with distilled water, followed by nanograde acetone. These precautions were necessary to insure that there was no interference from impurities.

Statistical Methods

Shrimp and fish mortality data were analyzed by the probit analysis method of Finney (1971) to determine LC_{50} values and their 95% confidence intervals. Oyster shell deposition data were analyzed by linear regression with probit transformation to determine an EC_{50} (the concentration of PCP effective in reducing shell deposition of exposed oysters to 50% of that of control oysters) and 95% confidence intervals.

TABLE 1. Acute Toxicity of Sodium Pentachlorophenate (Na-PCP)
to Several Estuarine Organisms

Species	Size[a] (x̄, mm)	96-hour EC_{50}[b], µg/l		Temp. (x̄, °C)	Salinity (x̄, ⁰/₀₀)
		Nominal	Measured		
Crassostrea virginica	45	104.0 (54-158)[c]	76.5 (37-116)	8.4	20.3
Penaeus aztecus	66	> 320.0	> 195.0	25.0	26.5
Palaemonetes pugio	18	> 560.0	> 515.0	24.8	24.3
Fundulus similis	42	> 560.0	> 306.0	24.4	22.9
Lagodon rhomboides	80	107.6 (93.7 - 122.0)	53.2 (42.4 - 65.4)	25.0	20.8
Mugil cephalus	58	221.6 (92.3 - 489.6)	112.1 (44.0 - 210.4)	24.7	25.5

[a]Size is height (umbo - distal valve edge) for oysters; rostrum - telson length for shrimp; standard length for fishes.
[b]Effect measured is shell deposition for oysters and mortality for shrimp and fishes.
[c]The 95% confidence intervals are in parenthesis.

Results of 96-hr Tests

Sodium pentachlorophenate was acutely toxic to oysters exposed for 192 hours (Table 1). The 192-hr EC_{50} was 76.5 µg/l. The EC_{50} value (measured) at the end of 96 hours approximately equalled that of the 192-hr EC_{50}, but due to the cold seawater temperature, control oysters deposited an average of only 1 mm shell/oyster. Therefore, we exposed the oysters for an additional 96 hours.

Exposure of brown shrimp to Na-PCP concentrations as high as 195 µg/l and the grass shrimp to concentrations up to 515 µg/l did not lead to significant mortality, or bioconcentration of the compound (Tables 1 and 2). Goodnight (1942) exposed various invertebrate species (including crustaceans) to 5 mg/l Na-PCP in static tests and reported that all but bloodworms were extremely resistant to the pesticide. Therefore, many invertebrate species appear to be relatively insensitive to Na-PCP compared to fishes.

Sodium pentachlorophenate at the concentrations tested, was acutely toxic to striped mullet and pinfish but not to longnose killifish (Table 1). Pinfish were the most sensitive species tested, the 96-hr LC_{50} being 53.2 µg/l. In comparison, Borthwick and Schimmel (this Symposium) reported a 96-hr LC_{50} of 38 µg/l for larval pinfish (< 48 hrs post-hatch) exposed to Na-PCP in a static test.

The lesser sensitivity of longnose killifish compared to pinfish in this test is similar to that shown by Schimmel *et al.* (1977a) with the insecticide toxaphene.

S.C. Schimmel *et al.*

TABLE 2. Uptake of Sodium Pentachlorophenate (Na-PCP)
by Several Estuarine Organisms

Species	Water conc. Measured (μg/l)	Whole-body residue (μg/kg wet wt)	Bioconc. factor
Penaeus aztecus	ND[a]	ND[a]	—
	14.0	10	—
	32.0	12	0.37[b]
	49.0	22	0.45
	92.0	11	0.12
	195.0	22	0.11
			$\bar{x} = 0.26$
Palaemonetes pugio	ND	ND	—
	32.0	50	1.6
	54.0	100	1.9
	76.0	230	3.0
	249.0	430	1.7
	515.0	240	0.5
			$\bar{x} = 1.7$
Fundulus similis	ND	ND	—
	36.0	1,100	31.0
	85.0	3,500	41.0
	287.9	3,300	12.0
	306.0	10,800	35.0
			$\bar{x} = 30.0$
Mugil cephalus	ND	ND	—
	26.0	230	8.8
	46.0	290	6.3
	85.0	6,700	79.0
	157.0	8,800	56.0
	308.0	—	$\bar{x} = 38.0$

[a]Control, non-detectable; = < 0.01 μg/l in water, < 10 μg/kg in tissue.
[b]Biconcentration factor = concentration of PCP in tissues divided by
the concentration of PCP in the external medium.

The reason for this range of sensitivity is unclear, but other investigators have reported a similar phenomenon when comparing susceptibility of freshwater fishes from various taxonomic groups to organochlorine insecticides (Macek and McAllister, 1970).

Our data on the toxicity of Na-PCP to estuarine fishes compare favorably with those for freshwater fishes. In static tests, the 96-hr LC_{50} value for bluegills *(Lepomis macrochirus)* was 20 µg/l; for goldfish *(Carassius auratus)*, 50 µg/l (Inglis and Davis, 1972). The 96-hr LC_{50} for the sockeye salmon *(Oncorhynchus nerka)* exposed in flow-through toxicity tests was 63 µg/l (Webb and Brett, 1973). For brook trout *(Salvelinus fontinalis),* fathead minnows *(Pimephales promelas),* goldfish, and bluegills exposed to Na-PCP in acute flow-through toxicity tests, a total mortality of all species was reported to occur at concentrations < 1.0 mg/l (Cardwell *et al.,* 1976).

Tissues of surviving longnose killifish and striped mullet from the 96-hr toxicity tests were analyzed for PCP. Results indicated that PCP was bioconcentrated by the killifish to an average of 30 times the amount measured in the exposure water and by mullet, 38 times (Table 2). These bioconcentration factors are extremely low compared to the bioconcentration (up to 1000×) by goldfish exposed to media containing PCP (Kobayashi and Akitake, 1975). Whether these contrasting results are attributable to possible differences in the rate of uptake of PCP among the fishes studied, or due to differences in the experimental procedures, remains uncertain.

The bioconcentration factors we observed were also extremely low when compared with many other organochlorine insecticides. For example, in other 96-hr studies, the average bioconcentration factor for pinfish exposed to toxaphene was 3,850 (Schimmel *et al.,* 1977a); for chlordane, 6,228 (Parrish, *et al.,* 1976) and for benzenehexachloride (BHC), 482 (Schimmel *et al.,* 1977b).

Uptake and Depuration of PCP by Oysters

Eastern oysters exposed to measured PCP concentrations of 25.0 and 2.5 µg/l for 28 days accumulated the chemical in their tissues an average of 41 and 78 times, respectively (Figure 1). Pentachlorophenol reached an apparent equilibrium in oysters within the first four days in both exposure concentrations and remained relatively constant throughout the uptake portion of the study. The mean concentration of PCP in oysters from the 25 µg/l aquarium was 1060 µg/kg; for the 2.5 µg/l aquarium, was 180 µg/kg. After the PCP delivery was discontinued, however, the oysters purged themselves of the pesticide within four days.

In summary, Na-PCP reduced shell deposition of eastern oysters at concentrations ≥ 34 µg/l during a 192-hr exposure. Concentrations of the pesticide as high as 195 and 515 µg/l, however, were neither toxic to brown shrimp and grass shrimp, respectively in 96-hr exposures, nor did the animals bioconcentrate the chemical. Pinfish were the most sensitive of all species tested (96-hr LC_{50} = 53 µg/l).

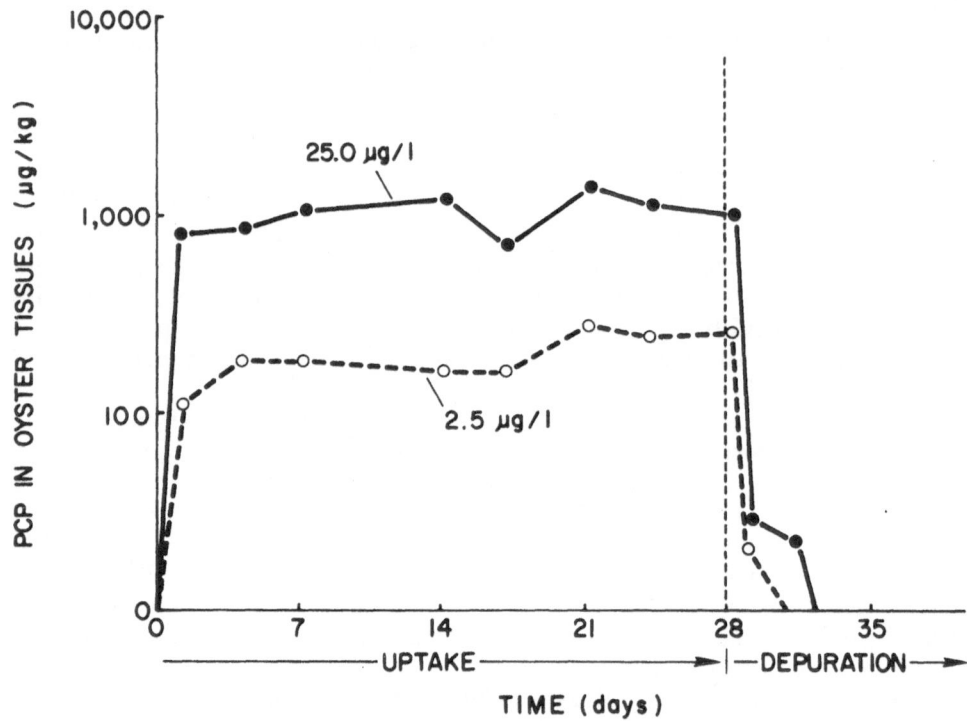

Figure 1. Uptake of pentachlorophenol (PCP) by eastern oysters *(Crassostrea virginica)* exposed 28 days, then allowed to depurate in PCP-free seawater.

Oysters bioconcentrated PCP 41 to 78 times the amount measured in water and when held in PCP-free seawater, depurated the chemical to nondetectable concentrations in four days. Compared with other organochlorine insecticides (such as toxaphene, chlordane and BHC) tested on estuarine animals, Na-PCP is relatively less toxic and is bioconcentrated to a lesser extent.

References

Abrams, F., D. Dericks, C.V. Fong, D.K. Guinan, and K.M. Slimak, 1975. Identification of organic compounds in effluents from industrial sources. Office of Toxic Substances, U.S. Environmental Protection Agency. EPA-560/3-75-002.

American Public Health Association, 1975. *Standard methods for the examination of water and waste water,* 14th ed., 1093 pp. American Public Health Association, Washington, D.C.

Bevenue, A., and H. Beckman, 1967. Pentachlorophenol: A discussion of its properties and its occurrence as a residue in human and animal tissues. *Residue Rev.,* **19:** 83-134.

Bevenue, A., J.W. Hylin, Y. Kawano, and T.W. Kelley, 1972. Organochlorine residues in water, sediment, algae, and fish, Hawaii - 1970-71. *Pesticide Monit. J.,* **6:** 56-64.

Boström, S.L., and R.G. Johansson, 1972. Effects of pentachlorophenol on enzymes involved in energy metabolism in the liver of the eel. *Comp. Biochem. Physiol.,* **41B:** 359-369.

Buhler, D.R., M.E. Rasmusson, and H.S. Nakaue, 1973. Occurrence of hexachlorophene and pentachlorophenol in sewage and water. *Environ. Sci. Tech.,* **7:** 929-934.

Cardwell, R.D., D.G. Foreman, T.R. Payne, and D.J. Wilbur, 1976. Acute toxicity of selected toxicants to six species of fish. U.S. Environmental Protection Agency. *Ecological Research Series.* EPA-600/3-76-008.

Finney, D.J., 1971. *Probit Analysis.* 3rd ed., 333 pp. Cambridge University Press, London.

Goodnight, C.J., 1942. Toxicity of sodium pentachlorophenate and pentachlorophenol to fish. *Industr. Eng. Chem.,* **34:** 886-872.

Holmberg, B., S. Jensen, A. Larsson, K. Lewander, and M. Olsson, 1972. Metabolic effects of technical pentachlorophenol (PCP) on the eel *Anguilla anguilla* L. *Comp. Biochem. Physiol.,* **43B:** 171-183.

Inglis, A., and E.L. Davis, 1972. Effects of water hardness on the toxicity of several organic and inorganic herbicides to fish. *U.S. Fish Wildlife Service, Bureau Sport Fisheries and Wildlife Tech. Paper* No. **67.**

Kobayashi, K., and H. Akitake, 1975. Studies on the metabolism of chlorophenols in fish — I. Absorption and excretion of PCP by goldfish. *Bull. Japan. Soc. Sci. Fish.,* **41:** 87-92.

Lowe, J.I., P.R. Parrish, J.M. Patrick, Jr., and J. Forester, 1972. Effects of the polychlorinated biphenyl Aroclor® 1254 on the American oyster, *Crassostrea virginica. Marine Biol.,* **17:** 209-214.

Macek, K.J., and W.A. McAllister, 1970. Insecticide susceptibility of some common fish family representatives. *Trans. Am. Fish. Soc.,* **99:** 20-27.

Parrish, P.R., S.C. Schimmel, D.J. Hansen, J.M. Patrick, Jr., and J. Forester, 1976. Chlordane: Effects on several estuarine organisms. *J. Toxicol. Environ. Hlth.,* **1:** 485-494.

Rivers, J.B. 1972. Gas chromatographic determination of PCP in human blood and urine. *Bull. Environ. Contam. Toxicol.,* **8:** 294-296.

Schimmel, S.C., J.M. Patrick, Jr., and J. Forester, 1977a. Uptake and toxicity of toxaphane in several estuarine organisms. *Arch. Environ. Contam. Toxicol.,* **5:** 353-367.

Schimmel, S.C., J.M. Patrick, Jr., and J. Forester, 1977b. Toxicity and bioconcentration of BHC and Lindane in selected estuarine animals. *Arch. Environ. Contam. Toxicol.,* **6:** 355-363.

Stanley, C.W., 1966. Derivatization of pesticide-related acids and phenols for gas chromatographic determination. *J. Agr. Food Chem.,* **14:** 321-323.

von Rümker, E.W. Lawless, A.F. Meiners, K.A. Lawrence, G.L. Kelso, and F. Horay, 1975. *Production, distribution, use and environmental impact potential of selected pesticides.* Office of Pesticide Programs, Office of Water and Hazardous Materials. U.S. Environmental Protection Agency. EPA 540/1-74-001.

Webb, P.W., and J.R. Brett, 1973. Effects of sublethal concentrations of sodium pentachlorophenate on growth rate, food conversion efficiency and swimming performance in underyearling sockeye salmon *(Oncorhynchus nerka). J. Fish. Res. Bd. Can.,* **30:** 499-507.

Zitko, V., O. Hutzinger, and P.M. K. Choi, 1974. Determination of pentachlorophenol and chlorobiphenyls in biological samples. *Bull. Environ. Contam. Toxicol.,* **12:** 649-653.

Effects of Dowicide® G-ST on Development
of Experimental Estuarine Macrobenthic Communities

M.E. TAGATZ, J.M. IVEY and M. TOBIA

Abstract—Aquaria containing clean sand received a continuous supply of flowing seawater from Santa Rosa Sound, Florida, mixed with known quantities of Dowicide® G-ST (79% sodium pentachlorophenate) for thirteen weeks. The measured concentrations of pentachlorophenol (PCP) in the aquaria were 1.8, 15.8 and 161 μg/l. At the end of the experiment, macrofauna established in control and experimental aquaria was examined. Mollusks, arthropods and annelids were numerically dominant among the macrofauna. Although exposure to 1.8 μg PCP/l had no effect, the higher concentrations of PCP caused marked reduction in the numbers of individuals and species. Mollusks were the most sensitive taxonomic group to PCP. These results and our previous studies on the effects of a nine-week exposure to PCP on the establishment of macrobenthic communities indicate that discharge of PCP into natural waters could alter the normal colonization by benthic animals and could impact various ecological relationships among localized populations.

Introduction

The extensive use of pentachlorophenol (PCP) as a wood preservative and as a biocide in a variety of industries threatens the aquatic environment. PCP has been

---

Contribution No. 352, Environmental Research Laboratory, Gulf Breeze.

Mention of commercial products does not constitute endorsement by the United States Environmental Protection Agency.

M.E. TAGATZ, J.M. IVEY and M. TOBIA • United States Environmental Protection Agency, Environmental Research Laboratory, Sabine Island, Gulf Breeze, Florida 32561, U.S.A.

TABLE 1. Animals Collected from Control Aquaria and from Aquaria Exposed to Dowicide® G-ST for 13 Weeks

Animal	Control		PCP concentration (μg/l)[a]					
			1.8		15.8		161	
Mollusca								
Enis minor	987	(10)[b]	1156	(10)	695	(10)	0	(0)
Mulina lateralis	45	(9)	59	(10)	26	(8)	0	(0)
Polinices duplicatus	6	(3)	5	(4)	4	(4)	0	(0)
Cyrtopleura costata	0	(0)	1	(1)	0	(0)	0	(0)
Lyonsia hyalina floridana	1	(1)	0	(0)	0	(0)	0	(0)
Doridella obscura	1	(1)	0	(0)	0	(0)	0	(0)
Diastoma varium	1	(1)	0	(0)	0	(0)	0	(0)
Arthropoda								
Balanus amphitrite niveus	352	(10)	301	(10)	275	(10)	0	(0)
Corophium acherusicum	26	(5)	53	(7)	15	(6)	0	(0)
Callinectes similis	1	(1)	0	(0)	3	(2)	2	(2)
Idoteidae	0	(0)	0	(0)	1	(1)	0	(0)
Annelida								
Polydora ligni	310	(10)	293	(10)	230	(10)	11	(4)
Neanthes succinea	22	(9)	31	(9)	18	(7)	24	(8)
Heteromastus filiformis	0	(0)	0	(0)	22	(6)	0	(0)
Spiophanes bombyx	7	(4)	5	(2)	3	(3)	0	(0)
Armandia agilis	1	(1)	2	(2)	2	(2)	0	(0)
Armandia maculata	2	(1)	1	(1)	0	(0)	0	(0)
Polydora websteri	4	(2)	2	(2)	1	(1)	0	(0)
Capitellides jonesi	0	(0)	0	(0)	1	(1)	1	(1)
Streblospio benedicti	1	(1)	0	(0)	2	(1)	0	(0)
Eumida sanguinea	2	(2)	0	(0)	2	(2)	0	(0)
Eteone heteropoda	1	(1)	0	(0)	0	(0)	0	(0)
Mediomastus californiensis	0	(0)	1	(1)	1	(1)	0	(0)
Polydora socialis	0	(0)	3	(2)	0	(0)	0	(0)
Polydora aggregata	0	(0)	0	(0)	1	(1)	0	(0)
Nemertea								
Nemertea	7	(4)	10	(4)	9	(6)	0	(0)
Coelenterata								
Actinaria	4	(2)	1	(1)	6	(3)	0	(0)

TABLE 1. continued

Animal	Control		PCP concentration (μg/l)[a]					
			1.8		15.8		161	
Chordata								
Atherinidae	2	(1)	0	(0)	0	(0)	0	(0)
Molgula manhattensis	0	(0)	1	(1)	0	(0)	0	(0)
Echinodermata								
Leptosynapta inhaerens	0	(0)	1	(1)	0	(0)	0	(0)
Sipunculoidea								
Sipunculoidea	0	(0)	1	(1)	0	(0)	0	(0)
Platyhelminthes								
Polycladida	0	(0)	1	(1)	0	(0)	0	(0)

[a]Measured concentration in the water
[b]Number of aquaria from which the animals were collected.

released into natural waters from industrial and agricultural operations and from runoff waters (Adelman *et al.*, 1976). There have been reports of toxic concentrations of PCP in aquatic environments (Fountaine *et al.*, 1976; van Dijk *et al.*, 1977). The available information indicates that PCP is toxic to a variety of aquatic organisms (Becker and Thatcher, 1973). However, relatively little is known of the effects of PCP and its salts on marine and estuarine organisms. Of particular concern is the use of PCP derivatives such as Dowicide® G-ST (79% sodium pentachlorophenate) in the expanding exploration for oil in the marine environment. These compounds are used as bactericides in drilling and completion fluids for oil-well drilling (Robichaux, 1975).

We initiated studies to determine the effects of PCP and its derivatives on the development of estuarine macrobenthic communities. We previously reported effects of a continuous nine-week exposure to concentrations of 7, 76 and 622 μg PCP/l on establishment of macrofauna in an experimental flow-through system (Tagatz *et al.*, 1977). We have since investigated the effects of exposure to Dowicide® G-ST for thirteen weeks on the establishment of macrobenthic communities, and results are presented here.

Experimental System

The experimental design (Hansen, 1974) consisted of 10 control aquaria and 10 aquaria for each of three concentrations of the toxicant. Aquaria (56 cm long, 9 cm wide, and 12 cm high) were filled to a depth of 6 cm with sand dredged from Santa Rosa Sound, Florida. The water levels were maintained at 9 cm (3 cm above substratum).

Unfiltered seawater with its constituent plankton was pumped from Santa Rosa Sound to overflow constant-head boxes, one for the control and one for each test concentration. Ten small standpipes within each of these boxes continually supplied water to the aquaria at a rate of 200 ml/min (maintained by adjusting the height of a 3.2 mm diameter hole in each standpipe). Dowicide® G-ST dissolved in deionized water was metered (by pump) into and mixed with the seawater in the constant-head boxes. Aquaria were exposed to the toxicant for thirteen weeks, December 27, 1976 - March 28, 1977. Water flowed from the aquaria through notched end-openings. Large animals, such as crabs, could escape through these openings before their number could drastically affect community structure.

At the end of exposure, the animals were collected by siphoning contents of the aquaria into a 1-mm mesh sieve, preserved and identified. Analysis of variance and the Student-Newman-Keul tests were used to test the statistical significance of differences in numbers and species of animals among control and exposed aquaria.

Concentrations of PCP in test water and sediment were determined by gas chromatography (Tagatz *et al.,* 1977; Schimmel *et al.,* this symposium) Samples of water were taken twice a week from the constant-head boxes, and sediment cores were taken from 3 of 10 aquaria per concentration at the end of exposure. Limits of detection were 0.2 μg PCP/l in seawater and 2.5 μg PCP/kg (dry weight) in sediment. Recoveries averaged greater than 80% for water and sediment samples.

The averages and ranges of PCP concentrations measured in water were 1.8 μg/l (0.5-4.1), 15.8 μg/l (8.9-22.0), and 161 μg/l (135-230). No PCP was detected in water samples from the control aquaria. The sediment samples from control aquaria and aquaria exposed to 1.8 μg PCP/l did not contain detectable amounts of PCP. A range of 3 to 6 μg PCP/kg was found in sediment from aquaria exposed to 15.8 μg PCP/l; 41 to 71 μg PCP/kg was found in sediment from aquaria exposed to 161 μg PCP/l.

Macrobenthic Communities

At the end of the 13-week period, the experimental and control aquaria contained a total of 5,066 animals, representing 32 species of 9 phyla (Table 1). The three most abundant species were the clam, *Enis minor* (2,838), the barnacle, *Balanus amphitrite niveus* (928), and the polychaete, *Polydora ligni* (844). The numerically dominant phyla were Mollusca, Arthropoda and Annelida.

The abundance of animals was affected by PCP (Figure 1). Total numbers of animals and species (Table 2) were significantly fewer ($P < 0.01$) in aquaria exposed to 161 μg/l than in control aquaria and aquaria exposed to lower concentrations of the toxicant. The number of animals in aquaria exposed to 15.8 μg/l also differed significantly from those in control aquaria and aquaria exposed to 1.8 μg PCP/l.

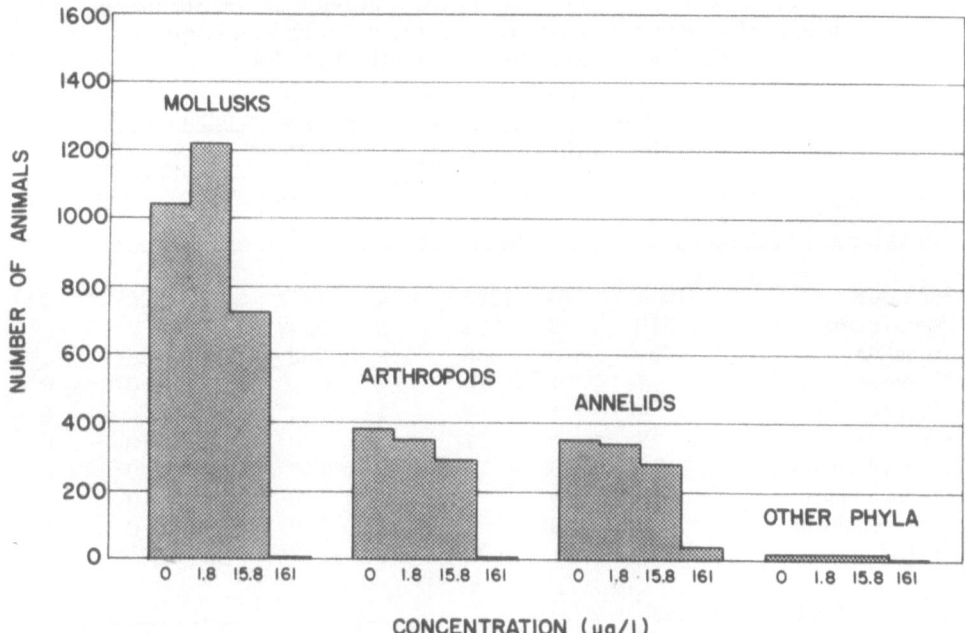

Figure 1. Number of animals, by phylum, collected from control aquaria and aquaria exposed for 13 weeks to Dowicide® G-ST. The average measured concentrations of PCP in the water from experimental aquaria were 1.8, 15.8 and 161 µg/l. The control aquaria did not contain detectable amounts (detection limit 0.2 µg/l) of PCP.

Mollusks were most affected by PCP. Although the number of individuals of arthropods and annelids decreased at 15.8 µg/l, only mollusks were significantly fewer than in control aquaria. This relative sensitivity of developing estuarine mollusks to PCP was also noted in our previous studies (Tagatz *et al.*, 1977). These results are consistent with the known molluscicidal applications of PCP (Bevenue and Beckman, 1967).

Exposure to concentrations of PCP such as 76 µg/l (Tagatz *et al.*, 1977), 161 µg/l (this study) and 622 µg/l (Tagatz *et al.*, 1977) caused a reduction not only in the number of individuals but also in the number of macrobenthic species in the experimental flow-through system. Changes in the relative abundance of species within a taxonomic group were also noted. For example, exposure to 161 µg PCP/l reduced the relative abundance of the polychaete worm, *Polydora ligni,* but not that of *Neanthes succinea.*

Conclusions

Discharges of PCP and its derivatives in nature, at levels tested in the laboratory, could disrupt stable ecological relationships among localized populations and have a direct adverse effect on abundance of various species normally

TABLE 2. Relative Abundance (Number of Individuals and Species) of Various Phylogenetic Groups from Control Aquaria and Aquaria Exposed to Dowicide® G-ST for 13 Weeks

Phylum	Control		PCP concentration (μg/liter)[a]					
	No.	Sp.	1.8		15.8		161	
			No.	Sp.	No.	Sp.	No.	Sp.
Mollusca	1041	6	1221	4	725[b]	3	0[b]	0[b]
Arthropoda	379	3	354	2	294	4	2[b]	1[b]
Annelida	350	9	338	8	283	11	36[b]	3[b]
Nemertea	7	1	10	1	9	1	0	0
Coelenterata	4	1	1	1	6	1	0	0
Chordata	2	1	1	1	0	0	0	0
Echinodermata	0	0	1	1	0	0	0	0
Sipunculoidea	0	0	1	1	0	0	0	0
Platyhelminthes	0	0	1	1	0	0	0	0
TOTAL	1783	21	1928	20	1317[b]	20	38[b]	4[b]

[a]Average measured concentration in the water
[b]Significantly less ($P < 0.01$) than control; significance is indicated only for the three most abundant phyla.

colonizing the substratum. For example, changes in the relative abundance of polychaete worms, a large group important in marine food webs, would impact many predator-prey interactions. Also, these compounds could decrease abundance of various commerically important species of arthropods and mollusks.

ACKNOWLEDGMENTS

The authors acknowledge the assistance of James Moore and Linda Faas who performed the chemical analysis. Dr. Jerry Oglesby, University of West Florida, Pensacola, Florida, provided assistance in statistical analyses.

References

Adelman, I.R., L.L. Smith, Jr., and G.D. Siesennop, 1976. Acute toxicity of sodium chloride, pentachlorophenol, Guthion®, and hexavalent chromium to fathead minnows *(Pimephales promelas)* and goldfish *(Carassius auratus), J. Fish. Res. Bd. Can.,* **33:** 203-208.

Becker, C.D., and T.O. Thatcher, 1973. Toxicity of power plant chemicals to aquatic life. Report No. WASH-1249, U.S. Atomic Energy Commission. Richland, Washington: Battelle Pacific Northwest Laboratories.

Bevenue, A., and H. Beckman, 1967. Pentachlorophenol: A discussion of its properties and its occurrence as a residue in human and animal tissues. *Residue Rev.,* **19:** 83-134.

Fountaine, J.E., P.B. Joshipura, and P.N. Keliher, 1976. Some observations regarding pentachlorophenol levels in Haverford Township, Pennsylvania. *Water Res.,* **10:** 185-188.

Hansen, D.J., 1974. Aroclor® 1254: Effect on composition of developing estuarine animal communities in the laboratory. *Contrib. Mar. Sci.,* **18:** 19-33.

Robichaux, T.J., 1975. Bactericides used in drilling and completion operations. In, *Environmental aspects of chemical use in well-drilling operations,* Conference Proceedings, May 1975, Houston, Texas. Report No. EPA-560/1-75-004, pp. 183-191, United States Environmental Protection Agency, Washington, D.C.

Tagatz, M.E. J.M. Ivey, J.C. Moore, and M. Tobia, 1977. Effects of pentachlorophenol on the development of estuarine communities. *J. Toxicol. Environ. Hlth.,* **3:** 501-506.

van Dijk, J.J., C. van der Meer, and M. Wijnans, 1977. The toxicity of sodium pentachlorophenolate for three species of decapod crustaceans and their larvae. *Bull. Environ. Contam. Toxicol.,* **17:** 622-630.

Effects of Pentachlorophenol on the Meiobenthic Nematodes in an Experimental System

FRANK R. CANTELMO and K. RANGA RAO

Abstract—Aquaria containing clean sand received a continuous supply of seawater from Santa Rosa Sound, Florida, mixed with known quantities of PCP for nine weeks (May 10 - July 12, 1976) for the first experiment and Dowicide® G-ST for thirteen weeks (December 27, 1976 - March 28, 1977) for the second experiment. The measured concentrations of PCP in the former experiment were 7, 76 and 622 μg/l while the concentrations of 1.8, 15.8 and 161 μg/l were used in the latter experiment. At the end of each experiment the meiofauna established in the control and experimental aquaria were examined. Nematodes were the dominant group and averaged 83% of all the meiofauna encountered. Concentrations of 1.8, 7 and 15.8 μg PCP/l did not affect the biomass and density of nematodes. An intermediate concentration of PCP (76 μg/l) caused an increase in biomass and density of nematodes compared to control aquaria. Higher concentrations of PCP (161 and 622 μg/l) caused a decrease in biomass and density of nematodes compared to control aquaria. Marked changes in nematode species composition and shifts in nematode feeding types were noticed in the aquaria exposed to 161 and 622 μg PCP/l. Nematodes classified as epistrate feeders were most abundant in the control aquaria and those exposed to 1.8, 7, 15.8 and 76 μg PCP/l. Deposit feeders were relatively abundant among the nematodes in aquaria exposed to 161 and 622 μg PCP/l. The alterations in nematodes observed in this investigation appeared to be due to the variations in macrobenthic fauna and food (algae) supply caused by the biocidal effects of PCP and also due to the toxic effects of PCP on meiofauna.

FRANK R. CANTELMO and K. RANGA RAO • Faculty of Biology, University of West Florida, Pensacola, Florida 32504, U.S.A.

Introduction

The toxic effects of pentachlorophenol on aquatic organisms are well documented (Goodnight, 1942; Crandall and Goodnight, 1959; Bevenue and Beckman, 1967; Yasuraoka and Hosaka, 1971; Shim and Self, 1973). Most of the previous investigations dealt with freshwater organisms and very little is known about the effects of PCP on estuarine benthic communities. Of all the benthic metazoans in marine sediments, meiofauna are generally the most numerous and may range in densities (numbers of individuals) from 55,000 to > 1,000,000 per square meter. Meiofauna may be defined as those organisms which pass through a 0.5 mm sieve and are retained on a sieve with mesh widths smaller than 0.1 mm. The permanent meiofauna includes the Nematoda, Copepoda, Ostracoda, Rotifera, Gastrotricha, Archiannelida, Tardigrada, Turbellaria, Mystococarida, Oligochaeta, Polychaeta, Gnathostomulida, Acarina and some specialized representatives of the Nemertinea, Hydrozoa, Bryozoa, Gastropoda, Holothuroidea, Tunicata, Priapulida, Soelenogastres and Sipunculoidea. Of these groups the Nematoda is generally the most common taxon in marine sediments.

It was not until the last decade that the significance of meiofauna for benthic communities had begun to be investigated. Although meiofauna biomass is about 1-4% of macrofauna biomass in sublittoral silty sand (Stripp, 1969), the average turnover rate of meiofauna is about five times greater than that for macrofauna. Thus, in terms of importance of food consumed and biomass provided for marine food web, the meiofauna may be of greater importance than the macrofauna in a given area. Meiofauna may also be important in the recycling of nutrients at lower trophic levels and in enhancing the mineralization of organic matter (Johannes, 1965; McIntyre, 1969, 1973; Marshall, 1970; Coull, 1973; Tenore et al., 1977). The concept of 'conveyor-belt detritivores' proposed by Rhoads (1974) for macrofauna may be expanded to include nematodes that act as conveyor-belt herbivores and bring oxidized material into the more reduced zones (Cantelmo, 1977).

Although the importance of meiofauna in estuarine systems continues to be documented (reviewed by McIntyre, 1969; Coull, 1973), there have been relatively few studies on the effects of pollution on nematode populations or meiofauna community structure (Marcotte and Coull, 1974; Gray, 1971; Ganapati and Sarma, 1973; Lorenzen, 1974; Tietjen, 1977). The present study was initiated to determine the effects of pentachlorophenol on laboratory-established meiobenthic communities.

Flow-through Aquaria

The effect of pentachlorophenol (PCP) on meiofauna was studied using an experimental system (Hansen, 1974) set up by M.E. Tagatz and his associates at the Gulf Breeze Environmental Research Laboratory. Each aquarium was 56 cm long, 9 cm wide and 12 cm high and was filled to a depth of 6 cm with a well-sorted (1.27) medium sand (0.41 mm grain size). All the sand was PCP free and thoroughly washed with water. Seawater levels were maintained at 9 cm (3 cm above the substrate) by seawater pumped from Santa Rosa Sound, Florida. A metered amount of PCP was mixed with the siphoned seawater flowing to the constant head boxes supplying the aquaria. Two flow-through experiments were conducted on PCP, one for nine weeks from May 10 - July 12, 1976 and another for 13 weeks from December 27, 1976 to March 28, 1977. During the nine week experiment, pentachlorophenol was used in average measured concentrations of 7, 76 and 622 μg/l PCP while Dowicide® G-ST was used in the 12-week experiment with average measured PCP concentrations of 1.8, 15.8 and 161 μg/l.

Meiofauna Samples

At the end of each experimental period, meiofauna cores were secured to a depth of 6 cm from control and experimental aquaria. Each core was preserved in buffered 5% formalin and Rose Bengal (0.3 g Rose Bengal/l of 5% formalin). The sediment was transferred to a 1000 ml Erlenmeyer flask and agitated with tap water. After allowing the heavier particles to settle for a few seconds, the supernatant was poured through a 55 μ nylon sieve. This decantation procedure which was repeated four times yielded better than 99% extraction efficiency for the groups considered in this study. Nematodes were transferred to a 3% aqueous glycerin solution for a week of infiltration before transferral to a mounting medium of anhydrous glycerin for the preparation of permanent glass slides.

Variations in Nematode Densities and Biomass

Nematoda accounted for greater than 80% of all the meiofauna encountered. Other meiofauna groups including the Copepoda, Foraminifera, Ostracoda, Rotifera, Polychaeta and Bivalvia were also represented in control and experimental aquaria but occurred in far fewer numbers (Cantelmo and Rao, 1978). The present report discusses the effects of PCP on Nematoda, the most abundant group encountered in this study. Although nematode densities in the control aquaria were lower (514 individuals 10 cm^{-2}) during the May-July experiment compared to the December-March experiment (3,502 individuals 10 cm^{-2}), the percent changes (relative to controls) in nematode densities and biomass in the aquaria exposed to PCP or Na-PCP revealed similar patterns in both experiments and will be discussed together. The difference in nematode densities between control aquaria from the two experiments may reflect seasonal differences in meiofauna in these flow-through systems in much the same manner as that reported in the natural environment (Jansson, 1966, 1968; Hopper and Meyers, 1967; Lasserre, 1971; Tietjen, 1969; Coull, 1970; McIntyre and Murrison, 1973; Elmgren, 1973; DeBovee and Soyer, 1974). Figure 1 shows changes in nematode biomass and density (expressed as % change compared to control) at various concentrations of PCP. The slight reductions in nematode biomass and densities noted in aquaria exposed to 1.8, 7.0 and 15.8 μg PCP/l were not statistically significant. However, the nematode biomass and densities exhibited a significant

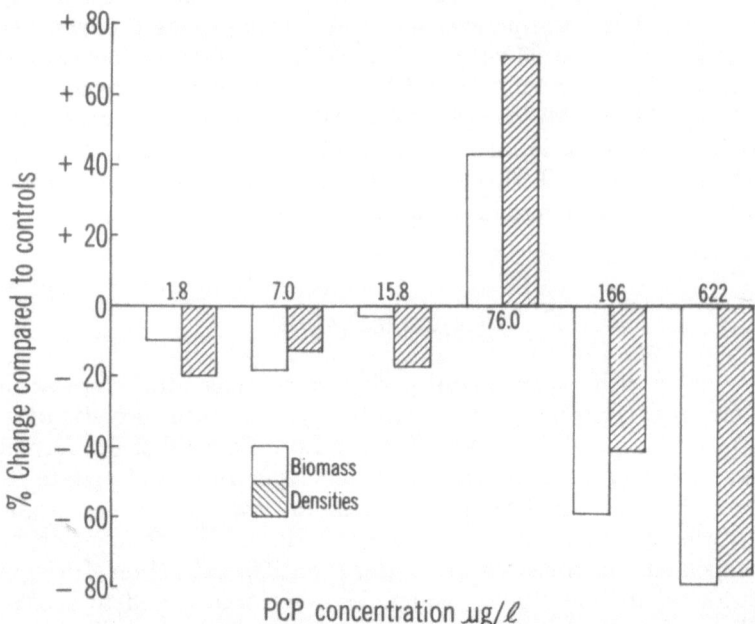

Figure 1. Changes in the densities and biomass of meiobenthic nematodes established in the experimental flow-through aquaria exposed either to pentachlorophenol for nine weeks or to Dowicide® G-ST for thirteen weeks. The measured concentrations of pentachlorophenol in the water are given.

increase (P < 0.05) in aquaria exposed to 76 µg PCP/l and a significant decrease (P < 0.05) in aquaria exposed to 161 and 622 µg PCP/l. At first one might expect a continued decline of nematode biomass and densities with increased concentrations of PCP because of its known biocidal effects. However, the observed increases in nematodes at 76 µg PCP/l (Cantelmo and Rao, 1978) may be partially explained by examining the distribution of macrofauna in these aquaria reported by Tagatz *et al.*, (1977). If we assume that macrofauna are in competition with the meiofauna for common food resources (Coull, 1973), the significantly fewer numbers of certain groups (Annelida) of macrofauna in the aquaria exposed to 76 µg PCP/l (Tagatz *et al.*, 1977) may offer the meiofauna reduced competition for food. Thus, reduced interspecific competition may cause the total population of nematodes to increase in the 76 µg PCP/l aquaria. At the higher concentrations tested (161 and 622 µg PCP/l), the toxic effects of PCP may have severely limited the macrofauna (Tagatz *et al.*, 1977) as well as the meiofauna.

Changes in Nematode Species Composition

In addition to changes in nematode biomass and density, there were changes in nematode species composition in aquaria exposed to higher concentrations of PCP.

A list of nematode species found in the May-July and/or December-March experiments is presented below.

Anoplostoma sp. 1
Anticoma sp. 1
Axonolaimus paraponticus (Hopper)
Chromadora nudicapitata Bastian
Chromadorina germanica Butschli
Chromadorita sp. 1
Chromadorita sp. 2
Cyatholaimus sp. 1
Diplolaimella punicea Timm
Eleutherolaimus stenosoma (De Man)
Eurystomina sp. 1
Microlaimus problematicus Allgen
Microlaimus sp. 1
Monhystera parva
Monhystera sp. 1
Monhystera sp. 2
Neochromadora sp. 1
Neonyx sp. 1
Oncholaimus domesticus (Chitwood and Chitwood)
Prochromadorella micoletzkyi Chitwood
Theristus sp. 1
Viscosia macramphida Chitwood

*Species found during May-July and December-March experiments.

Faunal similarities between control and experimental aquaria were examined using a matrix of Czekanowski coefficients, and affinity dendrograms were constructed (Fig. 2). A 100% similarity indicates that all the nematode species are in common. Among replicate cores, meiofauna percent similarity generally is 70-80% because of the patchiness of meiofauna. Thus, the percent similarity observed with controls and experimental aquaria exposed to lower concentrations of PCP (control *vs.* 1.8 μg PCP/l, control *vs.* 15.8 μg PCP/l, 1.8 *vs.* 15.8 μg PCP/l as shown in Fig. 2; control *vs.* 7 μg PCP/l; control *vs.* 76 μg PCP/l; 7 μg PCP/l *vs.* 76 μg PCP/l as reported by Cantelmo and Rao, 1978) is relatively high and does not indicate a shift in nematode species composition. However, the percent similarity of 44% noted with the nematodes in aquaria exposed to 161 μg PCP/l (Fig. 2) and 47% similarity noted with nematodes in aquaria exposed to 622 μg PCP/l (Cantelmo and Rao, 1978) indicates a change in nematode species composition.

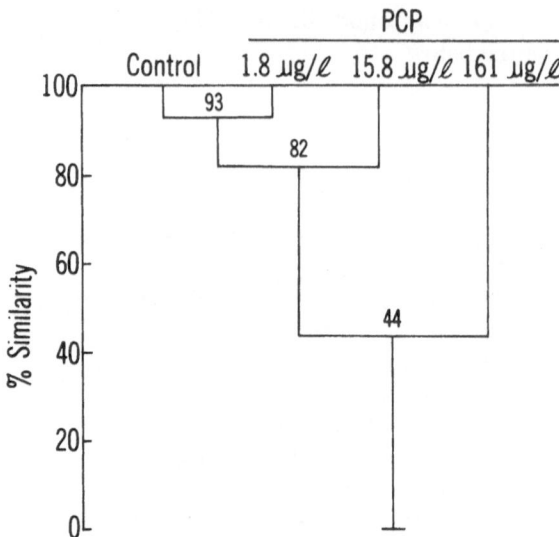

Figure 2. Affinity dendrogram showing the meiobenthic nematode faunal similarities between control aquaria and experimental aquaria exposed to Dowicide® G-ST for thirteen weeks. The measured concentrations of pentachlorophenol in the water are given.

Changes in Nematode Feeding Types

The changes in nematode species composition can also be clearly seen by examining the nematode feeding types. The identification of nematode feeding types followed the classification of Wieser (1953) and Boucher (1973).

Feeding Type	Characteristics
1A	with minute or no buccal cavity (selective deposit feeders)
1B	with a large unarmed buccal cavity (non-selective deposit feeder)
2A	buccal cavity provided with small armature (epistrate feeders)
2B	buccal cavity with big and powerful armature (predators and omnivores)

Non-selective deposit feeders and predator/omnivores were generally poorly represented in control aquaria and the aquaria exposed to 1.8, 7, 15.8, and 76 µg PCP/l (Fig. 3). However, epistrate feeders accounted for 69-82% of all nematode feeding types in control, 1.8, 7, 15.8, and 76 µg PCP/l aquaria and significantly declined in the 161 and 622 µg PCP/l aquaria. Selective deposit feeders showed the reverse trend with low percentages (1-19%) in the control aquaria and the experimental aquaria exposed to 1.8 to 76 µg PCP/l while their relative abundance among the nematodes in the aquaria exposed to 161 and 622 µg PCP/l was high (61-69%). Epistrate feeders *(Microlaimus problematicus, Prochromadorella micoletzkyi,* and *Chromadorita* sp.) and a selective deposit feeder *(Diplolaimella punicea)* were the main species responsible for the nematode changes recorded in the May-July experiment (Cantelmo and Rao, 1978) while epigrowth feeders *(Chromadorina germanica* and *Chromadora nudicapitata)* and a selective deposit feeder *(Monhystera parva)* were the main species responsible for the changes in relative abundance of feeding types recorded in the December-March experiment. At present, no additional data are available to explain these shifts, but some speculation is possible. Epistrate feeders are known to consume algae by scraping off the material from sedimentary particles, puncturing the cell wall and withdrawing the contents or ingesting whole algal cells. Although no data are available on benthic microflora in these flow-through systems, more benthic algae

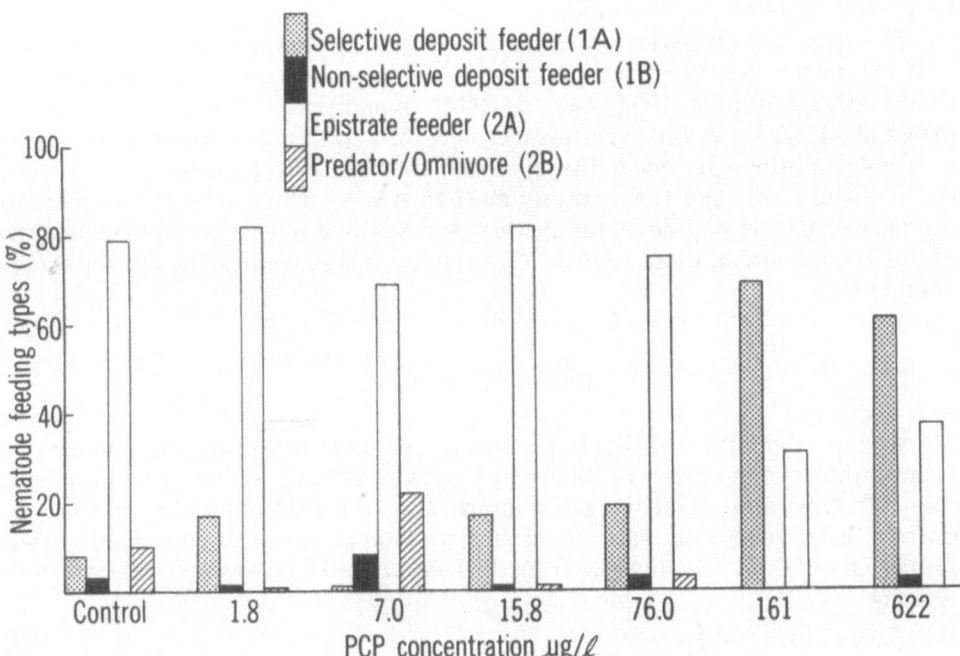

Figure 3. Distribution of nematodes, classified according to their feeding types, in control aquaria and experimental aquaria exposed to either Dowicide® G-ST for thirteen weeks or to pentachlorophenol for nine weeks. The measured concentrations of pentachlorophenol in the water are given.

were apparent (photographic documentation) in control, 1.8, 7.0, 15.8 and 76.0 μg PCP/l than in 161 and 622 μg PCP/l treated aquaria. The apparent reduction of benthic algae in aquaria with the higher concentrations of PCP may be due to the well-documented (Bevenue and Beckman, 1967; Mukherji, 1972) algicidal properties of PCP. Since nematodes have been shown to be selective in ingestion as well as digestion (Tietjen and Lee, 1973, 1977; Jennings and Deutsch, 1975), changes in the availability of food could account for a change in the relative abundance of epistrate and selective deposit feeders in these aquaria. The overall decline in nematode biomass and densities in aquaria exposed to 161 and 622 μg PCP/l may also be in part due to the toxic effects of PCP on meiofauna.

Conclusions

In summary, the highest concentrations of PCP (161 and 622 μg PCP/l) caused significant reductions in nematode densities, nematode biomass and changes in species composition. The observed effects may be due to one or more of the following factors: (a) alterations in food supply resulting from the algicidal properties of PCP, (b) possible differences in the sensitivity of nematode species to PCP, (c) variations in the competition between meiofauna and macrofauna caused by the toxicity of PCP to macrofauna.

The nematode densities established under these flow-through conditions are well within the ranges reported for sandy substrata in the natural environment (Ward, 1973; Lorenzen, 1974; Tietjen, 1977). In addition, the total meiofauna and nematode densities in the experimental flow-through aquaria appeared to exhibit seasonal variations in much the same manner as that expected in a natural environment. Although the dispersal mechanisms by which these nematodes are established in our experimental system remains unknown, the system appears useful in maintaining these animals for extended periods and evaluating the effects of pollutants.

ACKNOWLEDGMENTS

We are thankful to Mr. M.E. Tagatz for permitting us to examine the meiobenthic communities in the flow-through aquaria at the Environmental Research Laboratory, Gulf Breeze, Florida. Ms. Anita Brannon and Dr. Jerry L. Oglesby have generously assisted in statistical analyses. This investigation was supported by Grant R-804541-01 from the United States Environmental Protection Agency.

References

Bevenue, A., and H. Beckman, 1967. Pentachlorophenol: a discussion of its properties and its occurrence as a residue in human and animal tissues. *Residue Rev.*, **19**: 83-134.

Boucher, G., 1973. Premiers donees ecologiques sur les nematodes libres marins d'une station de vas côtière de Banyuls. *Vie Milieu,* **23B:** 69-100.

Cantelmo, F.C., 1977. *The Ecology of Sublittoral Meiofauna in a Shallow Marine Embayment.* Ph.D. Thesis, The City University of New York.

Cantelmo, F.C., and K.R. Rao, 1978. Effect of pentachlorophenol (PCP) on meiobenthic communities established in an experimental system. *Mar. Biol.,* In press.

Coull, B.C., 1970. Shallow water meiobenthos of the Bermuda platform. *Oecologia,* **4:** 325-327.

Coull, B.C., 1973. Estuarine meiofauna: a review: trophic relationships and microbial interactions. In, *Estuarine Microbial Ecology,* pp. 499-512. (H. Stevenson and R.R. Colwell, Eds.), University of South Carolina Press, Columbia.

Crandall, C.A., and C.T. Goodnight, 1959. The effect of various factors on the toxicity of sodium pentachlorophenate to fish. *Limnol. Oceanogr.,* **4:** 53-56.

De Boveé, F., and J. Soyer, 1974. Cycle annuel quantitatif du meiobenthos des vases terrigenes côtières distribution verticale. *Vie Milieu,* **24B:** 141-157.

Elmgren, R., 1973. Methods of sampling sublittoral soft bottom meiofauna. *Oikos,* **15:** 112-120.

Ganapati, P.N., and A.L.N. Sarma, 1973. The meiofauna and pollution in Visakhapatnam Harbour. *Current Sci.,* **42:** 724-725.

Goodnight, C.J., 1942. Toxicity of sodium pentachlorophenate and pentachlorophenol to fish. *Industr. Eng. Chem.,* **34:** 868-872.

Hansen, D.J., 1974. Aroclor® 1254: Effect on composition of developing estuarine animal communities in the laboratory. *Contrib. Mar. Sci.,* **18:** 19-33.

Hopper, B.E., and S.P. Meyers, 1967. Population studies on benthic nematodes within a sub-tropical sea grass community. *Mar. Biol.,* **1:** 85-96.

Jansson, B.O., 1966. Microdistribution of factors and fauna in marine sandy beaches. *Veroff. Inst. Meeresforsch. Bremerh.,* **2:** 77-86.

Jansson, B.O., 1968. Quantitative and experimental studies of the interstitial fauna in four Swedish sandy beaches. *Ophelia,* **5:** 1-71.

Jennings, J.B., and A. Deutsch, 1975. Occurrence and possible adaptive significance of β-glucuronidase and arylamidase (Leucine aminopeptidase) activity in two species of marine nematodes. *Comp. Biochem. Physiol.,* **52A:** 611-614.

Johannes, R.E., 1965. Influence of protozoa on nutrient regeneration. *Limnol. Oceanogr.,* **10:** 434-442.

Lasserre, P., 1971. Oligochaeta from the marine meiobenthos: taxonomy and ecology. *Smithsonian Contr. Zool.,* **76:** 71-86.

Lorenzen, S., 1974. Die Nematodenfauna der sublitoralen region der Deutschen Bucht, insbesondere in Titan-Abwassergebiet bei Helgoland. *Veroff. Inst. Meeresforsch. Bremerh.,* **14:** 305-327.

Marcotte, B.M., and B.C. Coull, 1974. Pollution, diversity and meiobenthic communities in the North Adriatic (Bay of Piran, Yugoslavia). *Vie Milieu,* **24B:** 281-300.

Marshall, N., 1970. Food transfer through the lower trophic levels of the benthic environment. In, *Marine Food Chains,* pp. 52-56. (J.H. Steele, Ed.), Oliver and Boyd, Edinburgh.

McIntyre, A.D., 1964. The ecology of marine meiobenthos. *Biol. Rev.,* **44:** 245-290.

McIntyre, A.D., and D.J. Murison, 1973. The meiofauna of a flatfish nursery ground. *J. Mar. Biol. Assoc. U.K.,* **53:** 93-118.

Mukherji, S.K., 1972. Use of pentachlorophenol as an algicide in paddy fields in West Bengal. *Weed Res.,* **12:** 389-390.

Rhoads, O.C., 1974. Organism-sediment relations on the muddy sea floor. *Ann. Rev. Oceanogr. Mar. Biol.,* pp. 263-300.

Shim, J.C., and L.S. Self, 1973. Toxicity of agricultural chemicals to larvivorous fish in Korean rice fields. *Trop. Med.,* **15:** 123-130.

Stripp, K., 1969. Das Verhaltnis von Makrofauna und Meiofauna in den Sedimenten der Helgolanden Bucht. *Veroff. Inst. Meeresforsch. Bremerh.,* **12:** 143-148.

Tagatz, M.E., J.M. Ivey, J.C. Moore and M. Tobia, 1977. Effects of pentachlorophenol on development of estuarine communities. *J. Toxicol. Environ. Hlth.,* **3:** 1-6.

Tenore, K.R., Tietjen, J.H. and Lee, J.J., 1977. Effect of meiofauna on incorporation of aged eelgrass, *Zostera marina,* detritus by the polychaete *Nephthys incisa. J. Fish. Res. Board Can.,* **34:** 563-567.

Tietjen, J.H., 1969. The ecology of shallow water meiofauna in two New England estuaries. *Oecologia,* **2:** 251-291.

Tietjen, J.H., 1977. Population distribution and structure of the free living nematodes of Long Island Sound. *Mar. Biol.,* in press.

Tietjen, J.H., and J.J Lee, 1973. Life history and feeding habits of the marine nematode *Chromadora macrolaimoides* Steiner. *Oecologia,* **12:** 303-314.

Tietjen, J.H., and J.J. Lee, 1977. Feeding behavior of marine nematodes. In, *Ecology of Marine Benthos,* pp. 23-36. (B.C. Coull, Ed.), University of South Carolina Press, Columbia.

Ward, R., 1973. Studies on the sublittoral free-living nematodes of Liverpool Bay. I. The structure and distribution of the nematode populations. *Mar. Biol.,* **22:** 53-66.

Wieser, W., 1953. Die Beziehung zwischen Mundhohlengestalt, Ernahrungsweise und Vorkommen bei freilebenden marinen Nematoden, *Ark. Zool.,* **4:** 439-484.

Yasuraoka, K., and Y. Hosaka, 1971. The problem of resistance of *Oncomelania* snail to sodium pentachlorophenate. *Japan J. Med. Sci. Biol.,* **24:** 393-394.

Effect of Sodium Pentachlorophenate on the
Feeding Activity of the Lugworm, *Arenicola cristata* Stimpson

NORMAN I. RUBINSTEIN

Abstract—A benthic bioassay utilizing time-lapse photography was used to measure the effect of four concentrations of Na-PCP (45, 80, 156 and 276 $\mu g/l$) on the feeding activity of *Arenicola cristata*. There was no marked effect on feeding activity at 45 $\mu g/l$. Na-PCP significantly affected feeding activity at concentrations of 80, 156, and 276 $\mu g/l$. As the lugworm feeds it mixes organic material and oxygenated water into the substrate. Inhibition of this activity could affect benthic community trophic structure and substrate-water column dynamics.

Introduction

Pentachlorophenol (PCP) and its sodium salt (Na-PCP) are the active ingredients of commercial bactericides used in drilling muds and completion fluids for oil-well drilling (Robichaux, 1975). Various avenues exist whereby these compounds enter the marine environment as a result of standard drilling procedures. Other inputs of PCP into the marine environment include effluent from wood processing and paper pulp operations, cooling water from steam electric plants and run-off from areas where PCP is used as an insecticide for control of termites and wood boring insects. Detectable levels of PCP in the aquatic environment have been reported by Bevenue *et al.* (1972), Byrd (1974) and Fountaine *et al.* (1976). Occasionally, PCP has been found at concentrations great enough to cause large fish kills (Fountaine *et al.*, 1976; Holmberg *et al.*, 1972).

Little is known of the effects of PCP on marine benthic invertebrates. Tagatz *et*

NORMAN I. RUBINSTEIN • Faculty of Biology, University of West Florida, Pensacola, Florida 32504, U.S.A.

al. (1977) found that PCP causes marked changes in the estuarine communities established under flow-through conditions. Although the acute toxicity of Na-PCP to several estuarine organisms has been determined (Borthwick and Schimmel; Schimmel *et al.,* this symposium) the behavioral toxicology of this compound has not been evaluated previously.

The aim of this investigation was to examine the effect of Na-PCP on the substrate reworking activity of an infaunal polychaete, *Arenicola cristata* Stimpson (commonly referred to as the lugworm). Lugworms are tube dwelling, deposit feeders found in most littoral habitats throughout the world. Their activities, which are somewhat analogous to those of the earthworm are responsible for bioturbations of the substrate to depths as great as 30 cm. The lugworm reworks sediment as a function of its feeding mode. As the lugworm feeds, it produces distinct topographical features in the form of funnel-shaped depressions on the substrate surface. Feeding is an integral part of an activity sequence which also incorporates excretion and peristaltic pumping of water through the burrow for respiration and ventilation. These combined activities comprise the Normal Cyclical Pattern which is believed to be controlled by internal pacemakers and is therefore relatively independent of normal environmental variables (Wells, 1966). A decrease in the formation of feeding funnels would indicate an interruption in this activity pattern.

Method for Testing Feeding Activity of Lugworms

A bioassay was developed that monitors the formation of surface features produced by lugworms (Rubinstein, 1977). Comparison of rates of feeding funnel formation between experimental and control lugworms forms the basis of this test. This test was employed to determine the effect of Na-PCP on lugworm activity.

Experiments were conducted in 125 l aquaria (0.25 m² surface area) containing 25 cm of clean beach sand (particle size 200-700 μm) and 75 l of filtered seawater (20 μm filter). All tests were conducted under static conditions; salinity ranged from 22 to 24°/∞ and temperatures fluctuated between 23 and 25 °C. Aquaria were aerated.

Six lugworms of similar size were introduced into each tank and allowed 48 hours to adapt to test conditions. Ground seagrass (70 g), predominantly *Thallassia testudium,* was added to each tank. The seagrass settled to the substrate surface forming a dark mat approximately 3 mm thick. This material served as a source of food for the lugworms and also provided photographic contrast against the white underlying sand in areas disturbed by the feeding animals. When the seagrass completely settled, the test compound was introduced into the experimental aquaria. At each dose level tested, Na-PCP was introduced into one aquarium while one aquarium served as the control. Four concentrations of Na-PCP (45, 80, 156 and 276 μg/l) were tested. Stock solutions were prepared from the commercial bactericide Dowicide® G-ST (79% Na-PCP). Seawater in the tanks was sampled one hour after introduction of the test compound. Concentrations of PCP were determined by gas chromatography (details of the method are described by Schimmel *et al.,* this symposium).

Changes in surface features produced by lugworms were monitored by time-lapse photography. A 35 mm single lens reflex camera was positioned above each tank. Cameras were equipped with automatic advance mechanisms, 24-hour timers and an automatic lighting system. Photographs of the substrate surface were taken at 12-hour intervals beginning 12 hours after introduction of the Na-PCP. The duration of each test was 144 hours. Upon completion of the test, photographs were analyzed to determine the surface area disturbed by the feeding lugworms. The total surface area turned under was plotted against time to give a substrate modification rate for both exposed and control animals. Rates were subjected to linear regression analysis and the slopes of the calculated lines were compared. Differences due to treatment were considered significant at $\alpha = 0.01$.

Effect of Na-PCP on Lugworm Feeding Activity

Sodium pentachlorophenate had no marked effect on lugworm feeding activity at 45 $\mu g/l$ (for a 144-hour period). Significant inhibition of feeding activity was

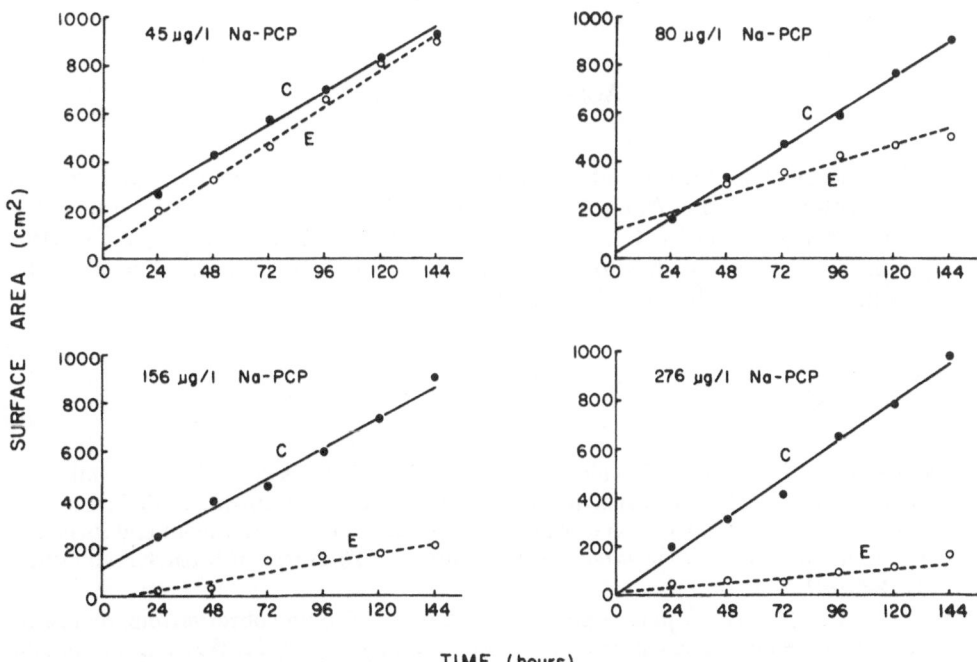

Figure 1. Comparison of the rates of sediment turned under by the lugworm *Arenicola cristata.* C: controls; E: experimental group exposed to sodium pentachlorophenate. Each group consisted of six lugworms.

observed at 80, 156 and 276 $\mu g/l$ (F $\geq$ 188, d.f. 2, 20; α = 0.01) (Fig. 1). Two of the six lugworms exposed to 276 $\mu g/l$ died, while no mortalities occurred at the lower concentrations tested.

Based on these results, a water concentration of Na-PCP between 45 and 80 $\mu g/l$ would interrupt the Normal Cyclical Pattern activity of the lugworm. The control lugworms and those exposed to 80 μg Na-PCP/l displayed similar rates of feeding funnel formation for the first 48 hours, however, from 60 to 144 hours significant inhibition of substrate reworking activity was observed with the latter worms (F = 446, d.f. 2, 12; α = 0.01). This latent effect suggests that lugworms may gradually accumulate Na-PCP until a threshold level is reached which then affects the Normal Cyclical Pattern.

Although the consequences of reduced lugworm activity are speculative at this time, it is possible that a suspension of substrate reworking by the lugworm and other infaunal organisms with similar deposit feeding habits could reduce the exchange of pollutants between water and sediment and thereby prolong the residence time of a pollutant in the water column. Garnas *et al.* (1977) showed that lugworms in a static system significantly affected the level of the insecticide methyl parathion remaining in the water column. Reduced feeding activity of the lugworm could also ultimately result in the death of these organisms. Such an event could affect the cycling of nutrients and toxicants through the benthic system as well as alter food chains of which the lugworm forms a part.

ACKNOWLEDGMENTS

This study was supported by Grant R-804458 from the United States Environmental Protection Agency. I would like to thank Ms. Lynn Faas for analysis of all water samples and Dr. C. N. D'Asaro for reviewing this manuscript. Facilities to complete this research were made available by Dr. Thomas Duke of the Environmental Research Laboratory, Gulf Breeze, Florida.

References

Bevenue, A., J.W. Hylin, Y. Kawano, and T.W. Kelly, 1972. Organochlorine residues in water, sediment, algae, and fish in Hawaii - 1970-71. *Pesticide Monit. J.,* **6**: 56-64.

Byrd, F.A., 1974. The distribution of pentachlorophenol in the environment and its effects as a pollutant. *Student-Oriented Studies Project, 1973. National Science Foundation.* **12**: 239-241.

Fountaine, J.E., P.B. Joshipura, and P.N. Keliher, 1976. Some observations regarding pentachlorophenol levels in Haverford Township, Pennsylvania. *Water Res.,* **10**: 185-188.

Garnas, R.L., C.N. D'Asaro, N.I. Rubinstein, and R.A. Dime, 1977. The fate of methyl parathion in a marine benthic microcosm. Paper #44 in Pesticide Chemistry Division, 173rd ACS meeting, New Orleans, Louisiana, March 20-25, 1977.

Holmberg, B., S. Jensen, A. Larson, K. Lewander, and M. Olsson, 1972. Metabolic effects of technical pentachlorophenol (PCP) on the eel *Anguilla anguilla*. L. *Comp. Biochem. Physiol.,* **43B:** 171-183.

Robichaux, T.J., 1975. Bactericides used in drilling and completion operations. pp. 183-191. In, *Environmental Aspects of Chemical Use in Well-Drilling Operations.* Conference Proceedings, May 1975, Houston, Texas. Report No. EPA-560/1-75-004, United States Environmental Protection Agency, Washington, D.C.

Rubinstein, N.I., 1977. A benthic bioassay utilizing time-lapse photography to measure the effect of toxicants on the feeding behavior of lugworms (Polychaeta: Arenicolidae). Submitted for publication.

Tagatz, M.E., J.M. Ivey, J.C. Moore, and M. Tobia, 1977. Effects of pentachlorophenol on development of estuarine communities. *J. Toxicol. Environ. Hlth.,* **3:** 1-6.

Wells, G.P., 1966. The lugworm *(Arenicola)* a study in adaptation. *Netherlands J. Sea Res.,* **3:** 294-313.

Toxicity of Sodium Pentachlorophenate to the Grass Shrimp, *Palaemonetes pugio,* in Relation to the Molt Cycle

PHILIP J. CONKLIN and K. RANGA RAO

Abstract—The toxicity of sodium pentachlorophenate (Na-PCP) to the grass shrimp, *Palaemonetes pugio,* was evaluated at different stages of the molt cycle. In 96-hour bioassays, the shrimp in later stages of the proecdysial period exhibited a greater sensitivity to Na-PCP than that exhibited by shrimp in the intermolt and early proecdysial stages of the molt cycle. The shrimp in later proecdysial stages generally molted (underwent ecdysis) during the 96-hour test period and died shortly after ecdysis. The 96-hour LC_{50} value obtained for these shrimp (0.436 ppm) is the lowest of all the LC_{50} values reported previously for adult crustaceans and is comparable to those for fish and larval crustaceans. The increased sensitivity to Na-PCP during the early postecdysial period was also apparent in a long-term (66 days) test. The observed postecdysial mortality of shrimp exposed to 1.0 ppm Na-PCP was not dependent on the duration of exposure of shrimp to Na-PCP during the proecdysial period. Studies with ^{14}C-PCP indicate that an abrupt increase in the uptake of PCP during the period shortly after ecdysis may cause increased mortalities during this period.

Introduction

The periodic shedding of the exoskeleton is one of the most spectacular events associated with the growth of crustaceans. Changes in the permeability of the cuticle occur in relation to the cyclic shedding, secretion and hardening of the

PHILIP J. CONKLIN and K. RANGA RAO • Faculty of Biology, University of West Florida, Pensacola, Florida 32504, U.S.A.

exoskeleton in crustaceans (Passano, 1960). During the period immediately following ecdysis (molt) the new thin cuticle is relatively more permeable and less protective than the thicker and calcified exoskeleton present during the intermolt period. Although there have been no studies to evaluate the toxicity of pesticides to crustaceans at different stages of the molt cycle, previous studies indicate that newly molted animals exhibit increased sensitivity to toxicants. Armstrong *et al.* (1976) observed increases in sensitivity of the Dungeness crab, *Cancer magister,* to methoxychlor during or soon after ecdysis. Similar increases in sensitivity have been suggested for the pink shrimp, *Penaeus duorarum* exposed to a polychlorinated biphenyl, Aroclor 1254® (Duke *et al.,* 1970; Nimmo *et al.,* 1971). Investigations of acute toxicity of copper to the crayfish, *Orconectes rusticus,* indicated that many of the test organisms died in the act of ecdysis (Hubschman, 1967).

Previous investigations indicate that adult crustaceans are more tolerant than fish to pentachlorophenol and sodium pentachlorophenate (Goodnight, 1942; Kaila and Saarikoski, 1977). The toxicity data available for crustaceans were derived from bioassays on animals whose physiological status in relation to the molt cycle was unknown. The aim of this investigation was to evaluate the toxicity of sodium pentachlorophenate (Na-PCP) to the grass shrimp, *Palaemonetes pugio,* at known stages of the molt cycle using standard 96-hour bioassays, long-term exposures and exposures for varying periods during the molt cycle.

Test Animals

Grass shrimp, *Palaemonetes pugio,* were collected from grass beds in Santa Rosa Sound, Gulf Breeze, Florida. Shrimp were used within two weeks after collection, as it has been shown that holding the animals in the laboratory for longer durations before use may lead to altered responses (Tatem *et al.,* 1976). The average length (rostrum-telson) of the shrimp used was 25 mm, with a range of 22 to 28 mm. Only non-gravid shrimp were used.

Test Conditions

The shrimp were maintained in sea water (salinity 10‰) at a temperature of 20 ± 1 °C under 12 hour light - 12 hour dark conditions. Shrimp were not fed during

---

Figure 1. The outer edge of a uropod (x 125) of an intermolt (Stage C) grass shrimp, *Palaemonetes pugio.* The epidermis shows no evidence of retraction from the old cuticle.

Figure 2. The outer edge of a uropod (x 125) of *Palaemonetes* in an early proecdysial (Stage D_0) stage of the molt cycle. Epidermal retraction can be seen.

Figure 3. The outer edge of a uropod (x 125) of *Palaemonetes* showing the initiation of neosetogenesis (Stage D_1).

Figure 4. The outer edge of a uropod (x 125) of *Palaemonetes* in late proecdysis (Stages D_3-D_4) with well developed new setae.

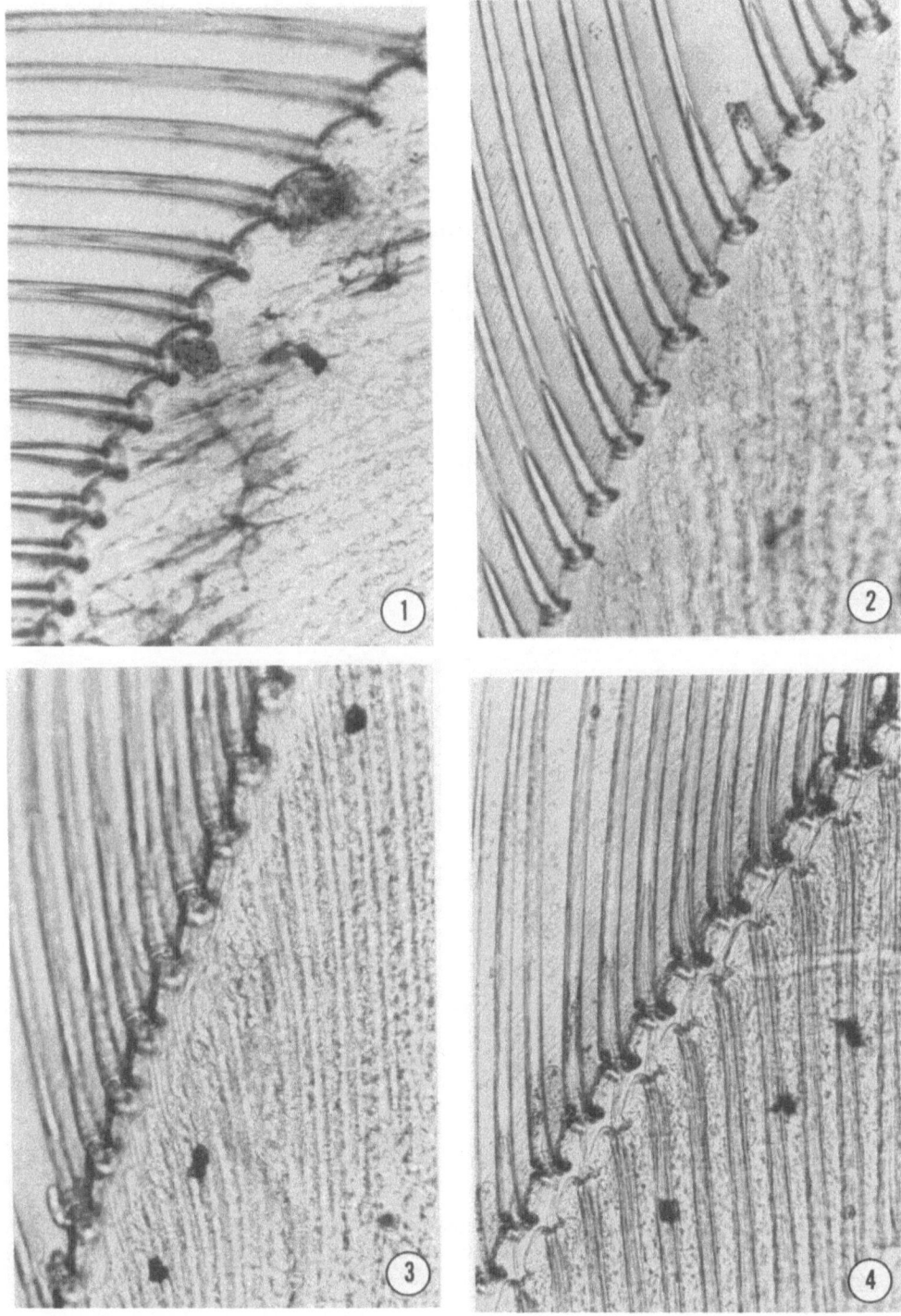

short-term (96-hour) bioassays. During long-term exposures the grass shrimp were fed live brine shrimp larvae on alternate days. The media of control and experimental shrimp were replaced daily with fresh solutions. Analytical grade Na-PCP (City Chemical Corp., New York; supplied by the Gulf Breeze Environmental Research Laboratory) was dissolved in filtered sea water. Solutions of Na-PCP were prepared daily, immediately prior to use, and diluted to the desired concentrations.

Identification of the Stages in the Molt Cycle

The crustacean molt cycle can be divided into five major stages (A through E) and several substages based on morphological criteria (Drach, 1939). One of the methods used in identifying the stages in the molt cycle involves an examination of the progress in formation of new setae during the proecdysial period (Drach and Tchernigovtzeff, 1967; Rao et al., 1973). Each shrimp was placed on a glass microscope slide, with a drop of sea water under the telson. The telson and uropods were gently spread with the help of a probe and the edges of the uropods were

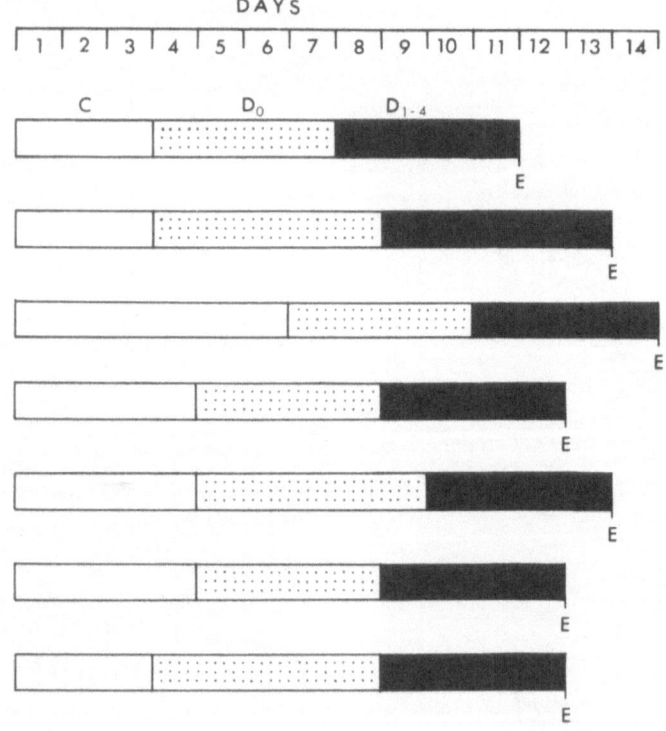

Figure 5. The duration of various stages in the molt cycle of the grass shrimp, *Palaemonetes pugio*. Although the relative durations of intermolt (Stage C) and early proecdysial (Stage D_0) stages are variable, the duration of Stages D_1 to D_4 remained constant (4 to 5 days). E: Ecdysis.

examined mircoscopically (100 × magnification) for signs of epidermal retraction (apolysis) and progress in the formation of new setae (neosetogenesis). Intermolt (stage C) shrimp show no evidence of epidermal retraction or neosetogenesis (Fig. 1). During early proecdysis (stage D_0) epidermal retraction is evident (Fig. 2). This is followed by neosetogenesis during stage D_1 (Fig. 3). Shrimp in advanced stages (D_3-D_4) of the proecdysial period can be easily identified by the extremely well developed new setae (Fig. 4). Finally, the old exoskeleton is cast off (stage E) and the early postecdysial stages (A and B) begin. Immediately after ecdysis the animal is protected by a thin new cuticle. During the postecdysial period additional layers of cuticle are secreted and calcified and these processes are completed in stage C (intermolt stage) of the molt cycle.

Duration of the Various Stages in the Molt Cycle

The length of the molt cycle varies with the size of the shrimp, season and other factors such as temperature and photoperiod. The average duration of the molt cycle for representative grass shrimp tested was 13 days. Although the duration of stages C and D_0 varied among individuals, the duration of the later proecdysial stages (D_1-D_4) appeared relatively constant (Fig. 5). Therefore, as the shrimp completed later proecdysial stages it was possible to predict reasonably well how soon ecdysis would occur.

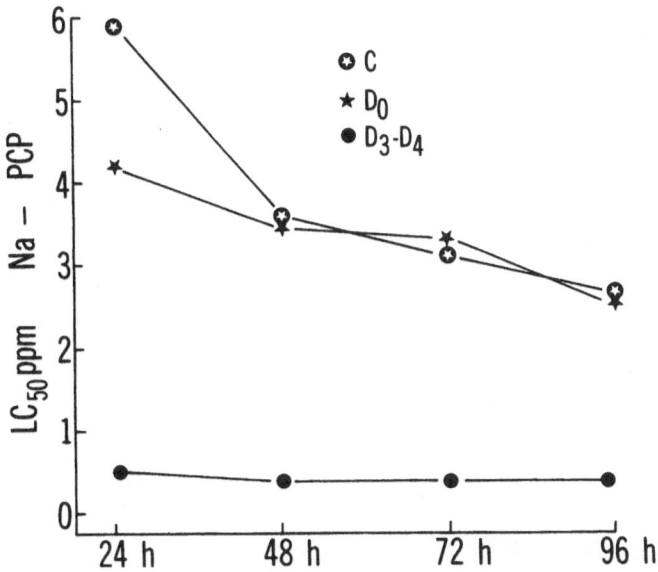

Figure 6. Short-term (96-hr) toxicity of sodium pentachlorophenate (Na-PCP) to *Palaemonetes pugio* at different stages of the molt cycle. The median lethal concentrations (LC_{50}) were computed based on the observed mortalities at 24, 48, 72 and 96 hours after exposure.

Short-term Toxicity Tests on Grass Shrimp
at Different Stages of the Molt Cycle

Standard 96-hour toxicity tests (American Public Health Association, 1975) were performed to determine the acute toxicity of Na-PCP to grass shrimp at different stages of the molt cycle (Conklin and Rao, 1977). Intermolt (stage C) shrimp and shrimp in early proecdysial (stage D_0) and late proecdysial (stages D_3-D_4) stages of the molt cycle were exposed to solutions of Na-PCP. The median lethal concentrations (LC_{50}; computed using probit analysis, Finney, 1971) of PCP for intermolt shrimp were not significantly different from those for shrimp in early proecdysis. However, shrimp in late proecdysial stages (D_3-D_4) exhibited increased sensitivity to Na-PCP as indicated by the substantially lower LC_{50} values (Fig. 6) than those obtained for shrimp in intermolt and early proecdysial stages. The shrimp that were in stages D_3-D_4 at the beginning of the test usually underwent ecdysis (molted) within the first 24 hours of the test period. Mortalities occurred shortly after ecdysis.

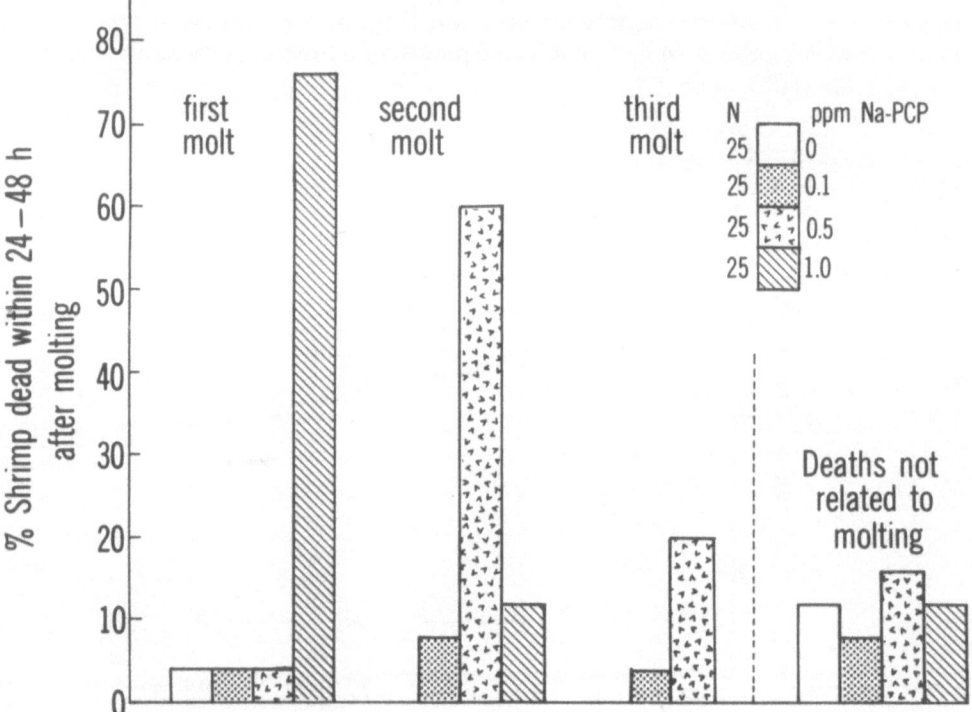

Figure 7. Results of a long-term (66 days) exposure of the grass shrimp, *Palaemonetes pugio,* to sodium pentachlorophenate (Na-PCP). The incidence of mortalities in relation to ecdysis (molt) is shown.

Long-term Exposure of Grass Shrimp to Na-PCP

A long-term experiment (66 days) was conducted to determine the toxicity of Na-PCP to the grass shrimp in relation to the molt cycle. One hundred intermolt (stage C) grass shrimp were divided into four equal groups. One group served as control while the others were exposed to 0.1, 0.5 and 1.0 ppm (mg/l) Na-PCP. The shrimp were kept individually in glass jars (250 ml solution). Each shrimp was examined microscopically on alternate days to assess the progress in the molt cycle. Daily examinations were made for evidence of ecdysis or mortality. Exposure to Na-PCP did not alter the relative durations of the stages in the molt cycle or the duration of the molt cycles (Rao *et al.,* this symposium). However, Na-PCP affected the survival of shrimp after ecdysis. A majority (63%) of the experimental shrimp died within 24 to 48 hours after ecdysis (Fig. 7). At the highest concentration of Na-PCP tested, 1.0 ppm, a majority (76%) of deaths occurred within 24 to 48 hours after the first ecdysis (molt). Although most of the shrimp (96%) exposed to 0.5 ppm

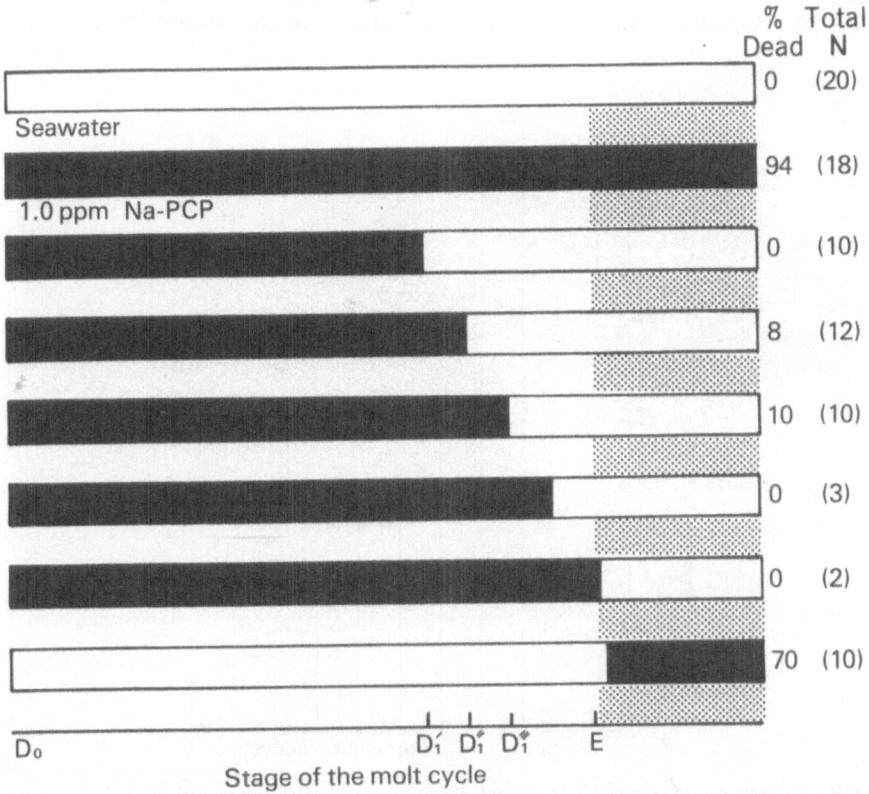

Figure 8. The critical period of the molt cycle for the toxicity of sodium pentachlorophenate (Na-PCP) to the grass shrimp, *Palaemonetes pugio.* Solid bar indicates the period of exposure to Na-PCP. Stipled area represents the postecdysial period. N represents the number of animals in each experimental group.

survived following the first ecdysis, a majority of the shrimp (60%) died within 24 to 48 hours after their second ecdysis. Exposure to 0.1 ppm Na-PCP did not affect the survival of shrimp.

Critical Period of the Molt Cycle

Further experiments were conducted to determine more precisely the time at which death occurred in relation to the molt cycle and to see if the toxic effects of Na-PCP were due to a cumulative uptake of the compound during the proecdysial period. Premolt (stage D_0) grass shrimp were exposed to 1.0 ppm Na-PCP. One group of shrimp was exposed to Na-PCP throughout the proecdysial period and allowed to complete ecdysis in this medium. As expected, all of these animals died following ecdysis (Fig. 8). When shrimp were exposed to Na-PCP for varying durations during proecdysis up to a period just prior to ecdysis and then transferred to sea water, very few mortalities occurred following ecdysis. This indicates that at the concentration tested (1.0 ppm), the duration of exposure to Na-PCP during the period of proecdysis was not the major factor affecting the postecdysial survival of

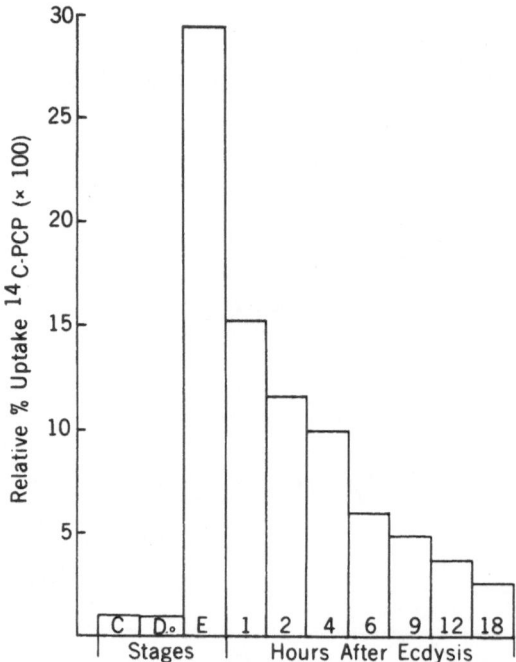

Figure 9. Relative uptake of ^{14}C-pentachlorophenol by the grass shrimp, *Palaemonetes pugio,* at different stages of the molt cycle. The uptake of ^{14}C-PCP by intermolt (Stage C) shrimp within one hour after exposure to a medium containing 2.0 ppm PCP is designated as 100% for comparison with other stages of the molt cycle. The intermolt shrimp exposed to 2.0 ppm PCP for one hour contained on an average 13 ppm PCP (13 mg/Kg wet weight).

grass shrimp. A group of shrimp was kept in sea water throughout the proecdysial period and allowed to complete ecdysis in this medium. As expected, no postecdysial mortalities occurred. However, when these shrimp were placed in 1.0 ppm Na-PCP within a few (1 to 3) hours after ecdysis, a majority of the shrimp died (Fig. 8). These results indicate that the period immediately following ecdysis is the most critical period for the toxic effects of Na-PCP on grass shrimp.

Uptake of ^{14}C-PCP in Relation to the Molt Cycle

The relative uptake of ^{14}C-PCP (Pathfinder Labs) by shrimp at different stages of the molt cycle was studied by Cantelmo, Conklin and Fox in this laboratory. Shrimp were exposed to a medium containing 2.0 ppm PCP for one hour and the total amount of PCP accumulated within this period by the grass shrimp was determined. Postecdysial shrimp exhibited substantially greater uptake of ^{14}C-PCP than intermolt (stage C) shrimp (Fig. 9). The greatest accumulation of PCP was noted in shrimp which were exposed to this medium immediately after ecdysis. A reduction in the uptake of ^{14}C-PCP was apparent with progress in the postecdysial secretion and hardening of the exoskeleton. The above studies show that an abrupt increase in the uptake of PCP shortly after ecdysis is responsible for the mortalities occurring during the postecdysial period.

A Survey of the Toxicity of PCP and Na-PCP to Crustaceans

As shown in Table 1, the 96-hour LC_{50} value (0.436 ppm; 95% C.I., 0.361-0.498) obtained for late proecdysial grass shrimp, *Palaemonetes pugio,* was the lowest of all the LC_{50} values reported for adult crustaceans. The LC_{50} value for adult *Palaemonetes pugio* is comparable to that reported by van Dijk *et al.* (1977) for the larvae of *Palaemonetes varians* (LC_{50}: 0.363 ppm; 95% C.I., 0.200-0.680) and is less than one-tenth of the value reported for adults of this species.

The generalization that crustaceans are less sensitive than fish to Na-PCP may not be valid. The 96-hour LC_{50} value obtained by us with the grass shrimp (0.436 ppm) and those obtained by van Dijk *et al.* (1977) for decapod crustacean larvae (0.084 to 0.363 ppm) are comparable to the range of LC_{50} values reported for fish (0.037 to 0.247 ppm; Davis and Hoos, 1975; Adelman *et al.*, 1976).

The present studies indicate that in evaluating the toxicity of chemicals to adult crustaceans, the physiological status of the animal in relation to the molt cycle should be considered. The relative toxicity at different stages of the molt cycle may vary depending on the type of chemical tested, the species examined, the relative thickness of the cuticle and frequency of ecdysis. In contrast to our observations on adult *Palaemonetes pugio,* the larvae of *Palaemonetes varians* did not show increased sensitivity to Na-PCP in relation to ecdysis (van Dijk *et al.*, 1977). The latter may be because the larval crustaceans generally undergo ecdysis more frequently and possess thinner cuticles compared to adults. The magnitude of changes in the permeability of the crustacean larval cuticle in relation to the molt cycle may not be as great as in the adult crustaceans.

TABLE 1. Toxicity of Pentachlorophenol and Sodium Pentachlorophenate to Crustaceans

Animal		Toxicity		Source
		Test	ppm	
ADULTS				
amphipod:	*Hyelella knickerbockerii*	72 hour toxicity	> 5.0	Goodnight, 1942
cladoceran:	*Daphnia magna*	24 hour toxicity	< 1.0	Weber, 1965
	Daphnia pulex	72 hour toxicity	> 5.0	Goodnight, 1942
crayfish:	*Cambarus (Orconectes) virilis*	72 hour toxicity	> 5.0	Goodnight, 1942
	Astacus fluviatilis	192 hour LC$_{50}$; pH 6.5	9.0	Kaila & Saarikoski, 1977
	Astacus fluviatilis	192 hour LC$_{50}$; pH 7.5	53.0	Kaila & Saarikoski, 1977
isopod:	*Asellus communis*	72 hour toxicity	> 5.0	Goodnight, 1942
prawns and	*Leander japonicus*	48 hour TLM	2.3	Tomiyama & Kawabe, 1962
shrimp:	*Palaemon elegans*	96 hour LC$_{50}$	10.39	van Dijk et al., 1977
	Crangon crangon	96 hour LC$_{50}$	1.79	van Dijk et al., 1977
	Palaemonetes varians	96 hour LC$_{50}$	5.09	van Dijk et al., 1977
	Palaemonetes pugio[a]	96 hour LC$_{50}$	0.436	Conklin & Rao, 1977
LARVAE	*Palaemonetes varians*	96 hour LC$_{50}$	0.363	van Dijk et al., 1977
	Palaemon elegans	96 hour LC$_{50}$	0.084	van Dijk et al., 1977
	Crangon crangon	96 hour LC$_{50}$	0.112	van Dijk et al., 1977

[a]Shrimp in later stages (D$_3$-D$_4$) of the molt cycle at the beginning of the test.

ACKNOWLEDGMENTS

This investigation was supported by Grant R-804541-01 from the United States Environmental Protection Agency. We thank Mr. Richard D. Summerall Jr. for his conscientious assistance in this study.

References

Adelman, I.R., L.L. Smith, Jr., and G.D. Siesennop, 1976. Effect of size or age of goldfish and fathead minnows on use of pentachlorophenol as a reference toxicant. *Water Res.,* **10:** 685-687.

American Public Health Association, 1975, *Standard methods for the examination of water and wastewater including bottom sediments and sludges.* 14th edition, 1193 pp., American Publ. Health Assn., Washington.

Armstrong, D.A., D.V. Buchanan, M.H. Mallon, R.S. Caldwell and R.E. Millemann, 1976. Toxicity of the insecticide methoxychlor to the Dungeness crab *Cancer magister. Mar. Biol.,* **38:** 239-252.

Conklin, P.J., and K.R. Rao, 1977. Toxicity of sodium pentachlorophenate (Na-PCP) to the grass shrimp, *Palaemonetes pugio,* at different stages of the molt cycle. *Bull. Environ. Contam. Toxicol.,* in press.

Davis, J.C., and R.A.W. Hoos, 1975. Use of sodium pentachlorophenate and dehydroabietic acids as reference toxicants for salmonid bioassays. *J. Fish. Res. Bd. Can.,* **32:** 411-416.

Drach, P., 1939. Mue et cycle d'intermue chez les Crustaces Decapodes. *Ann. Inst. Oceanogr., (Paris) (N.S.),* **19:** 103-391.

Drach, P., and C. Tchernigovtzeff, 1967. Sur la methode de determination des stages d'intermue et son application generale aux Crustaces. *Vie Millieu* **A18:** 595-610.

Duke, T.W., J.I. Lowe, and A.J. Wilson, Jr., 1970. A polychlorinated biphenyl (Aroclor 1254®) in the water, sediment, and biota of Escambia Bay, Florida. *Bull. Environ. Contam. Toxicol.,* **5:** 171-180.

Finney, D.J., 1971. *Probit Analysis.* 3rd edition, 333 pp. Cambridge University Press, London.

Goodnight, C.J., 1942. Toxicity of sodium pentachlorophenate and pentachlorophenol to fish. *Industr. Eng. Chem.,* **34:** 868-872.

Hubschman, J.H., 1967. Effects of copper on the crayfish *Orconectes rusticus* (Girard). *Crustaceana,* **12:** 33-42.

Kaila, K., and J. Saarikoski, 1977. Toxicity of pentachlorophenol and 2,3,6-trichlorophenol to the crayfish *(Astacus fluviatilis* L.). *Environ. Pollut.,* **12:** 119-123.

Nimmo, D.R., R.R. Blackman, A.J. Wilson, and J. Forester, 1971. Toxicity and distribution of Aroclor® 1254 in the pink shrimp *Penaeus duorarum. Mar. Biol.,* **11:** 191-197.

Passano, L.M., 1960. Molting and its control. In, *The Physiology of Crustacea* (T.H. Waterman, Ed.), Vol. **1:** pp. 473-536, Academic Press, New York.

Rao, K.R., S.W. Fingerman, and M. Fingerman, 1973. Effects of exogenous ecdysones on the molt cycles of fourth and fifth stage American lobsters, *Homarus americanus. Comp. Biochem. Physiol.,* **44A:** 1105-1120.

Tatem, H.E., J.W. Anderson, and J.M. Neff, 1976. Seasonal and laboratory variations in the health of grass shrimp *Palaemonetes pugio:* dodecyl sodium sulfate bioassay. *Bull. Environ. Contam. Toxicol.,* **16:** 368-375.

Tomiyama, T., and K. Kawabe, 1962. The toxic effect of pentachlorophenate, a herbicide, on fishery organisms in coastal waters. I. the effect on certain fishes and a shrimp. *Bull. Jap. Soc. Sci. Fish.*, **28:** 379-382.

van Dijk, J.J., C. van der Meer, and M. Wijnans, 1977. The toxicity of sodium pentachlorophenate for three species of decapod crustaceans and their larvae. *Bull. Environ. Contam. Toxicol.*, **17:** 622-630.

Weber, E., Einwirkung von Pentachlorophenolnatrium auf Fische und Fischnahrtiers. *Biol. Zentralbl.*, **84:** 81-93.

Inhibition of Limb Regeneration in the Grass Shrimp, *Palaemonetes pugio,* by Sodium Pentachlorophenate

K. RANGA RAO, PHILIP J. CONKLIN and ANITA C. BRANNON

Abstract—The initiation and progress of regeneration following the removal of the left fifth pereiopod were studied using the grass shrimp, *Palaemonetes pugio*. The regeneration patterns of 400 shrimp subjected to various treatments revealed that sodium pentachlorophenate (Na-PCP) affects the initiation and progress of limb regeneration. Depending on the concentration used, Na-PCP caused either a complete inhibition of regeneration, a delay of initiation of limb bud development, or a reduction of limb bud growth without altering the intermolt duration. By comparing the regeneration indices (R values) of control and experimental shrimp noted on specified days preceding ecdysis and on the day following ecdysis it was possible to determine the extent (%) of inhibition of regeneration in shrimp exposed to Na-PCP. EC_{50} values were computed using probit analysis. For example, the R values of shrimp nine days after limb removal yielded the following EC_{50} values with 95% confidence intervals shown in parenthesis: unfed shrimp, 0.473 ppm Na-PCP (0.306-0.670); fed shrimp, 0.565 ppm (0.452-0.706). The EC_{50} values based on postecdysial R values were: unfed shrimp, 0.615 ppm Na-PCP (0.451-0.852); fed shrimp, 0.637 ppm (0.485-0.850). The inhibitory effects of Na-PCP were more pronounced on the initial phases of limb regeneration (involving wound healing, cell division and dedifferentiation) than on the later phases of regeneration (involving further differentiation and cellular enlargement). Crustacean limb regeneration can be used as a sensitive bioassay for studying the effects of chemical pollutants.

K. RANGA RAO, PHILIP J. CONKLIN and ANITA C. BRANNON • Faculty of Biology, University of West Florida, Pensacola, Florida 32504, U.S.A.

Introduction

After a limb has been severed from the body of a crustacean, a new limb may grow to replace it. The initiation, progress and successful completion of limb regeneration depend on several factors, including the stage of the animal in the molt cycle, environmental conditions and the number of limbs lost (Bliss, 1960; Bliss and Boyer, 1964; Rao, 1966; Adiyodi, 1972; Skinner and Graham, 1972; Holland and Skinner, 1976; Weis, 1976a, 1977). Following limb removal, the process of wound healing begins. After wound healing is completed the processes of cell division and cellular dedifferentiation are initiated. During the initial stages of regeneration the limb bud appears as a tiny papilla. Two distinctive phases of regenerative growth, basal limb growth and proecdysial (premolt) limb growth can be distinguished in several crustaceans (Bliss, 1956, 1960). The basal limb growth which can occur independent of proecdysis establishes the primary organization of the limb which may be followed by a growth plateau (arrest in growth) if the environmental conditions are not suitable or the animal is not initiating proecdysial preparations. The next phase of regeneration, one of rapid growth and further development of existing tissues, is the proecdysial growth. This occurs in close association with the proecdysial stages of the molt cycle. The environmental and endocrine factors that accelerate the proecdysial stages may lead to precocious ecdysis and also cause an acceleration of regenerative limb growth (Bliss and Boyer, 1974; Passano and Jyssum, 1963; Rao, 1966).

Environmental pollutants such as insecticides (Weis and Mantel, 1976), and heavy metals (Weis, 1976b, 1977) have been shown to have marked effects on limb regeneration and molting in fiddler crabs. The present report deals with the effects of sodium pentachlorophenate on limb regeneration in the grass shrimp, *Palaemonetes pugio,* in relation to the molt cycle.

Limb Removal and Regeneration in Relation to the Molt Cycle

In order to determine the normal patterns of regenerative limb growth, an initial experiment was conducted with 50 grass shrimp. Each shrimp was microscopically examined to determine its stage in the molt cycle as described in the preceding paper (Conklin and Rao) in this symposium. The left fifth pereiopod was removed from each shrimp using fine tipped jeweler's forceps. Shrimp were maintained individually in glass jars containing 250 ml seawater (10‰) at $20 \pm 1\,°C$ under 12 hr light: 12 hr dark conditions. The shrimp were microscopically examined at two day intervals to assess the progress in limb regeneration and advances in the molt cycle. The regeneration index, R value, was determined following the method of Bliss (1956):

$$R \text{ value} = \frac{\text{length of limb bud}}{\text{carapace length}} \times 100$$

As shown in Figure 1, the incidence of limb regeneration, the rate of growth of the limb bud and the relative size of the new limb after ecdysis depended on (a) the

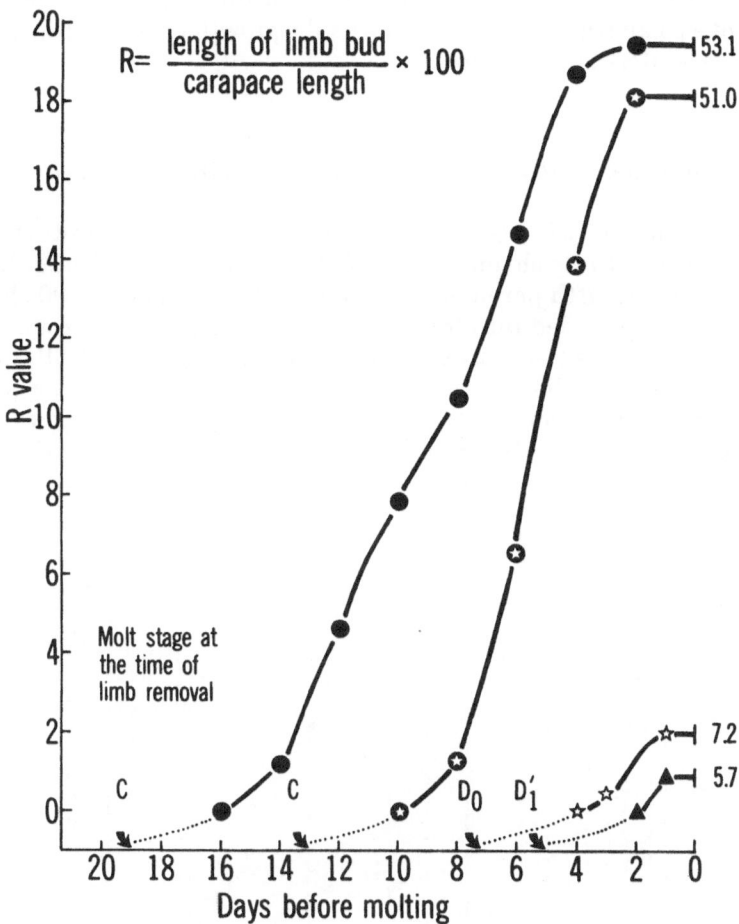

Figure 1. Representative patterns of limb regeneration in the grass shrimp *Palaemonetes pugio*. The left fifth pereiopod was removed (arrow) from shrimp in different stages of the molt cycle. The vertical line at the end of each curve indicates the incidence of ecdysis (molting) and the values given next the vertical bar represent postecdysial R values.

stage of molt cycle at which limb removal occurred and (b) the interval between limb removal and ecdysis. Limb removal during the intermolt stage (C) of the molt cycle led to successful completion of regeneration. No growth plateaus were observed during the progress of limb regeneration. At the time of ecdysis the limb bud stretched to a new dimension due to the uptake of water by the animal. The length of the new limb following ecdysis was usually 2.5 to 3 times greater than the length of the limb bud preceding ecdysis. The postecdysial size and functional morphology of the new limb were dependent on the degree of limb development during the proecdysial period. For example, in animals subjected to limb removal 5 to 7 days preceding ecdysis the proecdysial regeneration was limited to the formation of a small papilla. In these animals the growth of the limb bud continued during the next interecdysial period.

Limb Removal and Exposure of Grass Shrimp to Na-PCP

Two experiments were conducted using a total of 200 grass shrimp. The carapace length of these shrimp ranged between 8.8 and 11.3 mm. In the first experiment the left fifth pereiopod was removed from each of 100 shrimp. The shrimp were then divided into four equal groups. One of the groups served as control while the others were exposed to 0.1, 0.5 and 1.0 ppm Na-PCP from the day of limb removal until the animals completed at least one ecdysis or until the termination of the experiment (28 days). The media for control and experimental shrimp were replaced with fresh solutions daily. The shrimp were not fed during

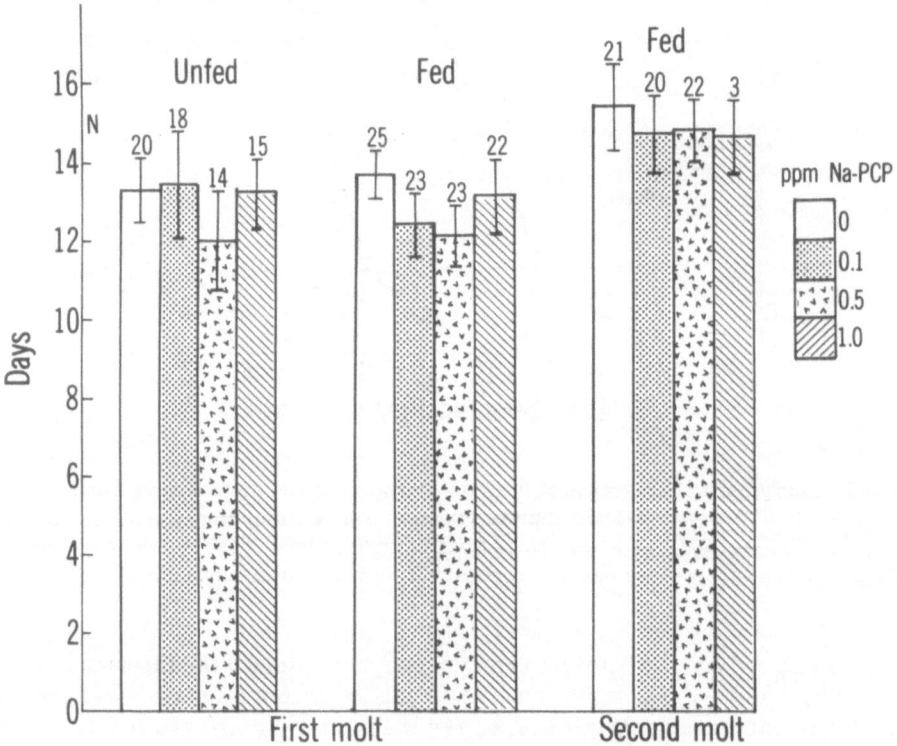

Figure 2. The durations of the molt cycles of unfed and fed grass shrimp, *Palaemonetes pugio,* exposed to 0, 0.1, 0.5 and 1.0 ppm sodium pentachlorophenate. Each bar represents the Mean while the vertical line represents the Standard Error of the Mean. N, number of shrimp which completed ecdysis.

this experiment. The second experiment utilizing 100 shrimp was conducted following the above procedure except that the shrimp were fed live brine shrimp larvae on alternate days for the duration of the experiment (66 days).

Lack of Effect of Na-PCP on the Duration of the Molt Cycle

Following limb removal, both the unfed and fed shrimp completed their first ecdysis. The interval between limb removal and the first ecdysis (molt) following this operation was not effected by exposure to Na-PCP or food (Fig. 2). The unfed animals were not followed through an additional molt cycle. Among the fed shrimp, several controls and those shrimp which survived the long-term exposure to Na-PCP were able to complete up to four ecdyses following limb removal. Since none of the shrimp exposed to 1.0 ppm survived beyond the second ecdysis, the data for subsequent ecdyses are not included in Figure 2. The interval between the first and

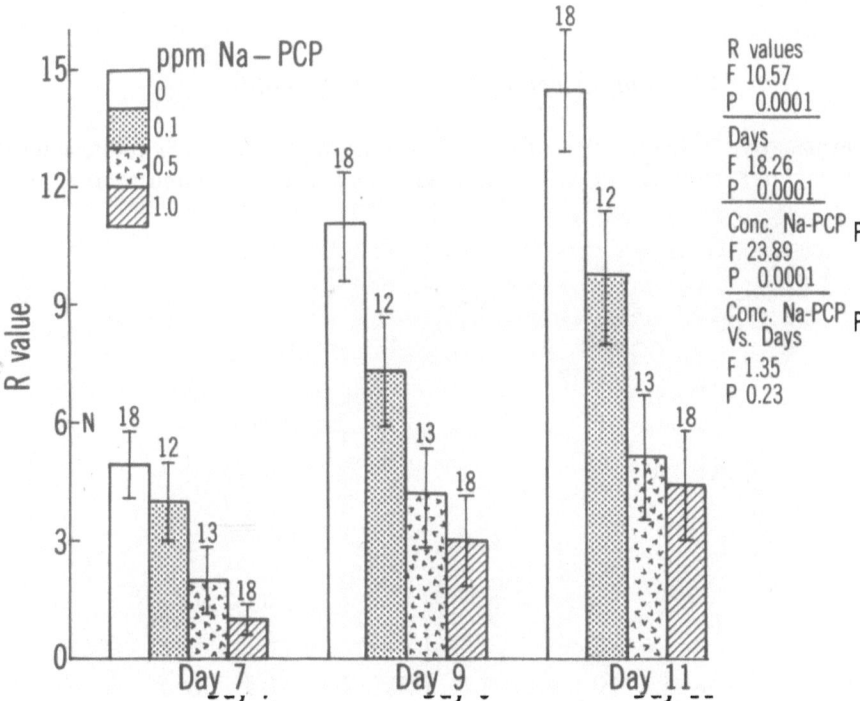

Figure 3. A comparison of the R values noted on the 7th, 9th and 11th day after removal of the left fifth pereiopod from the grass shrimp, *Palaemonetes pugio*. The control shrimp were maintained in seawater while the experimental shrimp were exposed to sodium pentachlorophenate beginning the day of limb removal until the completion of ecdysis. The shrimp were not fed during the experiment. The figure is based on Mean values and Standard Errors of the Mean.

second ecdysis in the control shrimp was not significantly different from that of the shrimp exposed to Na-PCP.

Inhibitory Effects of Na-PCP on Limb Regeneration

The effects of Na-PCP on limb regeneration in unfed shrimp are shown in Figure 3. An examination of the R values on the 7th, 9th and 11th days after limb removal shows the inhibitory effects of Na-PCP on limb regeneration. The degree of inhibition increased with the concentration of Na-PCP used. The inhibition of limb regeneration during the proecdysial period led to a marked reduction in the size of the new limb following ecdysis. The postecdysial (postmolt) R values of unfed and fed shrimp exposed to 0.1, 0.5 and 1.0 ppm Na-PCP from the day of limb removal up to the completion of the molt cycle are shown in Figure 4. It can be seen that the shrimp exposed to Na-PCP developed relatively shorter new limbs compared to the control shrimp. This inhibition by Na-PCP occurred in fed shrimp as well as in unfed shrimp.

Comparison of LC_{50} and EC_{50} Values

The effective concentration of Na-PCP required to cause 50% reduction (EC_{50}) in limb regeneration (R values) compared to controls was calculated. The EC_{50}

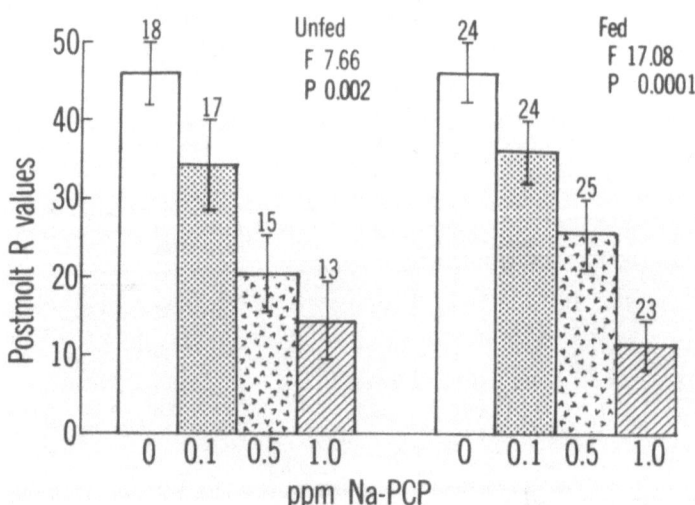

Figure 4. A comparison of the postecdysial (postmolt) R values of unfed and fed grass shrimp, *Palaemonetes pugio*, exposed to sodium pentachlorophenate. The R values are for the left fifth pereiopod. The figure is based on Mean values and Standard Errors of the Mean.

values obtained for fed shrimp were not significantly different from those of unfed shrimp, indicating that the inhibitory effects of Na-PCP on limb regeneration were independent of the shrimp's nutritional state. The EC_{50} values derived from the data on proecdysial R values are comparable to those obtained with postecdysial R values. This supports the earlier conclusion that inhibition of development of the limb bud during proecdysial period leads to a proportional reduction in the size of new limb following ecdysis.

During the intermolt (Stage C) and proecdysial (Stage D_0) periods the effective concentration of Na-PCP required to inhibit regeneration (EC_{50}) was much less than that required to cause death of the shrimp (LC_{50}), as can be seen in Figure 5. These results indicate that limb regeneration can be used as a sublethal bioassay. It should be noted that the EC_{50} values obtained for proecdysial (premolt) and postecdysial (postmolt) regeneration are comparable to the LC_{50} values obtained at the most sensitive stage of the molt cycle (shrimp in D_3-D_4 stages which molted during exposure to Na-PCP).

Figure 5. A comparison of the LC_{50} values (based on 96-hr toxicity tests) and EC_{50} values (based on the extent of inhibition of regenerative limb growth) for grass shrimp, *Palaemonetes pugio,* exposed to sodium pentachlorophenate. The LC_{50} values for shrimp at different stages of the molt cycle are presented.

Limb Regeneration in Relation to the Interval
Between Limb Removal and Exposure to Na-PCP

In order to determine the sensitivity of different phases of limb regeneration to Na-PCP, grass shrimp were exposed to Na-PCP beginning at varying times after limb removal. Two experiments were conducted in this series using 100 shrimp for each experiment. During the first experiment one group of shrimp served as controls, while the experimental groups were exposed to 1.0 ppm Na-PCP beginning the day of limb removal, two days after limb removal or four days after limb removal up to the completion of ecdysis. The postecdysial R values of these shrimp are shown in Figure 6. During the course of this experiment (January, 1977) the shrimp were exposed to relatively low temperatures (15-17 °C) due to a reduction in the heating of our laboratories. The latter was necessitated by a shortage in electric power supply during an unusually cold winter period. It is known that low temperatures usually cause a reduction in the growth of limb buds in crustaceans (Bliss, 1956; Passano, 1960; Rao, 1966; Miller and Vernberg, 1968). Consistent with these previous reports, we noted an overall reduction in limb regeneration in grass shrimp at 15-17 °C (Fig. 6) compared to the growth seen at 20-21 °C (Fig. 4).

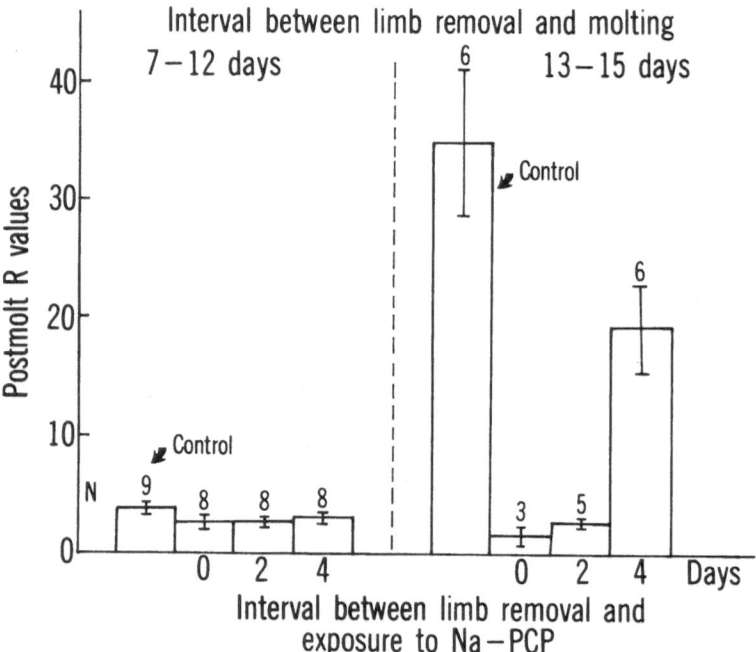

Figure 6. Postecdysial (postmolt) R values of grass shrimp, *Palaemonetes pugio*, exposed to 1.0 ppm sodium pentachlorophenate beginning the day of limb removal, two days after limb removal or four days after limb removal until the completion of ecdysis. The control shrimp were maintained in seawater throughout the experiment. All animals were fed live brine shrimp larvae on alternate days.

The control and experimental grass shrimp which underwent ecdysis within 7 to 12 days after limb removal exhibited very little limb bud growth. However, the control shrimp which underwent ecdysis 13 to 15 days after limb removal exhibited marked limb regeneration (Fig. 6). A pronounced inhibition of limb regeneration occurred in shrimp exposed to Na-PCP beginning either the day of limb removal or two days after limb removal. In shrimp exposed to Na-PCP beginning four days after limb removal, the extent of inhibition was relatively less.

An additional experiment was conducted to examine the effects of exposure to Na-PCP using grass shrimp with limb buds at different phases of development. The interval between the time of limb removal and exposure to Na-PCP, and the resulting regenerative limb growth are shown in Figure 7. The extent of inhibition decreased with increase in the interval between limb removal and the initiation of

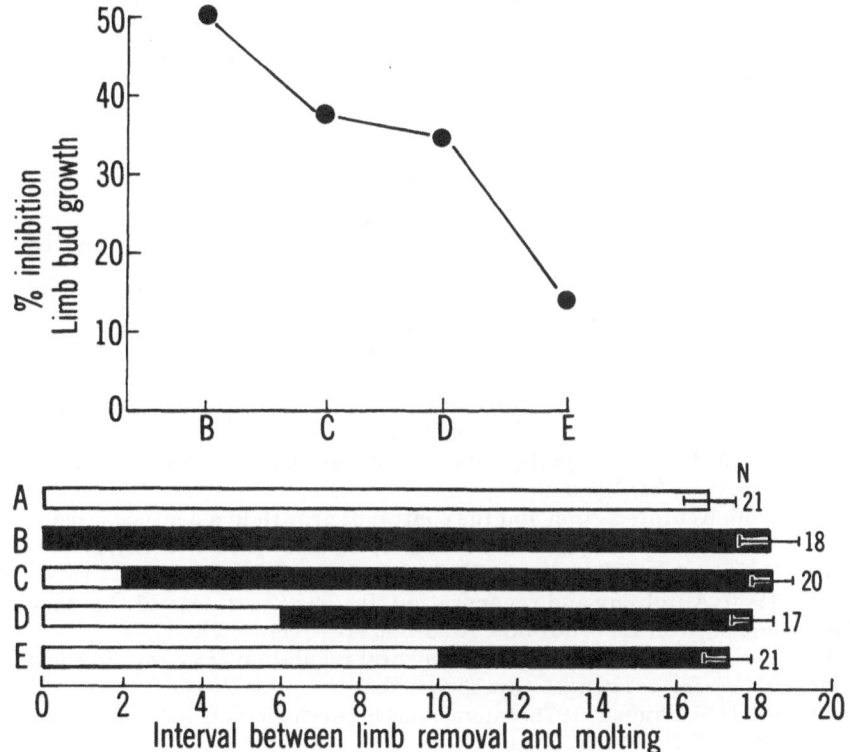

Figure 7. The relationship between the duration of exposure to sodium pentachlorophenate and the extent of inhibition (relative to controls) of regeneration of the left fifth pereiopod in the grass shrimp, *Palaemonetes pugio*. The control shrimp (A) were exposed to seawater throughout the experiment. B, shrimp exposed to 1.0 ppm Na-PCP beginning the day of limb removal through the completion of ecdysis. C, shrimp exposed to 1.0 ppm Na-PCP beginning two days after limb removal. D, shrimp exposed to 1.0 ppm Na-PCP beginning six days after limb removal. E, shrimp exposed to 1.0 ppm Na-PCP beginning ten days after limb removal.

exposure to Na-PCP. These results indicate that the early phases of regeneration (wound healing, mitosis and early dedifferentiation) are more sensitive to Na-PCP than later phases of limb regeneration. The inhibitory effects of PCP on mitotic activity has been previously noted with the flower buds and root cells of *Vicia fabra* (Amer and Ali, 1960).

A reduction in the inhibitory effects of Na-PCP on limb bud growth at later stages of development may also be related to changes in the permeability. It is known that a cuticle is secreted concurrent with the progress in the growth of crustacean limb buds.

Conclusions

Na-PCP inhibits the initiation and progress of limb regeneration in the grass shrimp without altering the duration of the molt cycle. Crustacean limb regeneration can be used as a sensitive bioassay for studying the effects of environmental pollutants.

ACKNOWLEDGMENTS

This investigation was supported by Grant R-804541-01 from the United States Environmental Protection Agency.

References

Adiyodi, R.G., 1972. Wound healing and regeneration in the crab *Paratelphusa hydrodromous*. *Intern. Rev. Cytol.*, **32:** 257-289.

Amer, S., and E.M. Ali, 1970. Cytological effects of pesticides. IV. Mitotic effects of some phenols. *Cytologia*, **34:** 533-540.

Bliss, D.E., 1956. Neurosecretion and the control of growth in a decapod crustacean. In *Bertil Hanström, Zoological papers in honor of his sixty-fifth birthday, November 20, 1956* (K.G. Wingstrand, Ed.), pp. 56-75, Zool. Inst. Lund, Sweden.

Bliss, D.E., 1960. Autotomy and regeneration. In: *The Physiology of Crustacea* (T.H. Waterman, Ed.), Vol. **1,** pp. 561-589. Academic Press, New York.

Bliss, D.E., and J.R. Boyer, 1964. Environmental regulation of growth in the decapod crustacean *Gecarcinus lateralis*. *Gen. Comp. Endocrinol.*, **4:** 15-41.

Holland, C., and D.M. Skinner, 1976. Interactions between molting and regeneration in the land crab. *Biol. Bull.*, **150:** 222-240.

Miller, D.C., and F.J. Vernberg, 1968. Some thermal requirements of fiddler crabs of the temperate and tropical zones and their influence on geographic distribution. *Amer. Zoologist*, **8:** 459-470.

Passano, L.M., 1960. Low temperature blockage of molting in *Uca pugnax*. Biol. Bull., **118:** 129-136.

Passano, L.M., and S. Jyssum, 1963. The role of the Y-organ in crab proecdysis and limb regeneration. *Comp. Biochem. Physiol.*, **9:** 195-213.

Rao, K.R., 1966. Studies on the influence of environmental factors on growth in the crab, *Ocypode macrocera*. H. Milne Edwards. *Crustaceana*, **11:** 257-276.

Skinner, D.M., and D.E. Graham, 1972. Loss of limbs as a stimulus to ecdysis in Brachyura (true crabs). *Biol. Bull.,* **143:** 222-233.

Weis, J.S., 1976a. Effects of environmental factors on regeneration and molting in fiddler crabs. *Biol. Bull.,* **150:** 152-162.

Weis, J.S., 1976b. Effects of mercury, cadmium and lead salts on limb regeneration and ecdysis in the fiddler crab, *Uca pugilator. U.S. Fish Wildl. Serv. Fish. Bull.,* **74:** 464-467.

Weis, J.S., 1977. Limb regeneration in fiddler crabs: species differences and effects of methyl mercury. *Biol. Bull.,* **152:** 263-274.

Weis, J.S., and L.H. Mantel, 1976. DDT as an accelerator of regeneration and molting in fiddler crabs. *Estuarine Coastal Mar. Sci.,* **4:** 461-466.

Effect of Sodium Pentachlorophenate on Exoskeletal Calcium in the Grass Shrimp, *Palaemonetes pugio*

ANITA C. BRANNON and PHILIP J. CONKLIN

Abstract—Exposure of the grass shrimp, *Palaemonetes pugio,* to media containing sodium pentachlorophenate (Na-PCP) led to an apparent increase in the dry weight of exuvia as well as an increase in the total quantity of calcium. The actual calcium concentration (μg Ca/mg dry exoskeleton) in exuvia did not vary significantly in relation to Na-PCP exposure. Whether the observed changes in exuvia from shrimp exposed to Na-PCP are due to a decrease in the resorption of the old exoskeleton preceding ecdysis remains to be clarified.

Introduction

Pentachlorophenol is an uncoupler of oxidative phosphorylation (Weinbach, 1954). It is also known to affect the activity of adenosine triphosphate phosphohydrolases (ATPases) and associated transport phenomena. Knudsen *et al.* (1974) noted a striking dose-related decrease of calcium deposits in the kidneys of rats fed pentachlorophenol. Fox and Rao (this symposium) reported that sodium pentachlorophenate (Na-PCP) and 2,4-dinitrophenol caused a dose-dependent inhibition of a calcium-activated ATPase in the microsomal fraction of the hepatopancreas of blue crabs. Cyclical exoskeletal deposition and resorption of calcium occur in relation to the crustacean molt cycle (Passano, 1960). The aim of this study was to determine the effect of Na-PCP on the calcium content of exuvia (cast exoskeleton) from the grass shrimp, *Palaemonetes pugio*.

ANITA C. BRANNON and PHILIP J. CONKLIN • Faculty of Biology, University of West Florida, Pensacola, Florida 32504, U.S.A.

Experimental Animals and Conditions

Two separate experiments were conducted using a total of 200 grass shrimp which were selected for uniformity of size (25 mm average rostrum-telson length). Initially, all shrimp were at the same stage of the molt cycle (Stage C). Animals were maintained in glass jars containing 250 ml of filtered seawater (10 %o) for controls or test solution for experimental animals. In the first experiment 100 shrimp were divided into four equal groups. One group served as the control while the remaining groups were exposed to 0.1, 0.5 or 1.0 ppm (mg/l) Na-PCP. The media for control and experimental shrimp were renewed daily. The shrimp were not fed during the first experiment.

The second experiment which also utilized 100 shrimp was conducted following the above procedure. However, for this experiment animals were fed live brine shrimp on alternate days.

Calcium Analysis

Cast exuvia were collected following ecdysis (molt) and stored in individual vials after rinsing with deionized, glass distilled water. Precaution was taken not to include exuvia which had been partially eaten by the molted shrimp. The exuvia were dried at 110 °C for 48 hr in individual preweighed 25 ml conical flasks. After obtaining dry weights, exuvia were ashed in approximately 10 ml of hot concentrated nitric acid.

A Perkin Elmer Model 290B atomic absorption spectrophotometer equipped with an Intensitron hollow cathode lamp was used for calcium determinations. Dilutions were made as necessary to be within linear working range for calcium. Lanthanum oxide was added to samples at a final concentration of approximately 0.05% to prevent phosphate interferences. The total amounts of calcium in the exuvia were calculated from a standard curve.

Effect of Na-PCP on Exuvial Weight and Exoskeletal Calcium

Control grass shrimp which were fed brine shrimp larvae appeared to have exuvia with higher dry weights than exuvia from unfed control shrimp (Fig. 1). Feeding also appeared to influence the total quantity of calcium in control exuvia (Fig. 2).

Unfed shrimp which were exposed to Na-PCP and completed ecdysis exhibited apparent higher exuvial dry weights (Fig. 1). For animals exposed to 1.0 ppm the dry weights of the exuvia were significantly higher ($P < 0.05$) than that of the controls. The total amount of exoskeletal calcium for shrimp exposed to Na-PCP also appeared to be greater (Fig. 2) especially at the highest concentration (1.0 ppm). The calculated concentrations of calcium in exuvia (μg Ca/mg dry exoskeleton) were not significantly affected by exposure to Na-PCP.

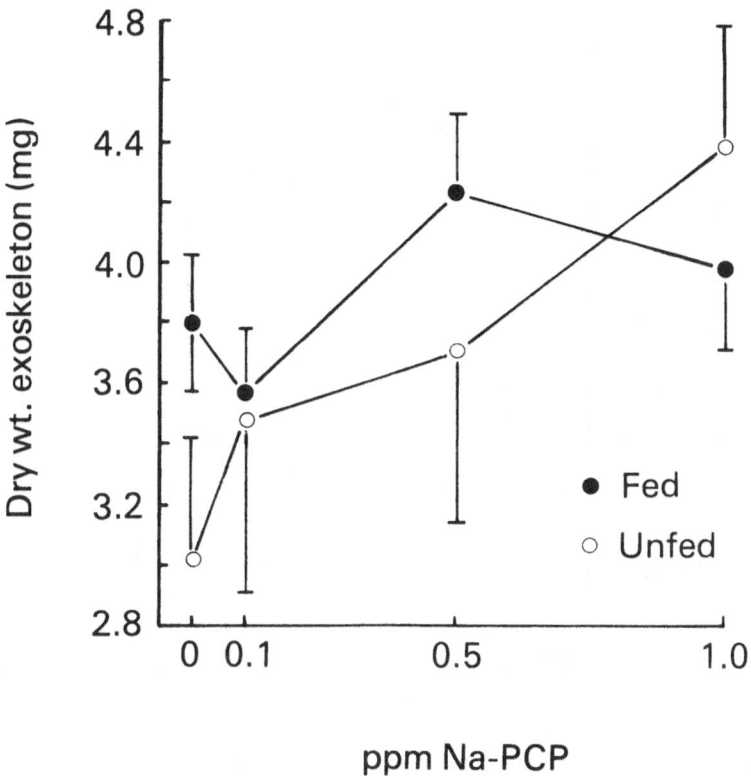

Figure 1. The dry weights of exuvia (cast exoskeleton) from fed and unfed grass shrimp, *Palaemonetes pugio,* exposed to 0, 0.1, 0.5 and 1.0 ppm Na-PCP. The shrimp were exposed to these solutions beginning from Stage C (intermolt) throughout the molt cycle and the exuvia at first ecdysis (molt) were dried and weighed. The data points are Mean ± Standard Error.

Fed shrimp were followed through a second ecdysis. However, almost all (> 90%) of the shrimp exposed to 1.0 ppm Na-PCP died immediately following the first ecdysis. Generally, exuvia from second ecdysis were lighter in dry weight and had less calcium than exuvia from the first ecdysis. Exuvia from shrimp exposed to 0.5 ppm Na-PCP which underwent a second ecdysis were heavier (Fig. 3) and had significantly higher ($P < 0.05$) amounts of exoskeletal calcium (Fig. 4) than controls. As noted for unfed animals, the calculated concentrations of calcium (μg Ca/mg dry exoskeleton) for fed experimentals were not significantly different from those of the controls.

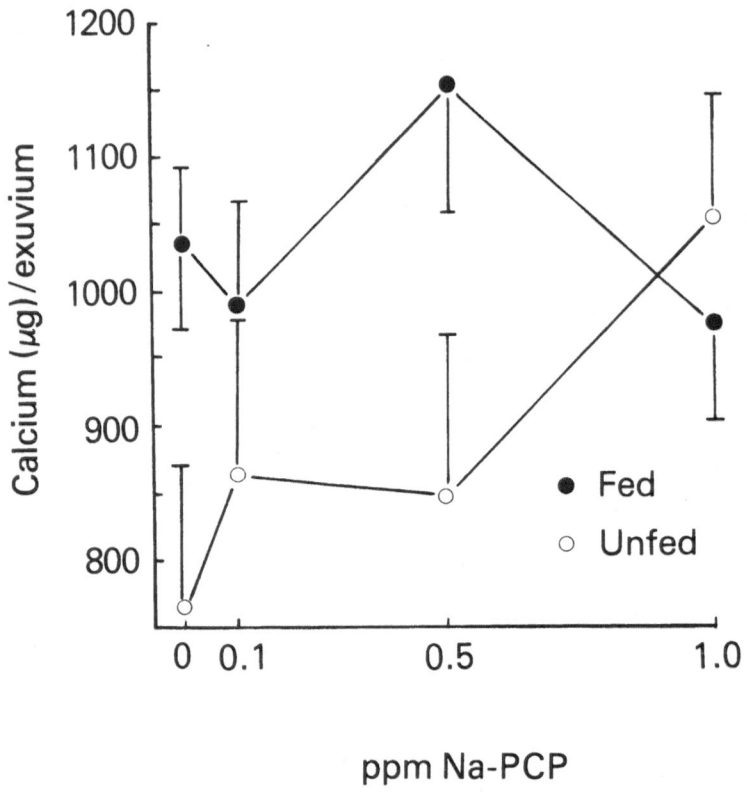

Figure 2. The total quantity of calcium in the exuvia (cast exoskeleton) from fed and unfed grass shrimp, *Palaemonetes pugio,* exposed to 0, 0.1, 0.5 and 1.0 ppm Na-PCP. The shrimp were exposed to these solutions beginning from Stage C (intermolt) throughout the molt cycle and the total quantity of calcium in the cast exuvia at first ecdysis (molt) was determined using atomic absorption spectrophotometry. The data points are Mean ± Standard Error.

Discussion and Conclusions

The weight of exuvia (cast exoskeleton) depends on factors such as the thickness of the cuticle, the degree of calcification and the extent of resorption occurring preceding ecdysis. The finding that the calcium concentration (μg Ca/mg dry exoskeleton) in the exuvia from control shrimp did not differ from that of shrimp exposed to Na-PCP indicates that this compound may not specifically affect the deposition or resorption of calcium.

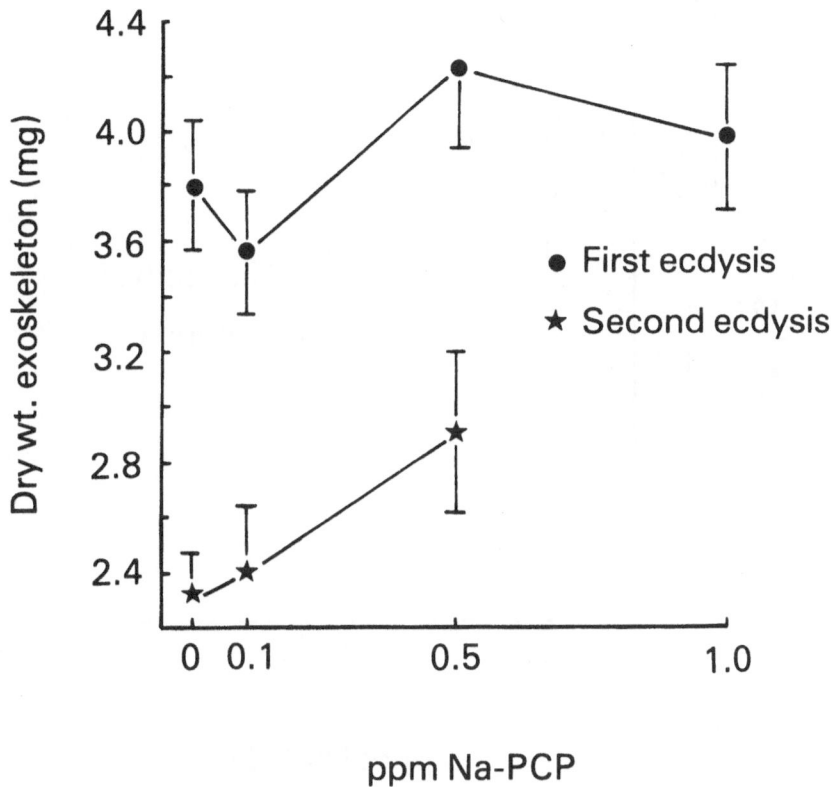

Figure 3. A comparison of the dry weights of exuvia collected at the first and second ecdyses (molts) occurring during the exposure of fed grass shrimp, *Palaemonetes pugio,* to 0, 0.1, 0.5 and 1.0 ppm Na-PCP. Since very few of the animals exposed to 1.0 ppm survived beyond the first ecdysis, data could not be obtained for the second ecdysis. All shrimp were exposed to appropriate media throughout the duration of the experiment. The data points are Mean ± Standard Error.

With the aid of secretion of the molting fluid, a partial resorption of the exoskeletal contents may be accomplished during the proecdysial stages of the molt cycle (Passano, 1960). The sequence of events leading to calcium resorption and the energy requirements of the transport processes that may be involved are not clearly established. Nevertheless, a reduction in the extent of resorption may lead to the shedding of a heavier exoskeleton. Whether such a reduction in exoskeletal resorption was responsible for the apparent increases in the total calcium and weight of the exuvia in some shrimp exposed to Na-PCP and in fed control shrimp (compared to unfed control shrimp) remains to be clarified.

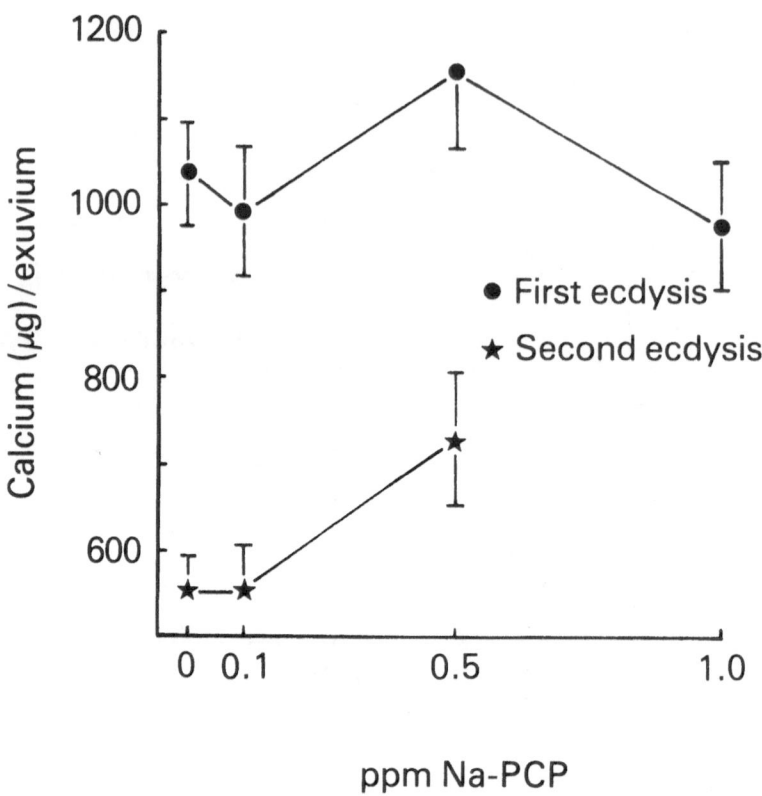

Figure 4. A comparison of the total quantities of calcium in exuvia collected during the exposure of fed grass shrimp, *Palaemonetes pugio,* to 0, 0.1, 0.5 and 1.0 ppm Na-PCP. The experimental conditions are the same as described in the legend for Fig. 3.

Knudsen *et al.* (1974) proposed that PCP acts indirectly on calcium metabolism and may therefore lower blood Ca^{++} levels in rats. Normally, crustaceans exhibit a rise in the level of Ca^{++} in the hemolymph during proecdysis (Passano, 1960). Further work is needed to determine the mechanism by which Na-PCP may affect the processes of exoskeletal deposition and resorption.

ACKNOWLEDGMENTS

This investigation was supported by Grant R-804541-01 from the United States Environmental Protection Agency.

References

Knudsen, I., H.G. Verschuuren, E.M. Den Tonkelaar, R. Kroes, and P.F.W. Helleman, 1974. Short-term toxicity of pentachlorophenol in rats. *Toxicology,* **2:** 141-152.

Passano, L.M. Molting and its control. In, *The Physiology of Crustacea* (T.H. Waterman, Ed.), Vol. **1:** pp. 473-536, Academic Press, New York.

Weinbach, E.E., 1954. The effect of pentachlorophenol on oxidative phosphorylation. *J. Biol. Chem.,* **210:** 545-550.

Ultrastructural Changes Induced by Sodium Pentachlorophenate in the Grass Shrimp, *Palaemonetes pugio,* in Relation to the Molt Cycle

DANIEL G. DOUGHTIE and K. RANGA RAO

Abstract—Intermolt (stage C) grass shrimp were exposed to 1.0 ppm Na-PCP for the duration of a molt cycle. Gills, hepatopancreas, midgut (portion of the digestive tract surrounded by hepatopancreas) and hindgut (portion of the digestive tract in the abdomen) from control and experimental shrimp at known stages of the molt cycle were examined at the ultrastructural level. Although signs of pathology were evident in late proecdysial shrimp, extensive pathological changes were not observed until after ecdysis. The extent of pathological changes varied with the tissue examined and the interval between ecdysis and the time of fixation for electron microscopy.

The following ultrastructural changes were seen in the gill epithelium of shrimp exposed to Na-PCP: formation of fluid filled invaginations of the intermicrovillar apical membrane, increase in lysosomal activity and eventual cytoplasmic and nuclear degeneration. The podocytes in the gill axis, the granular secretory cells and the tegumental gland cells also exhibited mitochondrial swelling, nuclear pyknosis and eventual cytoplasmic degeneration.

The cells lining the lumen of the midgut and hindgut of shrimp exposed to Na-PCP exhibited swelling of the apical membrane often accompanied by rupture, loss of microvilli from apical foci and increased lysosomal activity.

Pathological changes noted in the hepatopancreatic cells of the experimental shrimp were: high amplitude swelling of mitochondria including vesiculation of cristae, presence of myelin bodies within mitochondria and rough endoplasmic reticulum, increase of autophagic activity and loss of microvilli.

DANIEL G. DOUGHTIE and K. RANGA RAO • Faculty of Biology, University of West Florida, Pensacola, Florida 32504, U.S.A.

Introduction

Ultrastructural changes elicited by the *in vivo* or *in vitro* exposure of experimental animals to toxic concentrations of PCP have so far been reported for only one mammalian species (Weinbach, 1967). However, physiological and biochemical data have demonstrated the adverse effects PCP initiates in normal mitochondria through its ability to uncouple oxidative phosphorylation (Buffa *et al.,* 1963; Weinbach and Garbus, 1966; Weinbach *et al.,* 1967). The action of uncouplers of oxidative phosphorylation has been theorized on the basis of the chemical (Pressman, 1963) and chemiosmotic (Mitchell, 1961) interpretations. The former theory proposes that uncouplers promote conductivity of protons within mitochondrial membranes and subsequently prevent the formation of a gradient across the membrane. According to the latter theory the uncouplers promote the splitting of an energy rich intermediate prior to ATP production. A third examination (Weinbach, 1969) suggests that uncouplers traverse the lipid-protein layer of mitochondrial membranes and interact with protein groups that then undergo structural change. It is generally assumed that major changes in mitochondrial function are reflected in morphological alterations and that normal mitochondrial profiles are dependent on a continuing supply of high energy intermediates produced by oxidative phosphorylation. By an *in vitro* incubation of rat hepatic mitochondria with PCP in a standard phosphorylation medium, Weinbach (1967) has demonstrated a correlation between impaired functions indicated by a negligible P/O ratio, and altered morphology characterized by high amplitude swelling, flocculent appearing matrix and constricted or sparse cristae.

Conklin and Rao (this symposium) have shown that the grass shrimp, *Palaemonetes pugio,* is very sensitive to Na-PCP during the period immediately following ecdysis. This increased sensitivity of the post-ecdysial shrimp appeared to be due to a nearly 30-fold increase (relative to intermolt shrimp) in the uptake of the biocide following ecdysis. In the present communication, ultrastructural changes in the gills of grass shrimp exposed to 1.0 ppm Na-PCP are reported. In evaluating ultrastructural changes, gills from control and experimental shrimp at different stages of the molt cycle were examined. Portions of hindgut, midgut and hepatopancreas from experimental and control shrimp were also examined.

Experimental Animals and Electron Microscopy

Intermolt (stage C) grass shrimp were exposed to 1.0 ppm Na-PCP for the duration of a molt cycle. Selected on the basis of known stages of the molt cycle, shrimp from control and experimental groups were sacrificed for ultrastructural examination. Gills, midgut, hepatopancreas and hindgut were excised in cold 4.5% glutaraldehyde buffered with s-collidine and fixed at room temperature followed by fixation in cold s-collidine buffered 2% osmium tetroxide. After *en bloc* staining in 7% aqueous uranyl acetate, the tissue was dehydrated in a graded ethanol series and embedded in Epon 812. For light microscopy thick sections (1μ) of epon embedded material were cut with glass knives and stained with toluidine blue. Thin sections (60-80 nm) were made using glass knives on an LKB Ultrotome I and

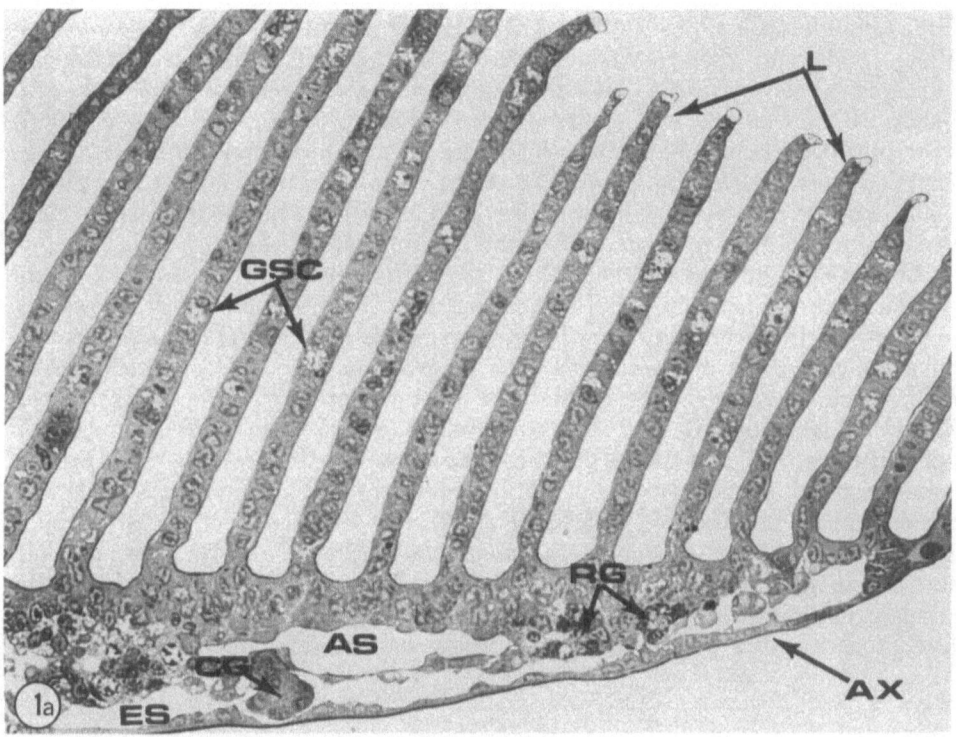

Figure 1a. Light micrograph of sagittal section of a pleurobranchiate gill of *Palaemonetes pugio* showing the association of lamella (L) with the main axis (AX). Note the affarent (AS) and efferent (ES) hemolymph sinuses and the close association between the clear gland (CG) and the efferent sinus. Note also the reticulate glands (RG) and the granular secretory cells (GSC).

stained in lead citrate. Grids were observed and photographed using a Phillips 201 electron microscope.

Ultrastructural and Functional Parameters of Normal Gills of the Grass Shrimp

Histological studies of crustacean gills have been conducted in *Astacus* (Huxley, 1896), *Carcinus* (Ali, 1966; Smyth, 1942), *Ocypode* (Flemister, 1959), *Palaemonetes* (Allen, 1893), and *Penaeus* (Young, 1959). More recently electron microscopic studies have been performed on gill tissues of *Artemia* (Copeland, 1967), *Astacus* (Fisher, 1972), *Callinectes* (Copeland and Fitzjarrel, 1968), *Carcinus* (Ali, 1966), *Gecarcinus* (Copeland, 1968), *Panulirus* (Strangways-Dixon and Smith, 1970) and *Penaeus* (Foster, 1976; Couch, 1977). Aside from its primary role as a site of respiration, the crustacean gill is also involved in osmoregulation (Copeland, 1967, 1968; Mantel, 1967, Koch *et al.,* 1954), and possibly detoxification.

As reported by Allen (1893), there are eight pairs of gills attached to the cephalothorax of *Palaemonetes.* These consist of podobranchiae borne by the second maxilliped, arthrobranchia borne by the third two, and pleurobranchiae situated above the five pairs of pereiopods. The pleurobranchiate gills of *Palaemonetes pugio,* which are studied here, consist of a main axis oriented dorso-ventrally with two rows of lamellae (Fig. 1a.). The two lamellar rows are separated by an angle of approximately 40° and slope dorso-laterally toward the inner surface of the cephalothoracic branchiostegite. Pertinent to the forthcoming discussion of pathological changes induced by Na-PCP are at least four major cell types found in the pleurobranchia of *Palaemonetes pugio;* (a) epithelial (or hypodermal) cells subjacent to the cuticle of both the axis and cuticle which are characterized by abundant infoldings of the apical plasma membrane; (b) cells which form the two types of tegumental glands, similar to those described in *Palaemonetes varians*

Figure 1b. Electron micrograph of a longitudinal section from a pleurobranchiate gill lamella of *Palaemonetes pugio.* Note the nucleus (N) of the epithelial cell, lateral extensions of epithelial perikaryon (EP), and numerous mitochondria (m) subjacent to the cuticle (c). Note also quiescent granular secretory cells (GSC), pillar regions (P), microtubule bundles (arrow) and hemolymph sinuses (H). × 3,400

Figure 2. Higher magnification of the gill epithelial perikaryon from control *Palaemonetes.* Note extensive infoldings of apical plasmalemma (arrow), cuticle (c) and mitochondria (m). × 9,000

Figure 3. A frontal section of apical infoldings which abut gill cuticle showing ramification of plasmalemma ridges. × 9,000

Figure 4. High magnification of the mitochondria from gill epithelium showing presence of matrix granules (arrows). Note golgi (g). × 12,600

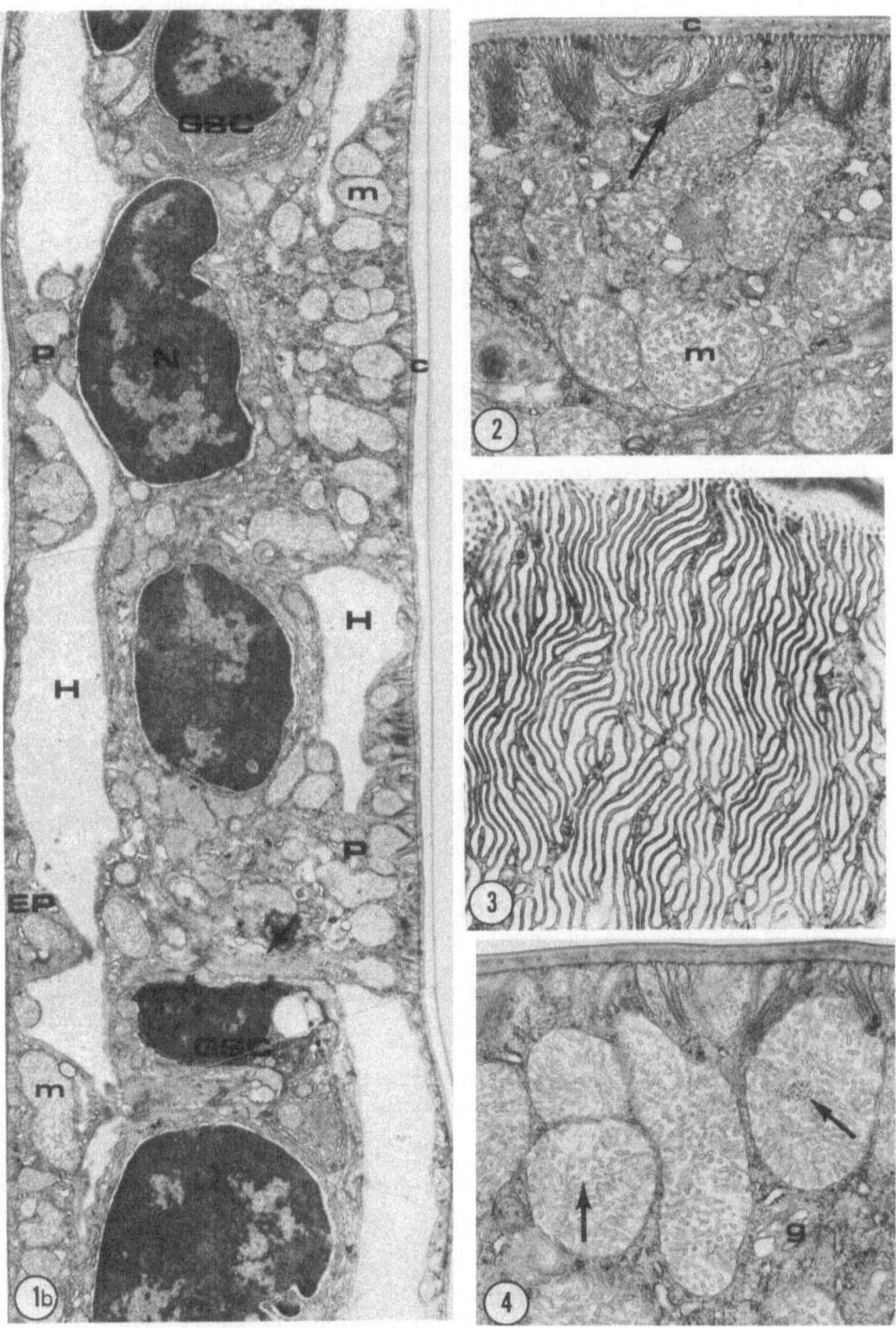

(Allen, 1893), which are located in the longitudinal axial septum that separates the efferent and afferent hemolymph sinuses; (c) Podocyte like cells resembling those found in the gills of *Panulirus* (Strangways-Dixon and Smith, 1970), *Hemigrapsus* and *Pachygrapsus* (Wright, 1964) which are primarily located around the efferent axial sinus and may function in the filtration and detoxification of hemolymph; and (d) granular secretory cells which are located in the epithelium (hypodermis) of the axis and lamella and contain unique spherical structures. A more detailed description of the normal ultrastructural and presumed functional characteristics of these cells are presented below.

Epithelial Cells of Normal Gills

The epithelial cells, while similar in subcellular organization in both the axis and lamella are cuboidal or columnar in the former. In the lamella, however, thin sheets of cytoplasm derived from cells whose nucleus is usually situated in the longitudinal medial septum pass between hemolymph sinuses and are apposed to the lamellar cuticle which varies in thickness from 0.2 - 0.6 μm (Fig. 1b). The cytoplasmic bridges formed by the epithelial cell perikaryon between the medial lamellar septum and the cuticle have been referred to as "pillar" regions (Smyth, 1942). Located within these pillar regions are numerous microtubules oriented perpendicular to the lamellar cuticle(Fig.1b)and it has been suggested that these are cytoskeletal in function and may provide support to the lamella during periods of changing hydrostatic pressures (Foster, 1976). These microtubules may also provide channels for the transport of cuticle precursor substances and/or products of cuticle degradation during appropriate stages of the molt cycle, as suggested by Locke (1969) for similar organelles in the epidermal cells of insects.

The apical plasmalemma of epithelial cells in the gill axis and lamellae of grass shrimp are characterized by abundant invaginations which give rise to microvilli-like ridges abutting the cuticle (Figs. 2,3). These ridges are much less pronounced or nonexistent during early proecdysial stage (D_0), but are more pronounced during later stages of proecdysis and the early postecdysial period associated with the secretion of new cuticle. The plasmalemmal invaginations and ridges, aside from participating in the secretion of new cuticle, are also morphologically

---

Figure 5. Podocyte cell in the efferent sinus (ES) of gill axis of *Palaemonetes*. Note mitochondria (m), pedicels and cisternae (arrow), endoplasmic reticulum (ER) and abundant cytoplasmic vesicles. × 7,700

Figure 6. A frontal section of podocyte surface showing interdigitating pedicels (P). × 18,000

Figure 7. High magnification micrograph of the pedicels, (P), subpedicellar cisternae (SPC). Coated vesicles can be seen forming from exocytosis of the cisternae. Note the diaphragms (arrows) joining the pedicels and also note basement membrane (BM). × 63,000

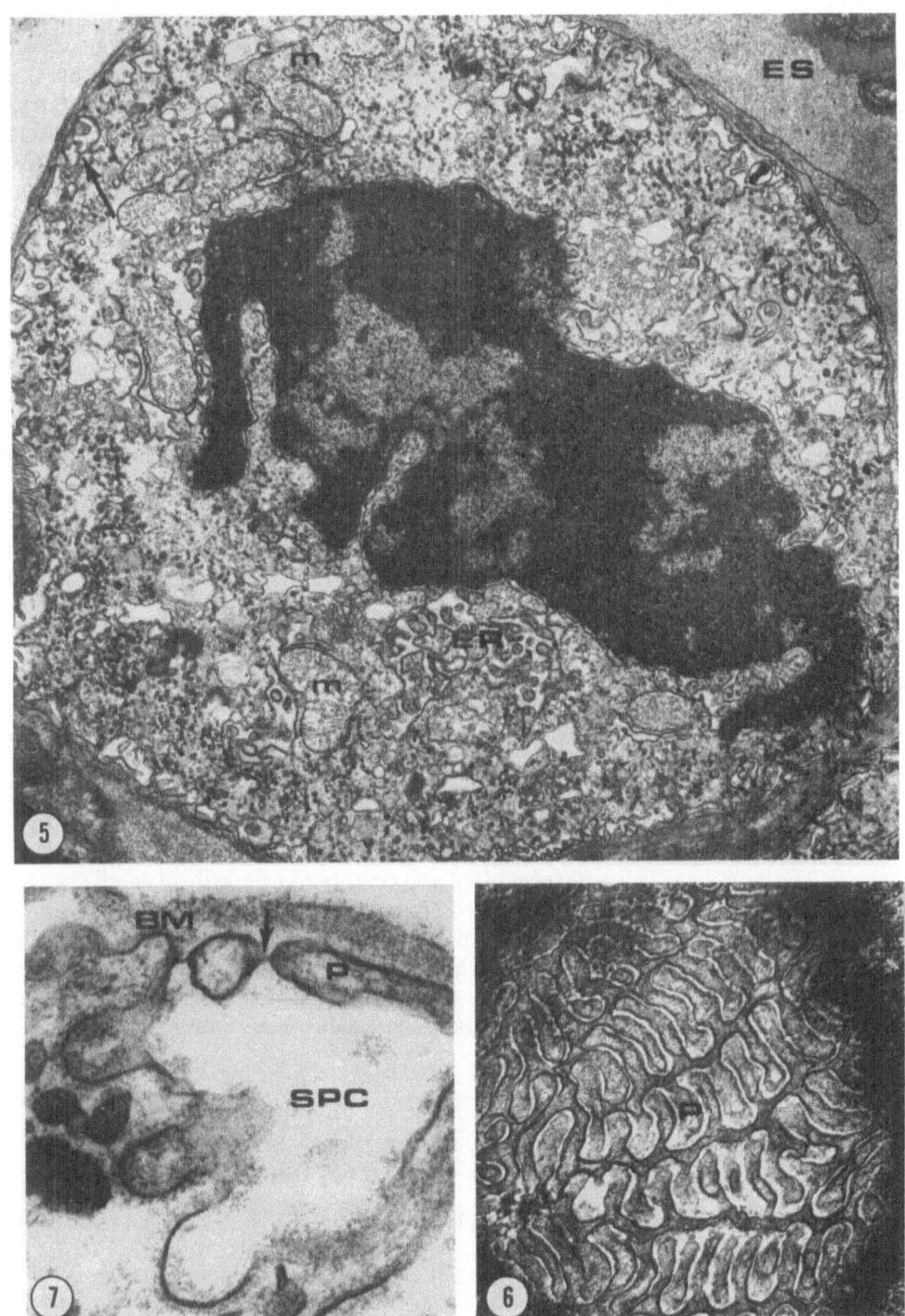

characteristic of transport epithelium due to the extensive infoldings of the apical membrane and interdigitations of the lateral membrane in the subcuticular region of the perikaryon. Energy for the transport of ions would be supplied by the numerous mitochondria associated with these regions. (see Diamond and Tormey, 1966a, 1966b; Diamond, 1969). An epithelium with suspected transport capabilities has been variously described for the gills of *Gammarus* (Milne and Ellis, 1973), *Astacus* (Bielawski, 1971) *Gecarcinus* (Copeland, 1968), *Callinectes* (Copeland and Fitzjarrel, 1968) and *Penaeus* (Foster, 1976; Couch, 1977).

The cytoplasm of gill epithelial cells varies considerably depending on the stage of the molting sequence, in terms of the relative abundance of cuticle precursor materials, mitochondria, microtubules, rough endoplasmic reticulum (RER) and ribosomes, etc., but the changes coincide closely with those of other arthropods studied in relation to the molt cycle as in *Panulirus* (Travis, 1955, 1957), *Calpodes* (Locke, 1969), and *Uca* (Green and Neff, 1972).

Typically, the epithelial mitochondria were elongated with a moderately electron dense matrix and the presence of matrix granules was occasionally noted (Fig. 4).

Podocytes of Normal Gills

The podocyte-like cells (Fig. 5) of *Palaemonetes* gill tissue are usually found protruding into the efferent axial sinus. The portion of the cell perimeter not embedded in or attached to the longitudinal septum and facing the hemocoel is characterized by a surface of rod-like interdigitating projections of the plasmalemma (Fig. 6) which lie above small subsurface cisternae. The plasmalemmal projections, called pedicels, are connected by thin diaphragms which reportedly act to filter out particles from the hemolymph (Strangways-Dixon and Smith, 1970). The basement membrane is composed of a thin mat of interwoven fibrin fibers and may provide both structural support and an additional filtration barrier. The cytoplasm is loosely organized and composed primarily of small vesicles of variable electron density (Fig. 5) which are derived from endocytotic activity of the subpedicellar cisternae (Fig. 7), tubular structures, and large intracellular spaces and vacuoles. Mitochondria and RER are scarce and are usually seen in close proximity to a central ovoid nucleus. The gill podocytes which bear a resemblance to those of the renal glomerulus described by Bloom and Fawcett (1968) may

---

Figure 8. Micrograph of reticulate gland complex in the gills of *Palaemonetes*. Note gland cell nuclei (GCN), duct cell nucleus (DCN) and apical vesicles (V). × 3,200.

Figure 9. Basal perimeter of reticulate gland cells showing extensive lamellar rough endoplasmic reticulum (RER). × 3,400

Figure 10. Central core of reticulate gland complex. Note duct cell and a portion of the duct (d) and the numerous apical vesicles (v) of the gland cells. × 3,400

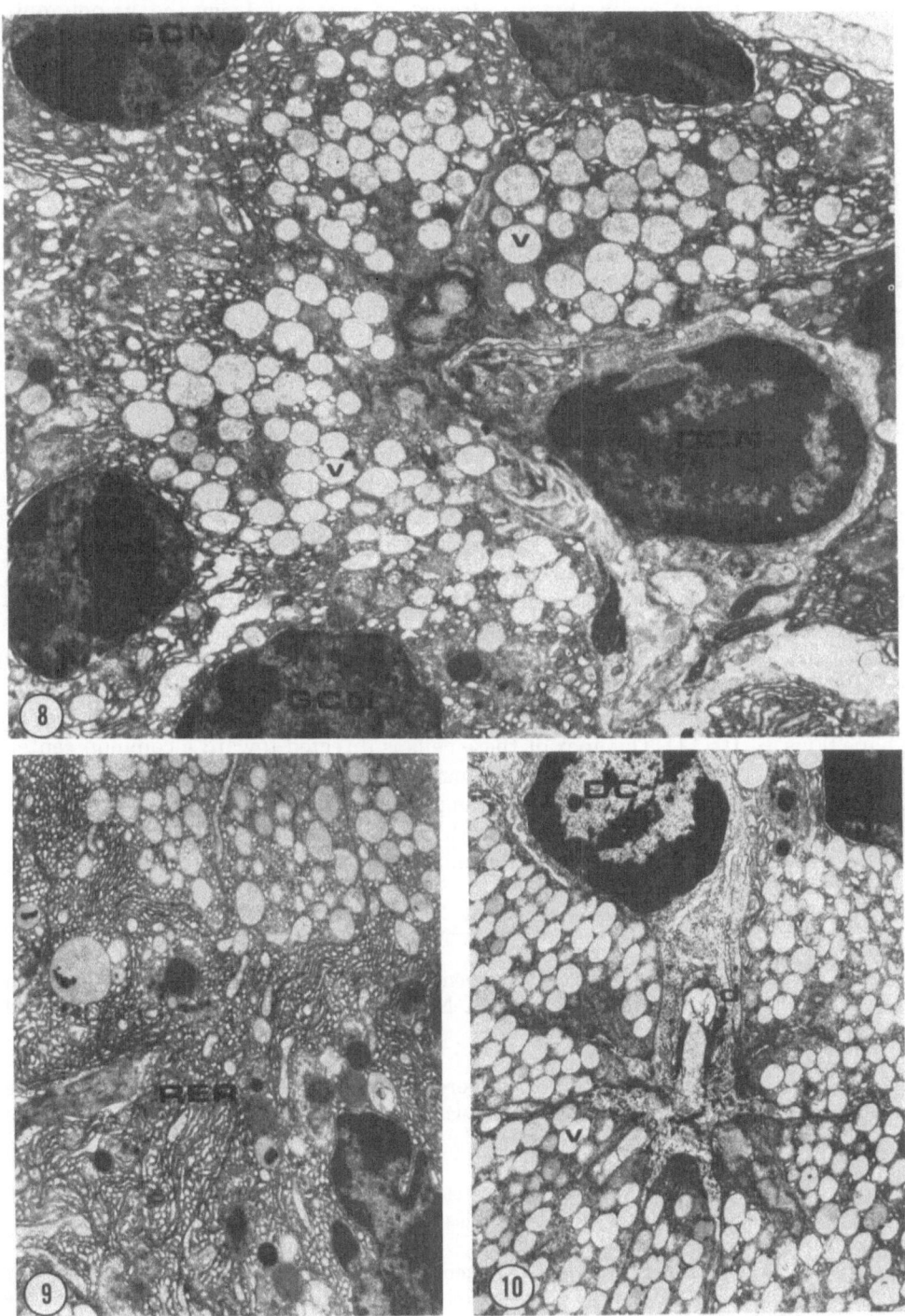

contribute to detoxification, by withdrawal of material from the hemolymph (Strangways-Dixon and Smith, 1970). Another possible role of these cells is one of hemolymph bacteriostatis; this has been attributed to the pericardial nephrocytes of *Calliphora* which secrete the bactericidal enzyme, lysozyme (Crossley, 1972).

Tegumental Gland Cells of Normal Gills

There have been no previous reports on the ultrastructure of crustacean tegumental glands, although histological studies are available concerning these glands in *Palaemonetes* (Allen, 1893), *Gecarcinus* (Skinner, 1962), and *Homarus* (Yonge, 1932). The exact role of tegumental glands is not well understood although it is generally assumed that they contribute to the formation of the new cuticle and/or production of exuvial fluid. A correlation between glandular recrudescence and phase of the molt cycle has been observed in *Homarus* (Yonge, 1932). However, Skinner (1962) was unable to show any correlation due to the presence of both replete and empty glands in *Gecarcinus* irrespective of the particular stage in the molting sequence. In his work on the gills of *Palaemonetes varians,* Allen (1893) described two distinct types of tegumental glands which on the basis of light microscopical observations are classified as being either clear or reticulate. Due to the gross similarities between these and the gland complexes found in *Palaemonetes pugio,* Allen's nomenclature will be employed in the following discussion. For the most part, the two tegumental glands studied here share several gross morphological characteristics. Both the clear and reticulate tegumental complexes are more or less spherical and are composed of six to ten cells which are conical in shape (Figs. 8, 10, 11). The apex of these cells is directed toward a common center where synthesized materials pass into a "cuticularized" duct formed by a centrally located cell (Figs. 10, 12). After traversing the interior of the gland complex, this duct assumes a straight or irregular path between and through hypodermal and epidermal cells where it eventually terminates in the cuticle. During the period of

---

Figure 11. Micrograph of the clear tegumental gland extending into the efferent axial sinus (ES) of the gill tissue of *Plaemonetes.* Note duct cell nucleus (DCN) and gland cell nucleus (GCN). × 3,000

Figure 12. High magnification of central core where clear gland cell and duct cell apices meet. Note the duct (d), numerous apical vesicles (v) and apparent release of material into the central lumen (L). × 9,000

Figure 13. High magnification of apical vesicles of a tegumental gland cell. Note the inclusion of cytoplasmic substance into vesicles by exotropy (arrows). × 27,000

Figure 14. Basal perimeter of clear tegumental gland cells. Note extensive infolding of plasmalemma (arrow) and the close association of mitochondria (m) with these intercellular spaces. × 8,200

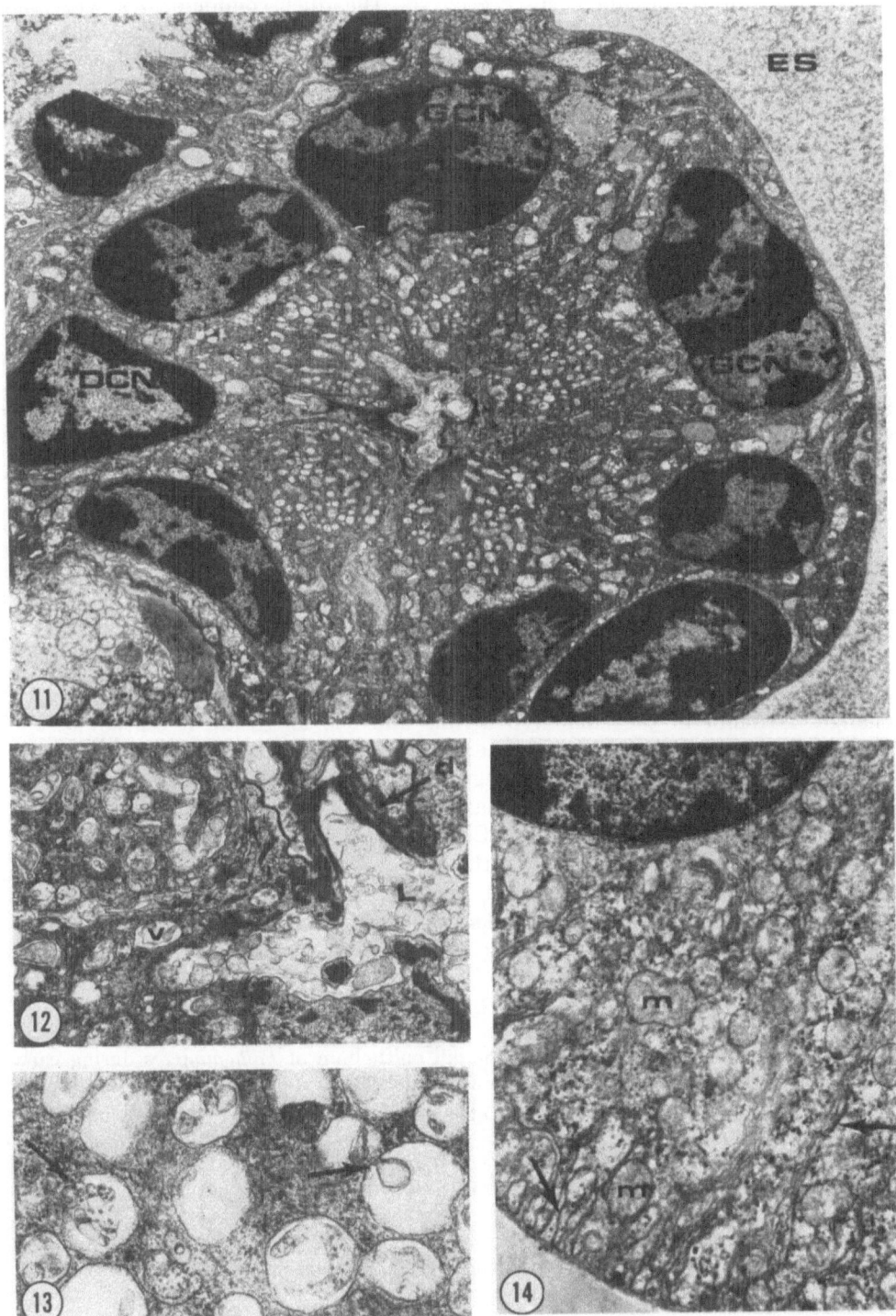

postecdysis, intermolt and early proecdysis, the duct communicates with the environment. However, during late proecdysis the duct communicates with and releases material into the exuvial space. Although tegumental ducts have occasionally been observed passing through the hypodermis subjacent to the lamella, they have never been observed entering or within the lamellar plates.

The reticulate glands, which are located at regular intervals within the longitudinal axial septum, are characterized by cells which contain numerous vesicles (in the median and apical cytoplasm) that vary both in size, electron density and texture depending on the stage of the molt cycle (Figs. 8, 9, 10). These vesicles appear to originate from the abundant parallel lamellar cisternae of the rough endoplasmic reticulum. Also present in the basal region of the reticulate cells are numerous polymorphic mitochondria, golgi with electron dense secretions and large vacuoles.

The clear glands are invariably located in close proximity to the efferent hemocoel, usually in secondary traverse efferent septa. They, like the reticulate gland, contain numerous albiet smaller vesicles in the median and apical cytoplasm (Fig. 11). The contents of the vesicles seem to be derived from exotropy of the vesicle membrane. Characteristically, the basal lamina is extensively infolded which gives rise to numerous intercellular pathways through the basal cytoplasm. Numerous mitochondria and polysomes are associated with these spaces (Fig. 14). The overall appearance of this region is suggestive of transport epithelium.

Granular Secretory Cells of Normal Gills

The fourth distinct cell type to be discussed in connection with pathological changes is found in the axial hypodermis and the longitudinal septum of the lamella. The granular secretory cell is unique both in its overall morphology and in the dearth of knowledge concerning its function. Cells similar to the granular secretory cell described briefly here have been reported in the integument of *Uca* (Green and Neff, 1972), *Palaemon* (Chassard-Bouchard, and Hubert 1973) and *Panulirus* (Travis, 1954) as well as in the gills of *Panulirus* (Smith, 1974), *Penaeus* (Foster, 1976; Couch, 1977) and *Callinectes* (Copeland, 1977, personal communication) and in the pericardial sacs of *Gecarcinus* (Copeland, 1968). In light of their presence in the integument of other crustaceans and their observed

---

Figure 15. Granular secretory cell in the gill epithelium of *Palaemonetes* during early proecdysis. Note the centrally located nucleus (N), golgi (g), mitochondria (m), rough endoplasmic reticulum (RER) and secretory granules (SGr). × 6,500

Figure 16. High magnification of mature secretory granules during late proecdysis. Note golgi (g) and rough endoplasmic reticulum. × 14,800

Figure 17. Duct region of granular secretory cell which appears to be instrumental in transporting substance of secretory granules to the exuvial space during proecdysis. × 12,600

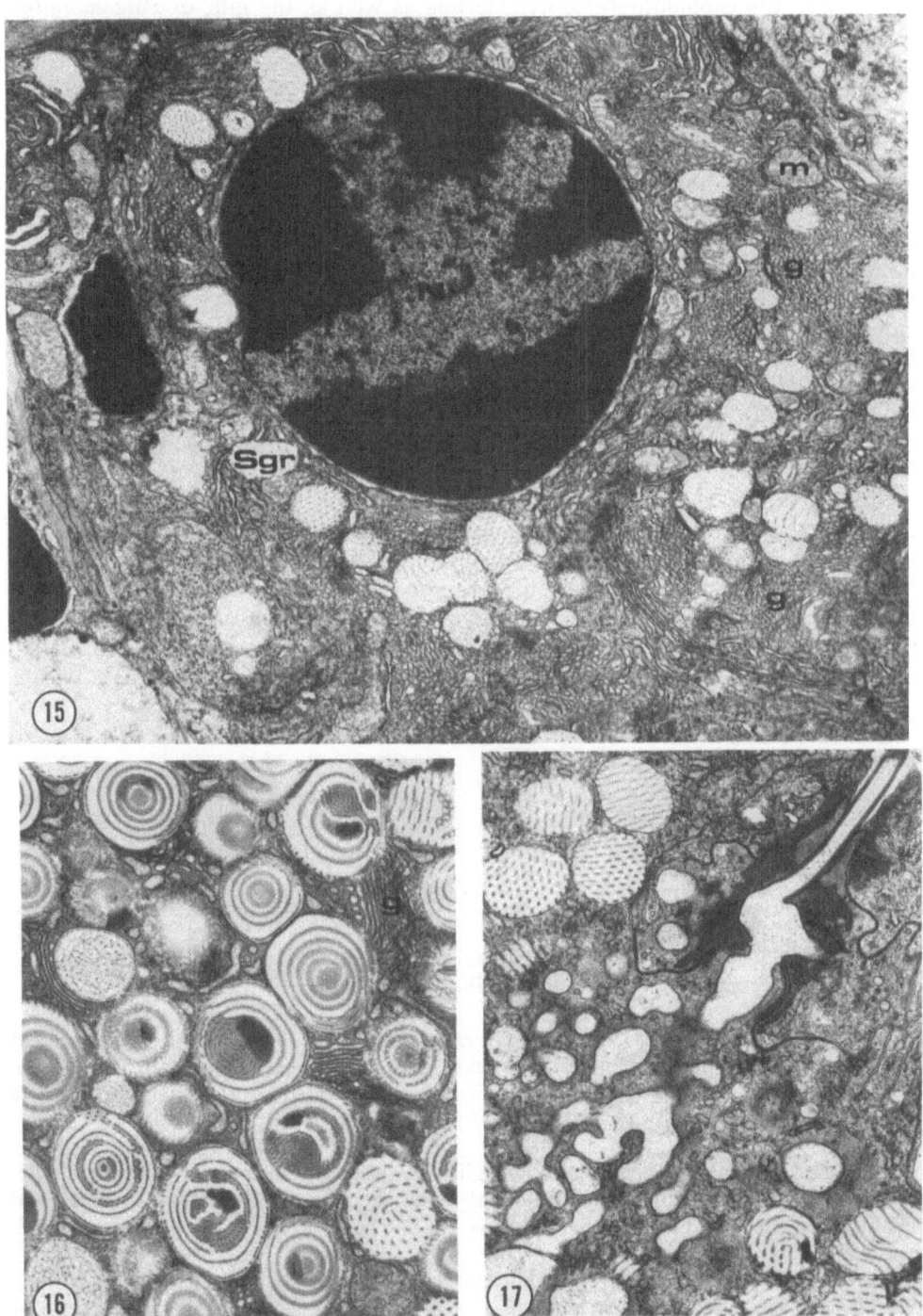

presence in the cephalothoracic hypodermis as well as the gills of *Palaemonetes pugio,* these cells appear to be closely associated with the crustacean cuticle. The ovoid or spherical nucleus (≈ 12.5 μm) is centrally located within a relatively dense cytoplasm (Fig. 15). This apparent density is derived from the normally abundant lamellar and vesicular cisternae of the rough endoplasmic reticulum, golgi and free ribosomes. During the proecdysial stages D_2 and D_3 the cytoplasm is replete with large, apparently laminated granules of unique morphology (Fig. 16) that are derived from the golgi and rough endoplasmic reticulum. Just prior to and presumably during ecdysis these granules appear to undergo dissolution and the product of dissolution exits the cell via a long thin cuticularized tubule (Fig. 17) which terminated in the exuvial space. It has been proposed by some authors that the contents of these granules, variously called asteroid bodies (Couch, 1977), maculated bodies (Foster, 1976), and op-art bodies (Steinbrecht, 1968) may be instrumental in the synthesis and secretion of cuticle (Chassard-Bouchaud, 1973; Green and Neff, 1972; Couch, 1977). Histochemical analysis of similar "reserve" cells in *Panulirus* (Travis, 1957) and in the cuticle synthesizing cells of *Penaeus* gills (Foster, 1976) implicates the granular secretory cells in muco- or glycoprotein synthesis. Travis (1963) has stated that the early stages of epicuticle deposition depend on the formation of a mucopolysaccharide matrix. In *Palaemonetes pugio* these cells exhibit periods of recrudescence and decline which closely coincide with the molting sequence. This as well as the observed release of material into the exuvial space implies their overall importance in the process of ecdysis. Copeland (1977, personal communication) suggested that the material in the granular secretory cells may assist in reducing friction and act as a lubricant while intricate organs such as the gills emerge from the old exuvium during ecdysis.

Hemocytes in Normal Gills

The predominant hemocyte found in the gills of *Palaemonetes pugio* is morphologically homologous with the granular hemocytes observed in the insects:

Figure 18. A hemocyte from the gill tissues of *Palaemonetes.* Note nucleus (N) and the numerous electron dense granules of the perikaryon. × 7,200

Figure 19. High magnification of the perikaryon of the hemocyte in the gill tissues of *Palaemonetes.* The dense granules (Gr) may be composed of carbohydrates. Note the fragmented cisternae of tubular endoplasmic reticulum (arrows) and normal mitochondria (m). × 16,900

Figure 20. Epithelial cell perikaryon in gills dissected 30 minutes after ecdysis from *Palaemonetes* exposed to 1.0 ppm Na-PCP. Note condensed mitochondrial matrix and slightly distended outer membrane (arrow). × 16,000

Figure 21. Micrograph of the epithelium in distal region of gill lamella dissected thirty minutes after ecdysis from *Palaemonetes* exposed to Na-PCP. Note the overall loss of cytoplasmic density (*) compared with the adjacent epithelial cell. × 12,200

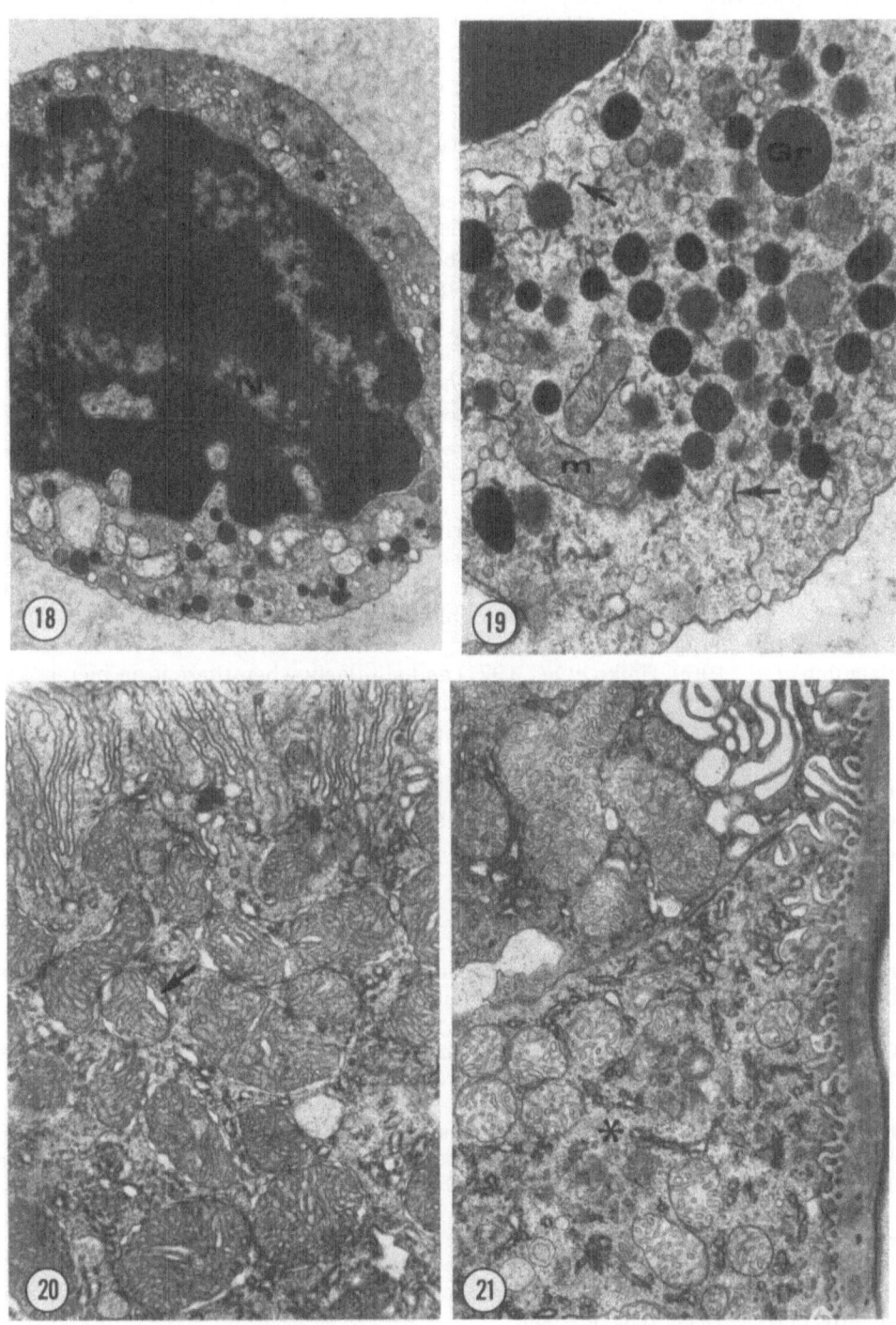

Galleria (Neuwirth, 1973), *Calpodes* (Lai-fook, 1973) *Leucophaea* (Hagopian, 1971), and *Antheraea* (Beaulaton, 1976), as well as in the crustaceans: *Eriocheir* (Bauchau, 1972), *Carcinus* (Johnstone *et al.*, 1973), and *Homarus* (Hearing and Vernich, 1967). These hemocytes are usually polymorphic with a multi-lobed nucleus (Fig. 18) and are characterized by the presence of conspicuous, electron-dense granules in the cytoplasm (Fig. 19). In some crustaceans it has been reported that the granules are composed of carbohydrates and protein (Hearing and Vernick, 1967; Johnstone *et al.*, 1973). Additionally, Hearing and Vernick (1967) have suggested that the granules may be microbodies or lysosomal in nature. Other cytoplasmic features of this hemocyte are numerous short lamellar cisternae of smooth endoplasmic reticulum, free ribosomes, and oblong mitochondria.

Ultrastructural Changes in the Gills of Palaemonetes pugio Associated With Exposure to Na-PCP (1 ppm)

Gill Epithelial Cells (15-30 Minutes after Ecdysis)

No change was observed in the majority of the cells. However, the mitochondria of some cells were observed in the condensed configuration (Fig. 20) characterized by a more dense matrix and distended outer membrane. It has been reported that mitochondrial condensation, a reversible phenomenon, results in cells where oxidative phosphorylation is stimulated or inhibited (Jasper and Bronk, 1968; Saladino *et al.*, 1969; Haiko *et al.*, 1971) and similarly when there is an increase in the ADP/ATP ratio (Trump and Arstilla, 1975). Therefore, in light of the ability of PCP to uncouple oxidative phosphorylation the observed condensed conformation is not unusual. Another abnormal change observed in the lamella was the apparent loss of cytoplasmic density, presumably due to swelling, in some of the epithelial cells in the most distal region of the lamella (Fig. 21). It has been suggested that the analogous region of the gill filaments of *Penaeus* are osmoregulatory in function

---

Figure 22. Pillar region of lamellar epithelium in gills dissected 30 minutes after ecdysis from *Palaemonetes* exposed to Na-PCP. Note abundant lysosomes (Ly). × 9,500

Figure 23. Podocyte of Na-PCP exposed *Palaemonetes* 30 minutes after ecdysis. Note condensed mitochondria characterized by a swollen outer membrane (arrow) × 11,700

Figure 24. Basal portion of a clear tegumental gland cell in gill tissues dissected 30 minutes after ecdysis from *Palaemonetes* exposed to Na-PCP. Note condensed mitochondria (m), slightly dispersed ribosomes (*). × 16,900

Figure 25. A granular secretory cell in the gill axis thirty minutes after ecdysis from *Palaemonetes* exposed to Na-PCP. Note the extreme loss of cytoplasmic density (*) fragmented RER (arrow) and granulated euchromatin (E). Compare with the structural integrity of the adjacent cells. × 3,600

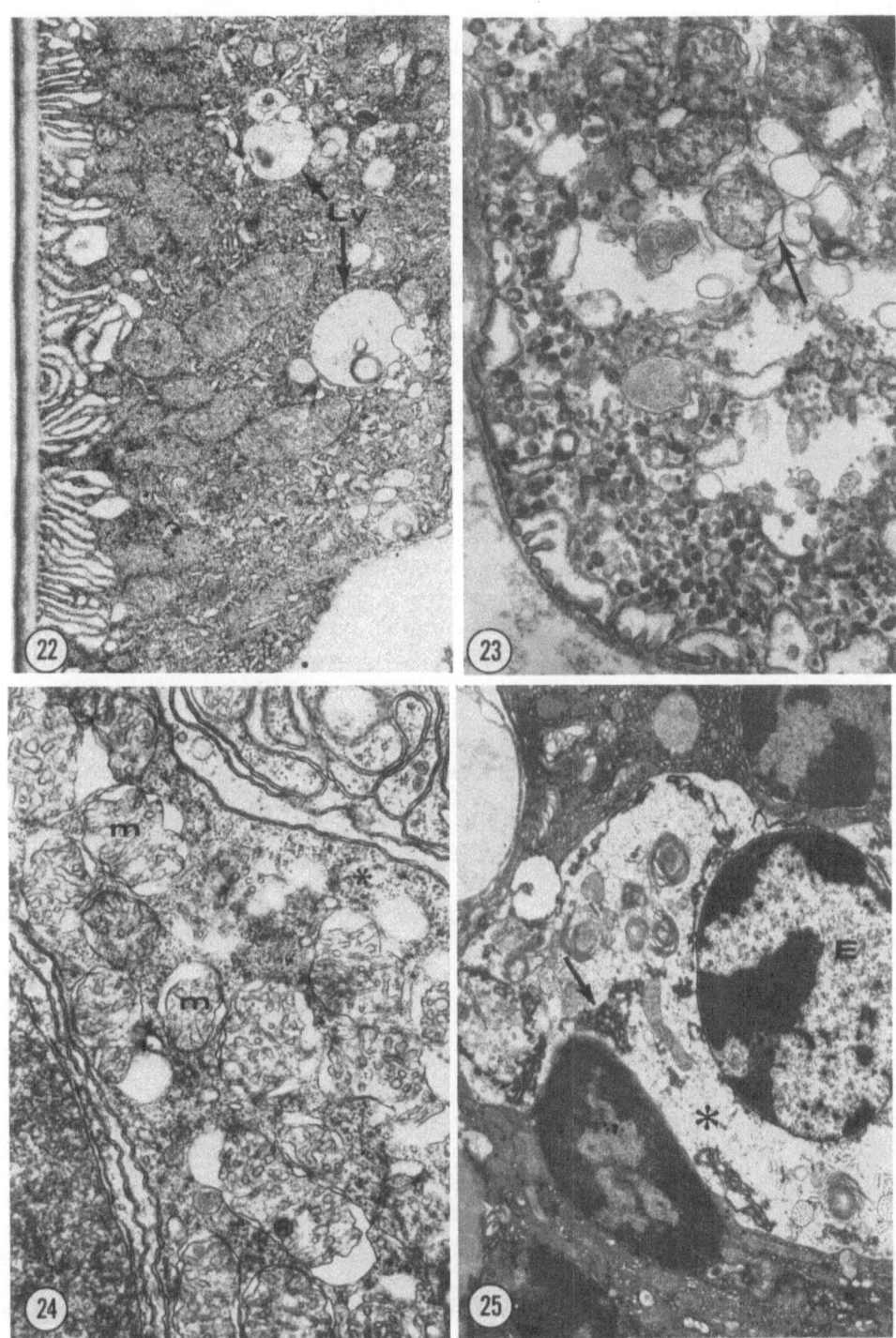

(Couch, 1977). In one animal there was a marked increase in a number of presumed lysosomes within the vicinity of the cytoplasmic bridges or pillar regions of the lamella (Fig. 22).

Podocytes (15-30 Minutes after Ecdysis)

In virtually all instances no abnormal changes were noticeable. However, in several cases condensation of mitochondria was observed (Fig. 23).

Tegumental Glands (15-30 Minutes after Ecdysis)

Condensation of mitochondria was observed in the basal region of some of the gland cells which make up the clear tegumental gland (Fig. 24). No abnormalities were noted in the reticulate glands.

Granular Secretory Cells (15-30 Minutes after Ecdysis)

No change was observed in most cells. In one animal, however, several of the cells exhibited considerable loss of cytoplasmic density and disorganization of contained organelles exemplified by fragmented RER dispersion of the euchromatin and absence of all normal structural integrity (Fig. 25). It is possible that the more sudden appearance of pathological deterioration in this cell type results from its direct communication with the environment and consequently Na-PCP, following ecdysis. (However, in spite of the similar communication noted in the tegumental glands, no commensurate increase in the rate of deterioration was noted in the associated cells. This could possibly relate to the length of the involved duct or tubule which is much longer in the tegumental glands.)

Hemocytes (15-30 Minutes after Ecdysis)

No pathological changes were noted in the hemocytes from animals at 15 to 30 minutes after ecdysis.

Figure 26. Longitudinal section of gill lamella 6 hours after ecdysis from *Palaemonetes* exposed to Na-PCP. Note the progressive deterioration of epithelial and septum cells. Note also the presence of large intracellular spaces (IS) associated with the epithelium and septum. Note the withdrawal of the basement membrane from the basal lamina (*) and the distension of the subcuticular apical infoldings (closed arrow). × 2,700

Figures 27 to 30. A series of high magnification micrographs showing the progressive distension of subcuticular apical infoldings in lamella of Na-PCP exposed *Palaemonetes* 6 hours after ecdysis. Note the included moderately electron dense subcuticular fluid. × 9,000

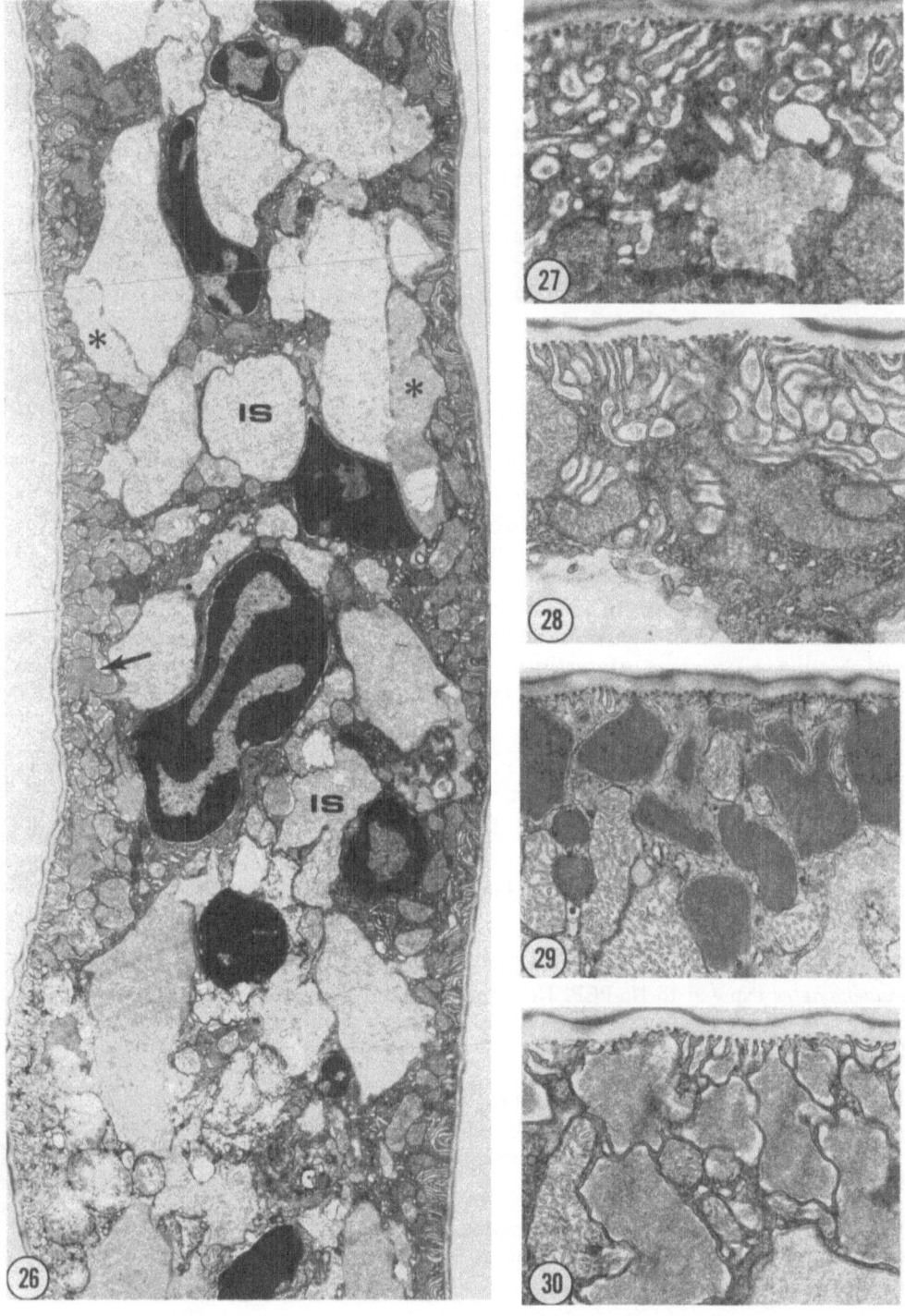

Gill Epithelial Cells (3-9 Hours after Ecdysis)

Extensive pathological changes were observed in both the lamellar and axial epithelial cells. Initially, it should be noted that the observed pathological degeneration leading to necrosis was not evident in all the lamella of exposed shrimp; on the contrary, a single gill would often exhibit lamellar and axial morphology ranging from normal to completely necrotic (Fig. 26). Often necrotic and normal lamella were situated side by side. The ramifications of this situation are discussed later. One of the more interesting changes observed, primarily in the lamella, was the progressive distension and swelling of the subcuticular invaginations of the perikaryon, which concurrently filled with a moderately electron dense fluid (Figs. 27-31). Cisternae of endoplasmic reticulum were characteristically swollen in degenerated cells. Large intracellular membrane bounded vacuoles containing a semi-lucent precipitate and occasional amorphous membranous structures were observed within the epithelial cells as well as in the cells comprising the lamellar septum and the axial hypodermis (Figs. 26, 31). Often, numerous vesicles derived from the endoplasmic reticulum and golgi were seen close to the vacuole, rupturing and releasing their contents (Fig. 33). The occurrence of intact and ruptured multivesicular bodies was common throughout the cytoplasm and may indicate an autolytic function. It has been reported that the occurrence of large clear vacuoles in injured cells may represent accumulations of watery fluid, usually isoosmotic with the hemolymph, in intracellular compartments such as the endoplasmic reticulum, lysosomes, mitochondria, nucleus and cell sap in a phenomenon referred to as hydropic degeneration (Trump and Arstilla, 1975). It is interesting to note that the contents of the vacuoles often possessed the same electron density and texture as the hemolymph (Fig. 26).

Figure 31. A low magnification micrograph of gill lamellar epithelium 9 hours after ecdysis from *Palaemonetes* exposed to Na-PCP. Note the distorted mitochondria (m), distended, fluid-filled subcuticular infoldings (sc) and the similarity of the subcuticular fluid and the hemolymph (H). × 5,700

Figure 32. Intracellular spaces (IS) in a granular secretory cell 6 hours after ecdysis from *Palaemonetes* exposed to Na-PCP. Note the abundant golgi and endoplasmic reticulum-derived vesicles entering the space (arrow) and subsequently rupturing indicative of a possible formative role. × 10,800

Figure 33. Intracellular space (IS) in a granular secretory cell 6 hours after ecdysis from *Palaemonetes* exposed to Na-PCP. Note the abundant golgi and endoplasmic reticulum-derived vesicles entering the space (arrow) and subsequently rupturing indicative of a possible formative role. × 10,800

Figure 34. Necrotic epithelial cell 6 hours after ecdysis in a gill lamella of *Palaemonetes* exposed to Na-PCP. This may contribute to the formation of the intercellular spaces. × 3,800

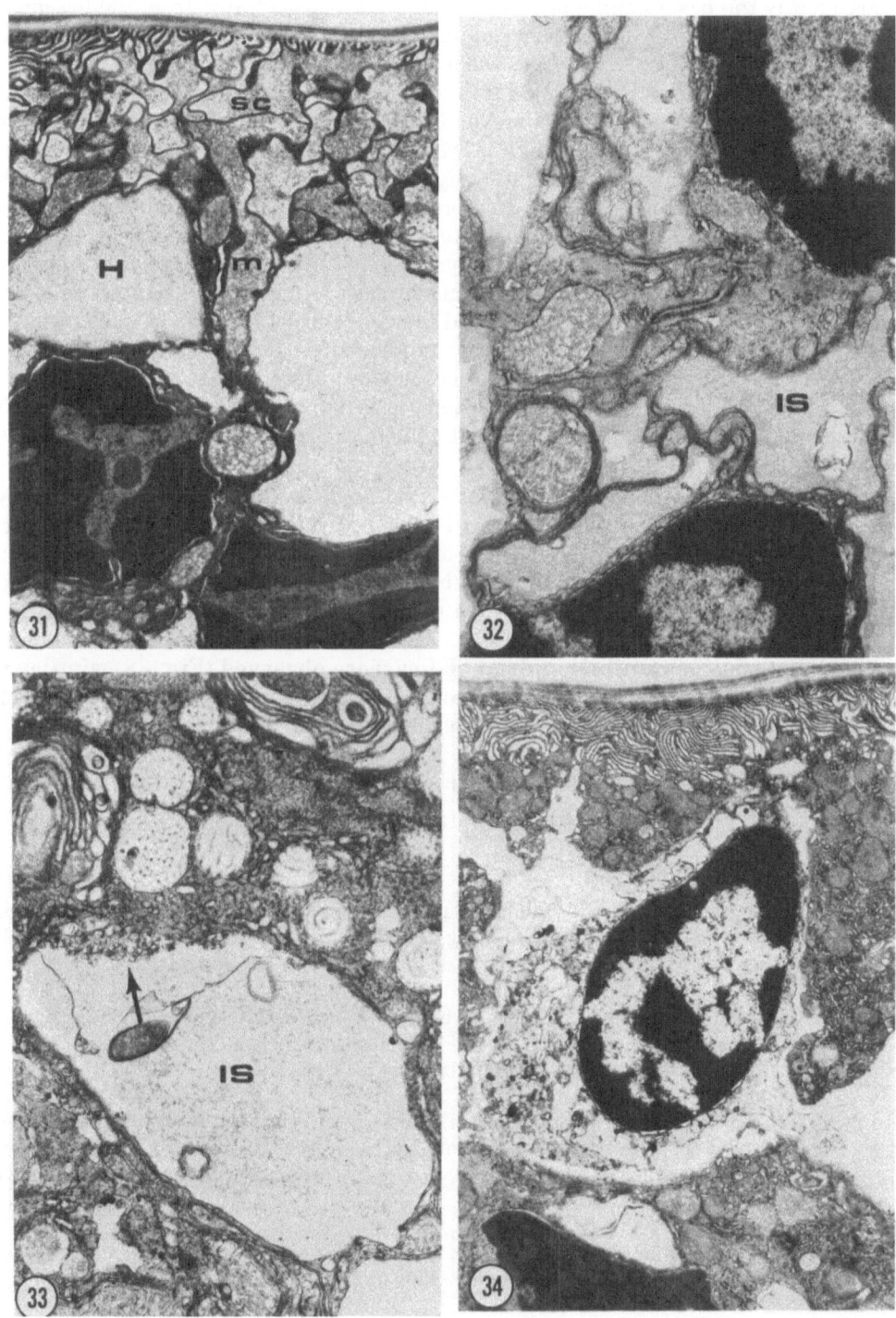

Frequently the basal lamina of the perikaryon was withdrawn from the basement membrane and the included space contained a fluid identical to the vacuolar fluid (Fig. 26). It is possible that some of the intercellular spaces observed in the lamellar medial septum were once occupied by viable cells as it was occasionally noted that entire, apparently necrotic, cells moved into the hemolymph sinuses (Fig. 34). The nuclei of degenerating cells exhibited various degrees of pathology, ranging from granulation of the euchromatin and slight distension of the nuclear envelope to extreme swelling of the nuclear membrane and vacuolation of intensely stained chromatin (Fig. 25). In all instances progressive pathological deterioration was accompanied by the dispersion of polyribosomes into their respective ribosome monomers indicative of inhibited protein synthesis. Autophagy, although rare, was observed in some instances and involved the convolution of endoplasmic reticulum around some injured organelle, usually mitochondria (Fig. 36) or portion of the cytoplasm. Presumably the infrequent occurrence of autophagic vacuoles is due to the energy required for their formation. It is known that by restricting ATP production in cells, the formation of autophagic vacuoles is inhibited (Shelburne *et al.*, 1970).

Mitochondria also exhibited various morphological responses to injury. Characteristically, various stages of high amplitude swelling coincident with vesiculation and fragmentation of cristae as well as the total absence of matrical granules were observed. The outer membrane was usually distended and polarity of the matrix was evident (Fig. 37). In necrotic cells, the outer mitochondrial membrane was often ruptured and small vesicular and sometimes flocculent electron dense intramitochondrial bodies were observed (Fig. 38), similar to inclusions reported in the renal mitochondria of rats exposed to mercuric chloride (McDowell *et al.*, 1976). These inclusions may represent abnormal accumulations of substances normally present in the matrix (e.g. calcium or ferritin). Less frequently, inclusions resembling myelin whorls were also observed in the lamellar

---

Figure 35. Nucleus in a necrotic gill lamellar epithelial cell as seen 9 hours after ecdysis in *Palaemonetes* exposed to Na-PCP exhibiting granulation of the euchromatin and intensely stained and vacuolated chromatin. Note the distended nuclear envelope (arrow). × 8,500

Figure 36. Autophagosome as seen 6 hours after ecdysis in lamellar septum of *Palaemonetes* exposed to Na-PCP. Note the convoluted membranes (arrow) surrounding a degenerate mitochondrion with vesiculated cristae (*). × 16,200

Figure 37. Epithelial mitochondria as seen 9 hours after ecdysis in *Palaemonetes* exposed to Na-PCP. Note polarization of fragmented cristae (cr), flocculent matrix (ma) and distended outer membrane. Note also the swollen endoplasmic reticulum (ER) and distorted apical infoldings (af). × 12,900

Figure 38. Epithelial mitochondria as seen 9 hours after ecdysis in *Palaemonetes* exposed to Na-PCP. Note absence of limiting membranes (arrows) and inclusion of electron dense intramitochondrial bodies. × 16,200

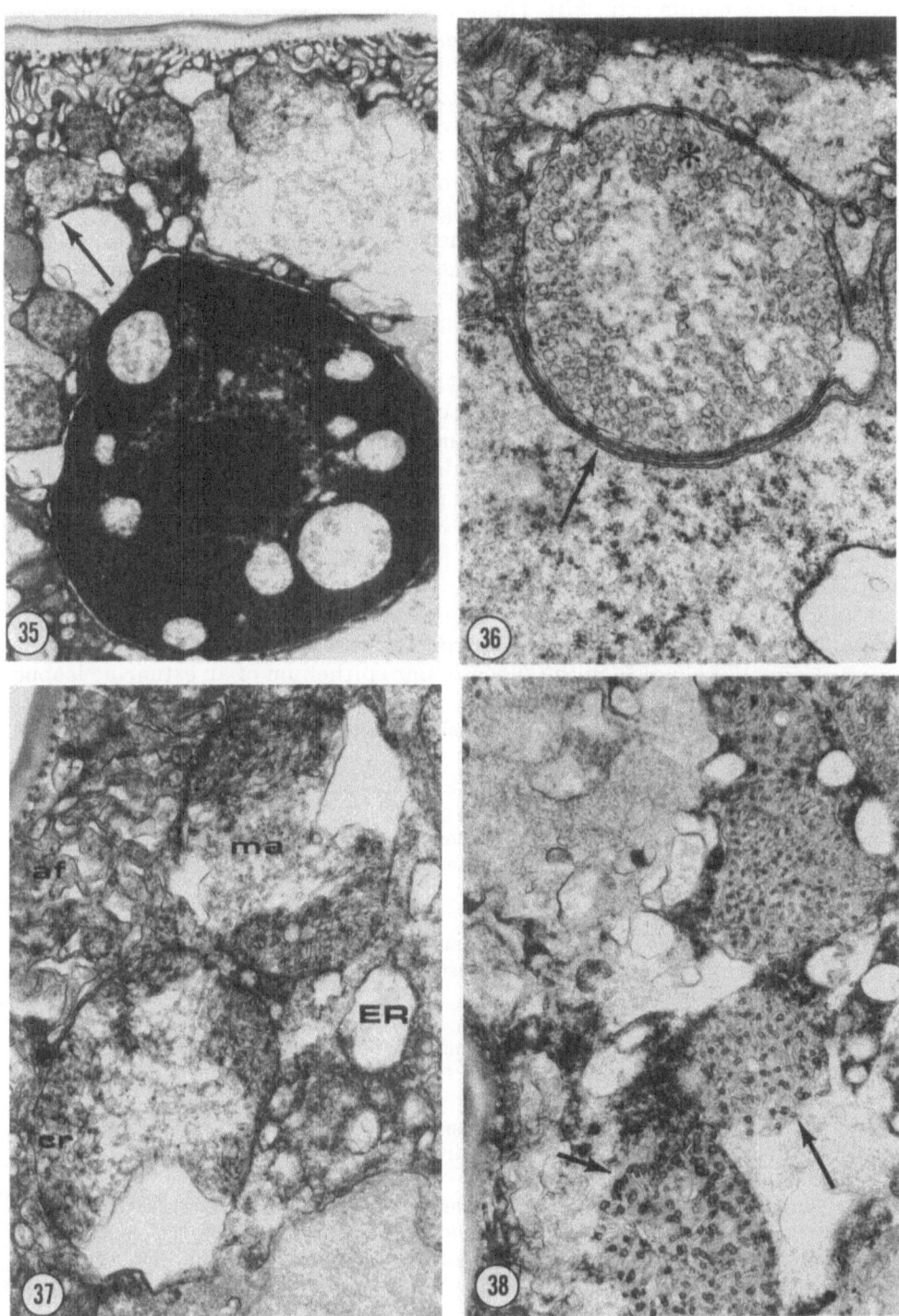

septum of some gills (Fig. 39). These myelin formations were also noted in degenerate hepatopancreatic tissue of *Palaemonetes pugio* (exposed to Na-PCP). The functional significance of these structures is not clear. Lamellar formations, as in hepatic mitochondria of pigs exposed to toxic concentrations of lead (Watrach, 1976), were often observed in the matrix which appeared to be continuous with the cristae membranes (Figs. 40, 41). These patterns may also be derived from the amorphous condensation of matrical proteins. Possibly due to the inability to effectively divide or abnormal fusion, huge megamitochondria were occasionally observed (Fig. 42). To explain this phenomenon in mitochondria of yeast exposed to toxic levels of copper, Keyhani (1973) suggested that high copper concentrations cause a hyperproduction of cytochromes, resulting in a change in the spatial arrangements of the respiratory components within the mitochondrial membranes which in turn leads to the fusion and subsequent formation of giant mitochondria. Finally, the lamella of some experimental animals exhibited cellular deterioration exemplified by an overall loss of structure and the formation of polymorphic electron-dense deposits around and within septum mitochondria (Fig. 43). Similar deposits have been observed in mitochondria of gills of penaeid shrimp exposed to cadmium (Couch, 1977). In general then, mitochondria were exemplified by high amplitude swelling, fragmentation and vesiculation of cristae, development of intramitochondrial bodies and lamellar formations, and a flocculent appearing matrix. In light of the widespread deterioration of mitochondria in some lamella, a possible explanation for the simultaneous occurrence of distended subcuticular invaginations in *Palaemonetes* will be proposed based on an earlier investigation reporting similar changes in the subcuticular epithelium of an estuarine isopod exposed to toxic concentrations of copper, mercury and cadmium (Bubel, 1976). As suggested by Green (1968), estuarine crustaceans possess the ability to maintain a

---

Figure 39. High magnification of the mitochondria in lamellar septum of Na-PCP exposed *Palaemonetes* 6 hours after ecdysis. Note presence of intramitochondrial myelin-form (My). × 31,000

Figure 40. Gill epithelial mitochondria 6 hours after ecdysis in *Palaemonetes* exposed to Na-PCP showing unusual conformation of cristae (cr) which may precede the formation of intramitochondrial lamellae. × 22,500

Figure 41. Intramitochondrial lamellar formations (arrow) seen 9 hours after ecdysis in *Palaemonetes* exposed to Na-PCP. × 18,900

Figure 42. Gill epithelial megamitochondrion seen 6 hours after ecdysis in *Palaemonetes* exposed to Na-PCP. May be indicative of impaired ability to effectively divide. × 7,200

Figure 43. Micrograph of a portion of lamellar septum of Na-PCP exposed *Palaemonetes* 9 hours after ecdysis. Note electron dense deposits within the cytoplasm and at the mitochondrial perimeter (apparently included between the inner and outer compartments). Note swollen endoplasmic reticulum (ER) and overall lack of cytoplasmic definition. × 16,900

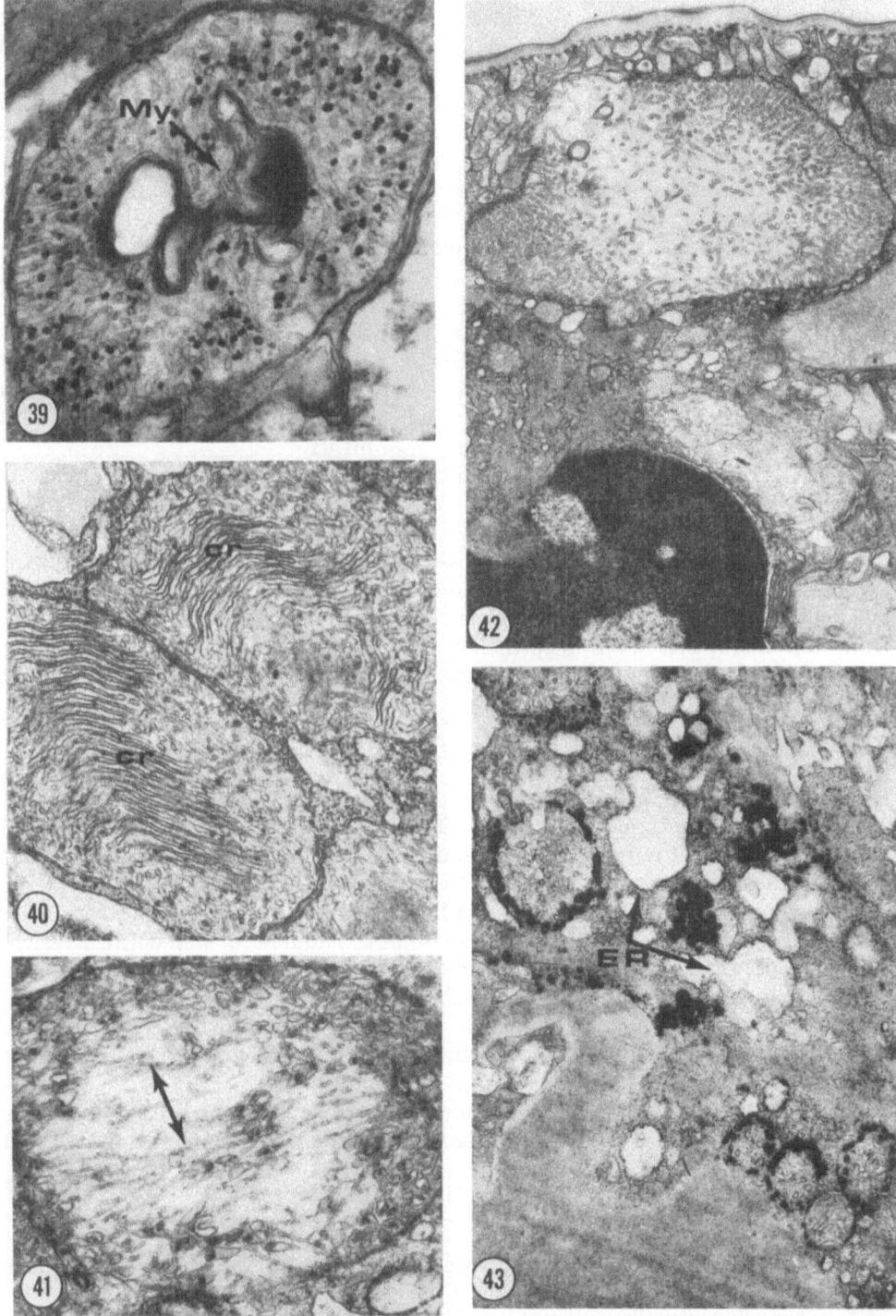

hyperosmotic hemolymph in low salinities (as employed in the present investigation 10 %) by the active transport of ions. If, as suggested earlier, the crustacean gill epithelium is capable of ion translocation through active transport mechanisms, mediated by a membrane ATPase system, then a breakdown of this system would necessarily create osmotic stress for the animal. Additionally it has been proposed that the major cause of osmotic swelling in injured cells is due to the colloid osmotic pressure of the macromolecules (Robinson, 1975). Under normal conditions this is balanced by the active extrusion of Na^+ into the extracellular fluid (Leaf, 1970). Therefore, due to the lack of metabolic energy supplied by mitochondria (possibly uncoupled by Na-PCP) sodium ions and subsequently water can enter the cells and bring about swelling.

In agreement with the findings of Lockwood *et al.*, (1973) the distended subcuticular invaginations observed in this investigation may be an attempt by the cells to establish a fluid-filled buffer zone between the bulk of the cell and the medium external to the cuticle. This would provide a longer pathway for the movement of water and thus decrease its rate of influx.

Finally, in reference to a statement made earlier regarding the lack of uniformly distributed pathological signs in both single as well as adjacent lamella the following proposal is made.

Often necrotic and normal lamella were situated side by side. In this regard, assuming that the distance between adjacent lamella is critical with respect to adequate respiratory function, the extreme expansion of necrotic lamella resulting from the disruption of the epithelial pillar region (Fig. 44) could instigate respiratory dysfunction in the adjacent normal lamella, thereby hastening organismic death.

Podocytes (3-9 Hours after Ecdysis)

As noted in the epithelial cells, podocytes exhibited all stages of cell injury ranging from almost normal, to minimal damage to necrosis. In intermediate states the subpedicellar cisternae were characterized by a marked decrease in endocytotic activity (Fig. 45). The cisternae were filled with a homogeneous electron lucent precipitate and there was a marked decrease in the number of well-defined vesicles in the subjacent cytoplasm. Mitochondria displayed some swelling, and fragmentation of cristae was evident. The nucleoplasm typically possessed granulated euchromatin and clumped or vacuolated chromatin. Some distension of the nuclear envelope as well as fragmentation of endoplasmic reticulum was also noted.

Other cells exhibiting more advanced necrosis invariably showed extreme swelling and rupture of the subpedicellar cisternae and the fine precipitate found

---

Figure 44. Longitudinal section of gill lamella of Na-PCP exposed *Palaemonetes* 9 hours after ecdysis. Note the extreme expansion of the lamella as a whole due to the widespread necrosis and subsequent dissolution of the pillar regions. Note the Lysosome (Ly). × 4,320

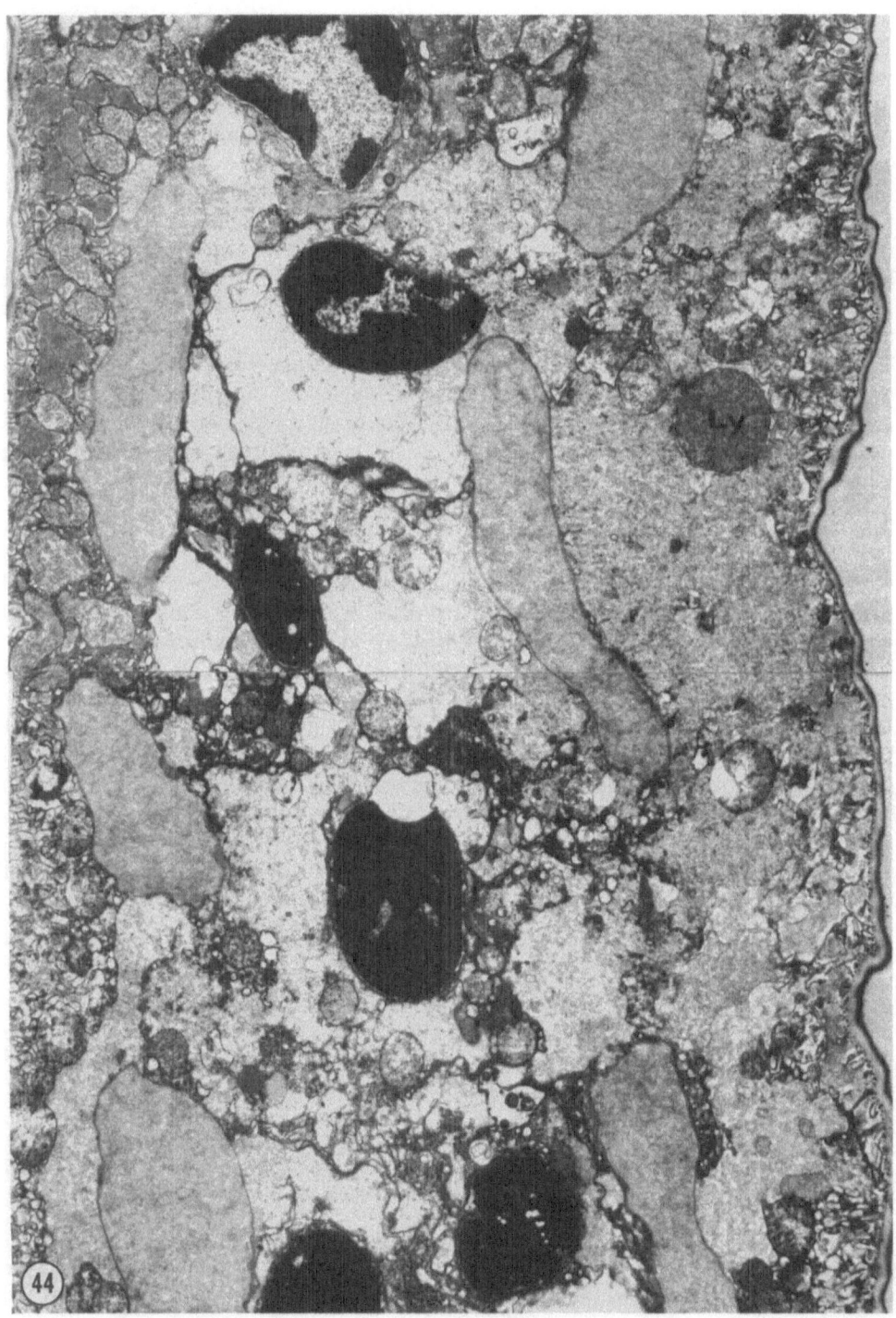

within the cisternae was common in the subjacent interstitial cytoplasm (Fig. 45). The well defined structural integrity of normal cells was absent and the cytoplasm was cloudy with widespread blurring of organelle profiles. This is possibly indicative of a change of intracellular protein configuration. Furthermore, the nuclear envelope was grossly distended and the nucleus exhibited margination of chromatin and pyknosis (Fig. 45). Extreme vesiculation and polarization of mitochondrial cristae was evident. The central matrical region contained a flocculent material and some lamellar formations. Vesicular-electron opaque intramitochondrial bodies were occasionally observed. In some instances, mitochondria characterized by high amplitude swelling, were observed entering the large central vacuole (Fig. 46) which presumably is capable of autolytic processes (Smith, 1968).

Finally, in the most extreme cases, cells were characterized by an almost total absence of cytoplasmic organelles. Only a degenerate nucleus and vestigial structures were present, suspended in an electron lucent precipitate (Fig. 47) not unlike that observed in the large intracellular vacuoles observed in the degenerating lamellar epithelium.

Tegumental Glands (3-9 Hours after Ecdysis)

Many tegumental glands of both the clear and reticulate variety displayed symptoms of injury and degeneration. In intermediate clear glands mitochondrial condensation was persistent and there was a marked dispersion of polysomes (Figs. 24 and 48) indicative of impaired protein synthesis. Loss of cytoplasmic density was observed in both the gland cells (Fig. 48) and the cells responsible for duct formation (Fig. 49). Pronounced granulation of the euchromatin was also apparent in the latter. High amplitude swelling and sparse cristae were commonly observed in the peripheral mitochondria (Fig. 50). The apical vesicles of the gland cells often exhibited electron dense deposits at their periphery.

The reticulate tegumental glands typically displayed progressive deterioration. As in the clear glands, the duct-forming cell was the first to show signs of deterioration (Fig. 51). The basal rough endoplasmic reticulum was often

---

Figure 45. Podocyte cell from gill axis of Na-PCP exposed *Palaemonetes* 6 hours after ecdysis. Note the nuclear pyknosis as well as the extreme dilation of the nuclear envelope (ne). Note the ruptured subpedicellar cisternae (arrow) and the included precipitated fluid (*) throughout the poorly defined cytoplasm. × 6,300

Figure 46. Podocyte in the gill axis of Na-PCP exposed *Palaemonetes* 6 hours after ecdysis. Note the apparent movement of swollen mitochondria (m) into the intracellular vacuole (IV). This vacuole may carry out autolytic activity in normal cells. × 10,800

Figure 47. Necrotic podocyte from gill of Na-PCP exposed *Palaemonetes* 9 hours after ecdysis. Note the absence of all but traces of the normal cytoplasmic constituents. Note the omnipresence of precipitated fluid (*), ruptured subpedicellar cisternae (closed arrow) and distended nuclear envelope (ne). × 7,600

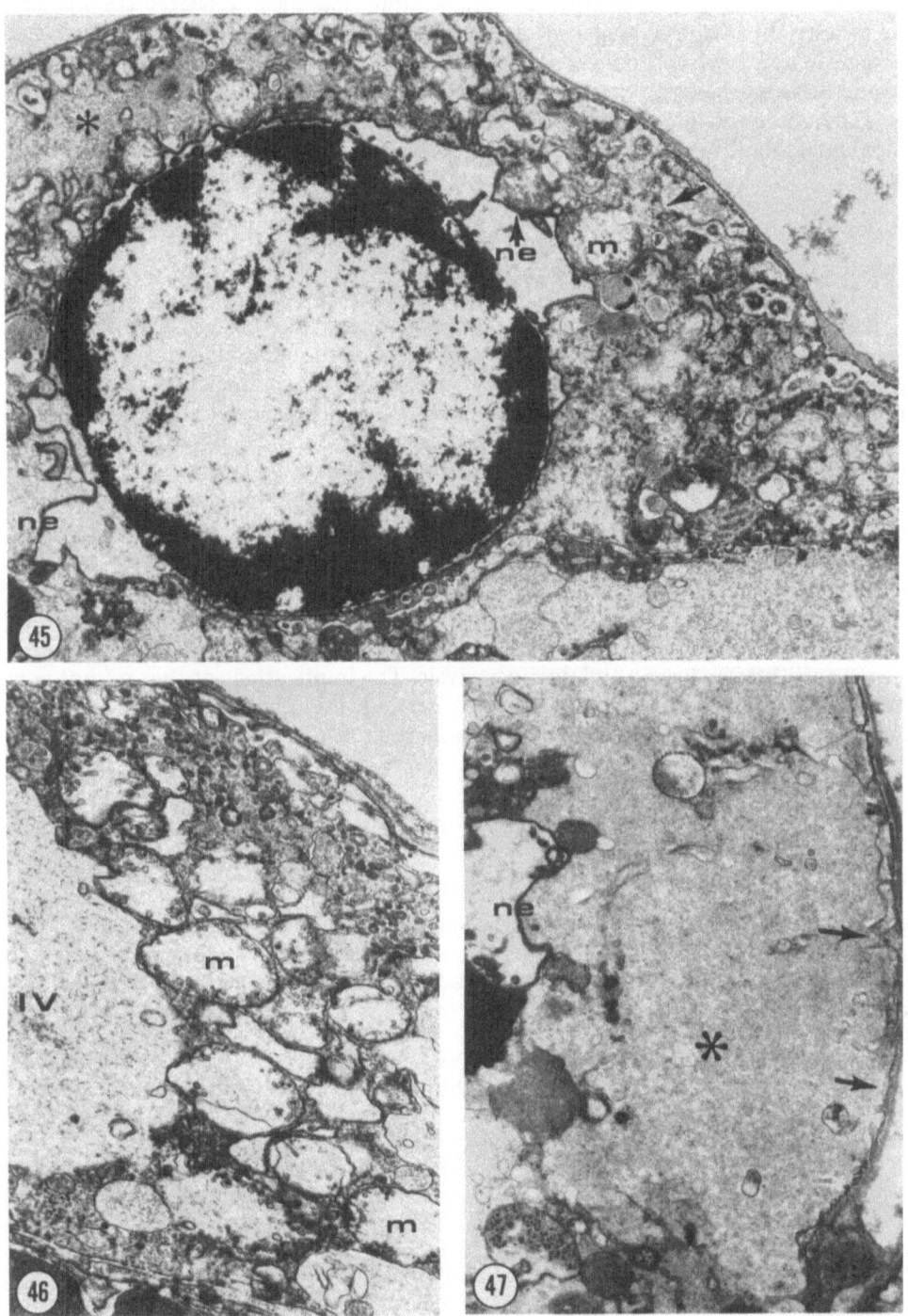

proliferated in complex configurations and in other instances appeared fragmented and swollen. In advanced stages of injury the nucleus was characterized by clumped or marginated heterochromatin (Fig. 52). An interesting phenomenon was the degenerative morphology exhibited by some mitochondria. In these, large regions of the matrix were devoid of cristae (Fig. 53). A similar response was observed in the spinal mitochondria of a primate exposed to tricresylphosphate (Ahmed and Glees, 1977).

Granular Secretory Cells (3-9 Hours after Ecdysis)

Degradative reactions in many cells were severe. The observed deterioration was not restricted to either the lamella or the axis. As noted earlier in the other cell types, the pathological changes were progressive and included fragmented endoplasmic reticulum with detached ribosomes, loss of cytoplasmic density, formation of intracellular spaces and extreme distension and rupture of the nuclear envelope (Figs. 54 and 55). It is interesting to note that while many granular secretory cells appeared normal there was a considerable discrepancy, compared with equivalent controls, in the relative abundance of laminated granules found within the cytoplasm. In control animals sacrificed 3-24 hours after ecdysis the presence of the granules was subdued or negligible. However, in comparable Na-PCP exposed animals, the cells contained numerous fully formed granules. This suggests that some as yet unknown factor inhibited the dissolution and release of the granular matrix before and during ecdysis. The implications of this are not now understood. More research into the chemical or mineral composition of these structures is needed so as to adequately assess their role in the molting process.

Figure 48. Apical cytoplasm of clear tegumental gland cell from gill axis of Na-PCP exposed *Palaemonetes* 9 hours after ecdysis. Note the low cytoplasmic density (*), condensed mitochondria (m) with vesiculated cristae and distended outer membrane (arrow). × 8,200

Figure 49. Clear tegumental gland of gill axis, 6 hours after ecdysis in *Palaemonetes* exposed to Na-PCP. Note the low density of the gland cell nuclei (GCN) and granulation of the duct cell nucleus (DCN). Note also the duct (arrow). × 3,500

Figure 50. Basal portion of clear tegumental gland cell from Na-PCP exposed *Palaemonetes* 6 hours after ecdysis. Note dispersed ribosomes (*) condensed mitochondria (m) and swollen mitochondria (sm) with sparse cristae and flocculent matrix. × 16,900

Figure 51. Reticulate gland complex from gill axis of Na-PCP exposed *Palaemonetes* 6 hours after ecdysis. Note loss of cytoplasmic density (*) in duct cell (DC) as well as the granulation of nuclear euchromatin and margination of chromatin. × 3,400

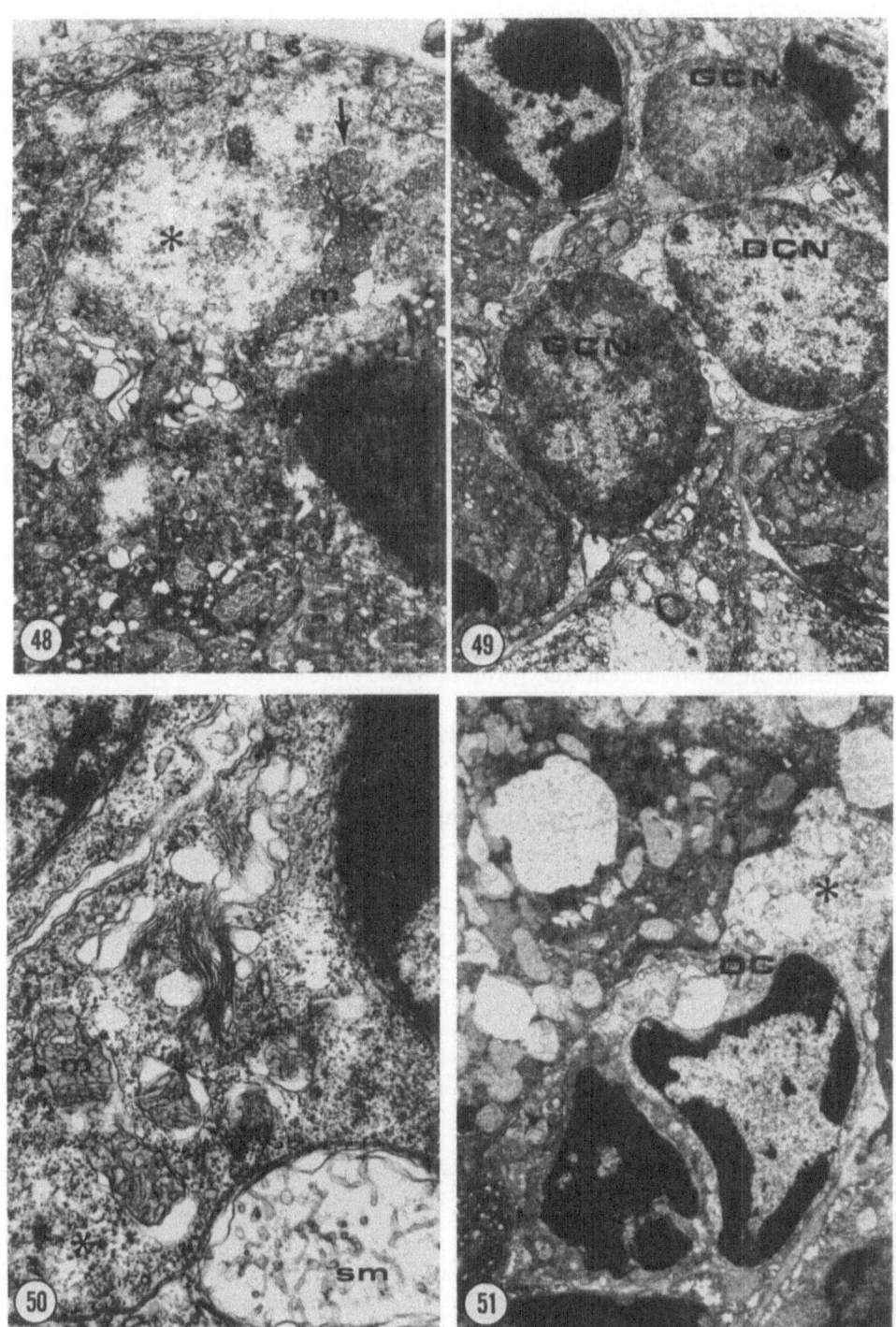

Hemocytes (3-9 Hours after Ecdysis)

Approximately one half of the granular hemocytes exhibited mitochondrial degeneration ranging from slight condensation of the inner compartment to extreme dilation of the outer compartment coupled with matrix polarization (Figs. 56 and 57). The cytoplasm of deteriorated cells displayed a lack of density, due to swelling (Fig. 57). In a number of hemocytes electron opaque deposits, which resembled myelin whorls, were observed within both the cytoplasm and the mitochondria (Fig. 58).

Ultrastructural Changes in the Midgut, Hindgut and Hepatopancreas

As part of this study, an investigation has been made concerning abnormal changes in the midgut, hindgut and hepatopancreas of grass shrimp exposed to Na-PCP but due to the limited available space in this symposium proceedings these results will be published elsewhere. However, a brief description of the pathological changes observed in these tissues will be presented in the following paragraphs.

The midgut and hindgut epithelium was the only tissue which exhibited abnormal changes during the late proecdysial stages (D_3-D_4) of the molting sequence. These changes were manifested in swollen, fragmented cisternae of RER and a distended nuclear envelope. Withdrawal of the basal lamina from the basement membrane was also noted. Shortly after ecdysis (15-30 min), condensation of mitochondria, proliferation and swelling of tubular smooth endoplasmic reticulum and abnormal vesiculation of the apical microvilli were observed. More extensive pathological deterioration was noted in these tissues from shrimp sacrificed 3-9 hours after ecdysis. These changes included extreme swelling and rupture of the apical cell foci and a concurrent loss of microvilli. Mitochondria displayed various pathological symptoms exemplified by high amplitude swelling,

---

Figure 52. Reticulate gland cell from the gill axis of Na-PCP exposed *Palaemonetes* 9 hours after ecdysis. Note the pyknotic nucleus (GCN), fragmented endoplasmic reticulum (arrow) and overall loss of cytoplasmic organization. × 3,400

Figure 53. Median cytoplasm of reticulate tegumental gland cell from the gill axis of Na-PCP exposed *Palaemonetes* 6 hours after ecdysis. Note presence of voids in mitochondrial matrix (arrows). × 7,900

Figure 54. Granular secretory cell 6 hours after ecdysis from *Palaemonetes* exposed to Na-PCP. Note granulated euchromatin (E), ruptured nuclear envelope (arrow) and deteriorated cytoplasmic organization. × 5,700

Figure 55. Granular secretory cell 9 hours after ecdysis from *Palaemonetes* exposed to Na-PCP. Note condensed mitochondria (m) with distended outer membrane and vesiculated cristae. Note precipitated interstitial fluid (*). × 7,900

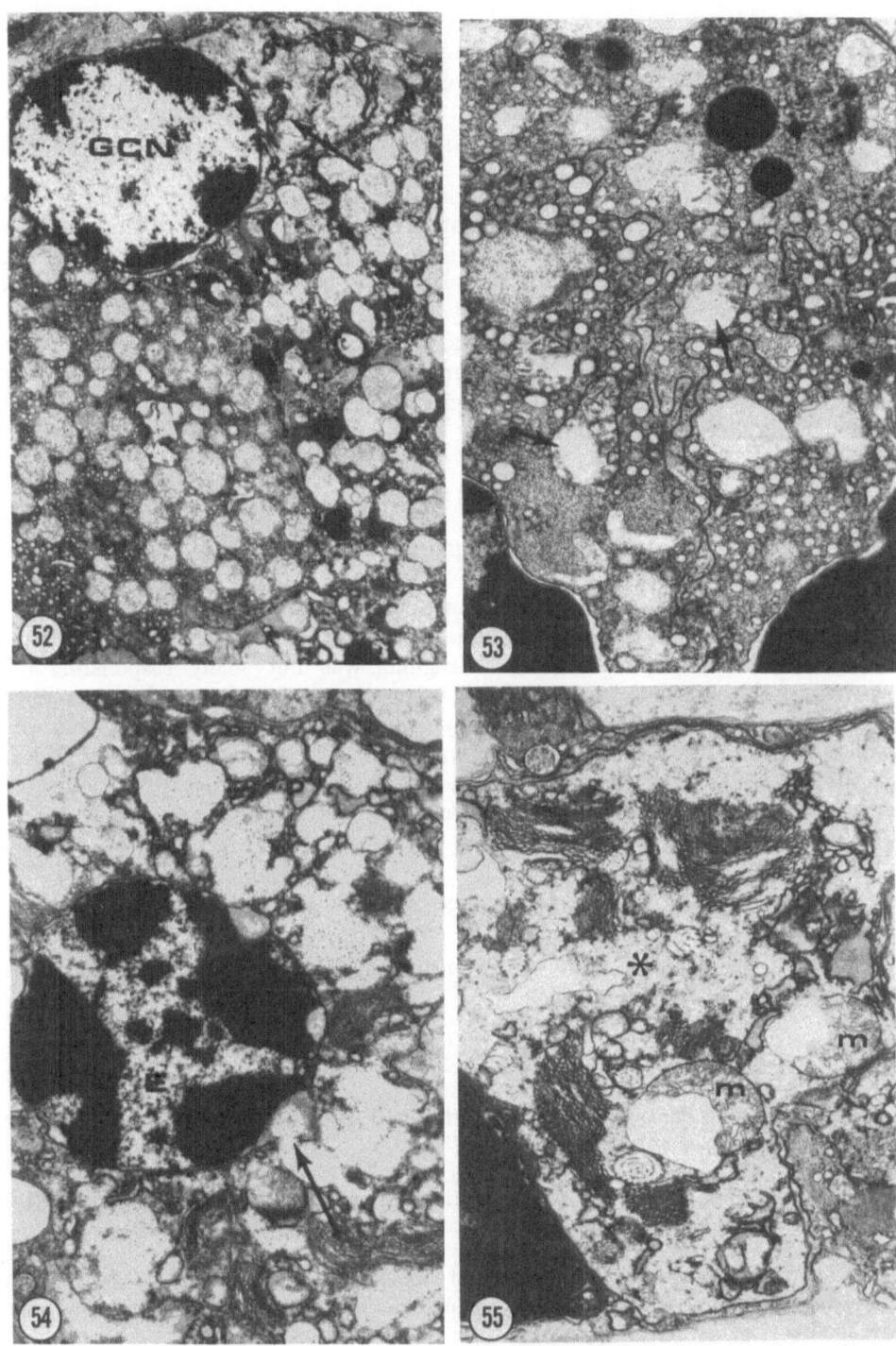

loss of matrical granules and sparse cristae. In several animals, the mitochondria of the basal cytoplasm were proliferated and exhibited compartmentalization of the matrix coincident with differential swelling. Typically the apical cytoplasm possessed numerous distended golgi, ruptured multivesicular bodies, intracellular spaces and myeloid bodies. The lack of cytoplasmic density in this region is indicative of a low viscosity due to swelling.

Similar to changes observed in the gut epithelium some cells of the hepatopancreas showed abnormal vesiculation of microvilli within 30 minutes after ecdysis. At this stage, fragmented E.R. and swollen nuclear envelopes were also observed. Three to nine hours after ecdysis all cells exhibited various stages of progressive cytoplasmic deterioration. Myeloid figures were observed in the rough endoplasmic reticulum and mitochondria. Other mitochondria were swollen and possessed sparse constricted cristae in a flocculent appearing matrix. The widespread occurrence of swollen golgi, dispersed polysomes and increased autophagic activity were also noted.

Conclusions

In light of the apparently simultaneous deterioration of the midgut, hindgut, hepatopancreas and gill tissues of shrimp exposed to Na-PCP, it seems clear that focal cell death in any one of these regions was not solely responsible for perpetrating the eventual death of the organism. Due to the increased permeability of the new untanned cuticle immediately after ecdysis it seems likely that the peripheral epithelium became inundated with Na-PCP. The subsequent entry of Na-PCP into the hemolymph and its transport into other major organs may account for the concurrent degeneration of the gut and hepatopancreatic tissues.

ACKNOWLEDGMENTS

This investigation was supported by Grant R-804541-01 from the United States Environmental Protection Agency.

Figure 56. Granular hemocyte from a gill of Na-PCP exposed *Palaemonetes* 6 hours after ecdysis. Note condensed mitochondria (m). × 25,600

Figure 57. Granular hemocyte from a gill of Na-PCP exposed *Palaemonetes* 9 hours after ecdysis. Note distorted mitochondria (m) and loss of cytoplasmic density. × 17,100

Figure 58. Granular hemocyte from a gill of *Palaemonetes* exposed to Na-PCP, 6 hours after ecdysis. Note the inclusion of electron dense bodies within both the cytoplasm and the mitochondria (arrows). × 16,200

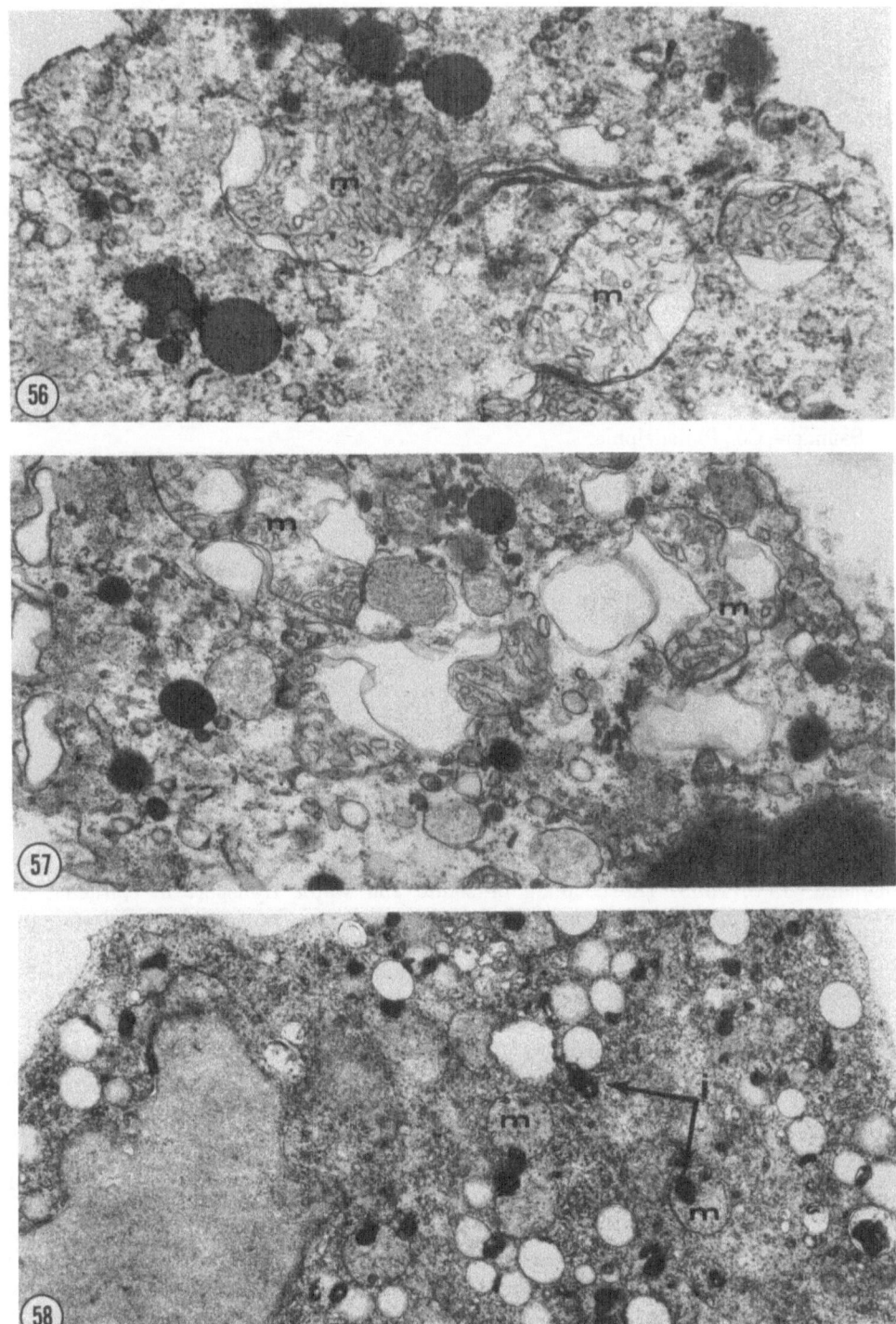

References

Ahmed, M.M., and P. Glees, 1977. Mitochondrial degeneration after organic phosphate poisoning in prosimian primates. *Cell Tiss. Res.,* **175:** 459-465.

Ali, M., 1966. The histology and fine structure of the gills of *Carcinus maenas* L. and other decapod crustacea. Ph.D. thesis, University of Newcastle upon Tyne.

Allen, E.J., 1893. On the minute structure of the gills of *Palaemonetes varians. Quart. J. Microscop. Sci.,* **34:** 75-84.

Bauchau, A.G. and M. DeBrouwer, 1972. Ultrastructure des hemocytes d'*Eriocheir sinensis* (Crustace, decapode, brachyoure). *J. Microscopie,* **15:** 171-180.

Beaulaton, J., and M. Monpeyssin, 1976. Ultrastructure et cytochimie des hemocytes d'*Antheraea pernyi* Guer. (lepidoptera, Attacidea) au cours du cinquieme age larvaire. *J. Ultrastr. Res.,* **55:** 143-156.

Bielawski, J., 1971. Ultrastructure and ion transport in gill epithelium of the crayfish *Astacus leptodactylus* Esch. *Protoplasma,* **73:** 177-190.

Bloom, W., and D.W. Fawcett, 1968. *A Textbook of Histology.* (9th ed.), pp. 652-684. W.B. Saunders, Co., Philadelphia.

Bubel, A., 1976. Histological and electron microscopical observations on the effects of different salinities and heavy metal ions, on the gills of *Jaera nordmanni* (Rathke) (Crustacea, Isopoda). *Cell Tiss. Res.,* **167:** 65-95.

Chassard-Bouchaud, C., and M. Hubert, 1973. Etude ultrastructurale du tegument des crustaces decapodes en fonction du cycle d'intermue. *J. Microscopie,* **16:** 75-86.

Copeland, D.E., 1967. A study of salt secreting cells in the brine shrimp *(Artemia salina). Protoplasma,* **63:** 363-384.

Copeland, D.E., 1968. Fine structure of salt and water uptake in the land crab, *Gecarcinus lateralis. Am. Zoologist,* **8:** 417-432.

Copeland, D.E., and A.T. Fitzjarrel, 1968. The salt absorbing cells in the gills of the blue crab *(Callinectes sapidus* Rathbun) with notes on modified mitochondria. *Z. Zellforsch.,* **92:** 1-22.

Couch, J.A., 1977. Ultrastructural study of lesions in gills of a marine shrimp exposed to cadmium. *J. Invert. Pathol.,* **29:** 267-288.

Crossley, A.C., 1972. The ultrastructure and function of pericardial cells and other nephrocytes in an insect *Calliphora erythrocephala. Tissue Cell,* **4:** 529-560.

Diamond, J.M., 1969. The coupling of solute and water transport in epithelia. In, *Proceedings of the International Symposium on Renal Transport and Diuretics* (K. Thurau and J. Jahrmarker, Ed.), pp. 77-97, Springer-Verlag, Berlin.

Diamond, J.M., and J. McD. Tormey, 1966a. Role of long extracellular channels in fluid transport across epithelia. *Nature,* **210:** 817-820.

Diamond, J.M., and J. McD. Tormey, 1966b. Studies on the structural basis of water transport across epithelial membranes. *Fed. Proc.,* **25:** 1458-1463.

Fisher, J.M., 1972. Fine structural observations on the gill filaments of the freshwater crayfish, *Astacus pallipes* (Lereboullet). *Tissue Cell,* **4:** 287-299.

Flemister, S.C., 1959. Histophysiology of gill and kidney of crab *Ocypode albicans. Biol. Bull.,* **116:** 37-48.

Foster, C.A., 1976. Morphology of the gill of the brown shrimp, *Penaeus aztecus* (Ives). Masters thesis, University of Southern Mississippi, Hattiesburg, Mississippi.

Huxley, T.H., 1896. *The Crayfish. An Introduction to the Study of Zoology.* The International Scientific series. Paul, Trench and Trubner, London.

Green, J., 1968. *The Biology of Estuarine Animals,* 401 pp., Sedgewich and Jackson, London.

Green, J.P., and M.R. Neff, 1972. A survey of the fine structure of the integument of the fiddler crab. *Tissue Cell,* **4:** 137-171.

Hagopian, M., 1971. Unique structures in the insect granular hemocytes. *J. Ultrastr. Res.,* **36:** 646-658.

Hearing, V., and S.H. Vernick, 1967. Fine structure of the blood cells of the lobster *Homarus americanus. Chesapeake Sci.,* **8:** 170-186.

Jasper, D.K., and J.R. Bronk, 1968. Studies on the physiological and structural characteristics of rat intestinal mucosa. Mitochondrial structural changes during amino acid absorption. *J. Cell Biol.,* **38:** 277-291.

Johnstone, M.A., H.Y. Elder, and P.S. Davies, 1973. Cytology of *Carcinus* Haemocytes and their function in carbohydrate metabolism. *Comp. Biochem. Physiol.,* **46A:** 569-581.

Keyhani, E., 1973. Morphological changes in yeast cell mitochondria grown at various copper concentrations. *Exper. Cell Res.,* **81:** 73-78.

Koch, H.J., J. Evans, and E. Schicks, 1954. The active absorption of ions by the isolated gills of the crab *Eriocheir sinensis* (M. Edw.) *Mededel. Kon. VI Acad. Kl. Wet.,* **16:** 1-16.

Lai-fook, J., 1973. The structure of the hemocytes of *Calpodes ethlius* (Lepidoptera). *J. Morph.,* **139:** 79-104.

Laiho, K.V., J.D. Shelburne, and B.F. Trump, 1971. Observations on cell volume, ultrastructure, mitochondrial conformation and vital dye uptake in Erlich ascites tumor cells: effects of inhibiting energy production and function of the plasma membrane. *Amer. J. Path.,* **65:** 203-222.

Locke, M., 1969. The structure of an epidermal cell during development of the protein epicuticle and the uptake of molting fluid in an insect. *J. Morph.,* **127:** 7-40.

Mantel, L.H., 1967. Asymmetry potentials, metabolism and sodium fluxes in gills of the blue crab, *Callinectes sapidus. Comp. Biochem. Physiol.,* **20:** 743-753.

McDowell, E.M., R.B. Nagle, R.C. Zalme, J.C. McNeil, W. Flamenbaum, and B.F. Trump, 1976. Studies on the pathophysiology of acute renal failure. I. Correlation of ultrastructure and function in the proximal tubule of the rat following administration of mercuric chloride. *Virchows Arch., B Cell Path.,* **22:** 173-196.

Milne, D.J., and R.A. Ellis, 1973. The affect of salinity acclimation on the ultrastructure of the gills of *Gammarus oceanicus* (Segerstrale, 1947) (Crustacea, amphipoda). *Z. Zellforsch,* **139:** 311-318.

Mitchell, P., 1961. Coupling of phosphorylation to electron and hydrogen transfer by a chemi-osmotic type of mechanism. *Nature,* **191:** 144-148.

Neuwirth, M., 1973. The structure of the hemocytes of *Galleria mellonella* (Lepidoptera). *J. Morph.,* **139:** 105-124.

Pressman, B.C., 1963. In, *Energy Linked Functions of Mitochondria* (B. Chance, Ed.), pp. 188-199, Academic Press, New York.

Robinson, J.R., 1975. Colloid osmotic pressure as a cause of pathological swelling of cells. In, *Pathobiology of Cell Membranes* (B.F. Trump and A.V. Arstilla, Ed.), pp. 173-189, Academic Press, London.

Saladino, A.J., P.J. Bentley, and B.F. Trump, 1969. Ion movements in cell injury. Effect of amphotericin B on the ultrastructure and function of the epithelial cells of the toad bladder. *Amer. J. Path.,* **54:** 421-466.

Shelburne, J.D., A.V. Arstilla, and B.F. Trump, 1970. *Amer. J. Path.,* **59:** 106a.

Skinner, D.M., 1962. The structure and metabolism of a crustacean integumentary tissue during a molt cycle. *Biol. Bull.,* **123:** 635-647.

Smith, M., 1974. A preliminary report on complex fibrillar aggregates in the gill epithelium of *Panulirus interruptus* (Randall). *Tissue Cell,* **6:** 515-519.

Smyth, J.D., 1942. A note on the morphology and cytology of the branchiae of *Carcinus maenus. Proc. Roy. Irish Acad.,* **48B:** 105-199.

Steinbrecht, R.A., 1968. Op-art-bodies. A new class of periodically layered cell inclusions observed in crayfish epidermis cells. 4th Europ. Reg. Conf. Electron Microscopy (D. Steve-Bocciarelli, Ed.) Tipografia Vaticana, Rome.

Stangways-Dixon, J., and D.S. Smith, 1970. The fine structure of gill podocytes in *Panulirus argus* (Crustacea). *Tissue Cell,* **2:** 611-624.

Travis, D.F., 1954. The molting cycle of the spiny lobster, *Panulirus argus* (Latreille). I. Molting and growth in laboratory maintained individuals. *Biol. Bull.,* **107:** 433-450.

Travis, D.F., 1955. The molting cycle of the spiny lobster *Panulirus argus* Latreille. II. Pre-ecdysial histological and histochemical changes in the hepatopancreas and integumental tissues. *Biol. Bull.,* **108:** 88-112.

Travis, D.F., 1957. The molting cycle of the spiny lobster *Panulirus argus* Latreille. IV. Postecdysial histological and histochemical changes in the hepatopancreas and integumental tissues. *Biol. Bull.,* **131:** 451-479.

Travis, D.F., 1963. Structural features of mineralization from tissue to macromolecular levels of organization in the decapod crustacea. *Annals New York Acad. Sci.,* **109:** 177-245.

Trump, B.F., and A.V. Arstilla, 1975. Cellular reaction to injury. In, *Principles of Pathobiology* (M.F. LaVia and R.B. Hill, Jr., Ed.), pp. 9-96, Oxford University Press, New York.

Wienbach, E.C., and J. Garbus, 1966. Restoration by albumin of oxidative phosphorylation and related reactions. *J. Biol. Chem.,* **241:** 169-175.

Weinbach, E.C., J. Garbus, and H.G. Sheffield, 1967. Morphology of mitochondria in the coupled, uncoupled and recoupled states. *Exper. Cell Res.,* **46:** 129-143.

Weinbach, E.C., and J. Garbus, 1969. Mechanism of action of reagents that uncouple oxidative phosphorylation. *Nature,* **221:** 1016-1018.

Wright, K.A., 1964. The fine structure of the nephrocyte of the gills of two marine decapods. *J. Ultrastr. Res.,* **10:** 1-13.

Yonge, C.M., 1932. On the nature and permeability of chitin. I. The chitin lining the foregut of decapod Crustacea and the function of the tegumental glands. *Proc. Roy. Soc. London,* **B111:** 298-329.

Young, J.H., 1959. Morphology of the white shrimp *Penaeus setiferus* (Linnaeus 1758). *Fish Bull. U.S. Fish Wildl. Serv.,* **59:** 145.

Effects of Sodium Pentachlorophenate and 2,4-Dinitrophenol on Respiration in Crustaceans

ANGELA C. CANTELMO, PHILIP J. CONKLIN, FERRIS R. FOX
and K. RANGA RAO

Abstract—The oxygen consumption of the grass shrimp, *Palaemonetes pugio,* was determined at different stages of the molt cycle. At each stage of the molt cycle, the oxygen consumption varied in relation to periods of activity. In order to minimize the errors in establishing basal (control) rates of oxygen consumption, measurements were made over extended periods (18 to 24 hours). In contrast to the previous reports of progressive increases in oxygen consumption during proecdysial stages in other crustaceans, we noted significant increases in oxygen consumption just prior to and during the actual shedding of exoskeleton (ecdysis) in grass shrimp. The effects of sodium pentachlorophenate (Na-PCP) on oxygen consumption varied depending on the stage of the molt cycle, concentration of Na-PCP and extent of pre-exposure of shrimp to Na-PCP. At concentrations of 1.5 and 5.0 ppm, Na-PCP did not alter the oxygen consumption of shrimp in intermolt and proecdysial stages of the molt cycle. Late proecdysial shrimp exposed to 5.0 ppm Na-PCP exhibited an increase in oxygen consumption in relation to ecdysis to the same level as that of control shrimp. However, following ecdysis, the shrimp exposed to 5.0 ppm Na-PCP exhibited a dramatic decline in oxygen consumption and died within three hours. This increased sensitivity during the early postecdysial period appeared to be related to an increase in the uptake of Na-PCP at this stage compared to intermolt and proecdysial stages. A decline in oxygen consumption as noted above could be induced in intermolt shrimp by using higher concentrations of Na-PCP. Exposure of shrimp to 10 or 20 ppm Na-PCP, or to 5 ppm followed by 20 ppm Na-PCP caused an initial increase in oxygen consumption and a subsequent decline leading to death. The survival time of intermolt shrimp pretreated with 5 ppm Na-PCP was longer

ANGELA C. CANTELMO, PHILIP J. CONKLIN, FERRIS R. FOX and K. RANGA RAO • Faculty of Biology, University of West Florida, Pensacola, Florida 32504, U.S.A.

than that of shrimp exposed directly to 10 or 20 ppm Na-PCP. Although 20 ppm 2,4-dinitrophenol (DNP) caused an initial increase in oxygen consumption in intermolt shrimp, this was not followed by any decline in oxygen consumption or death during a 24-hour exposure.

The effects of Na-PCP and DNP on tissue respiration *in vitro* were studied using the blue crab, *Callinectes sapidus*. At concentrations of 1×10^{-6} M and 5×10^{-5} M, these compounds did not alter the oxygen consumption of the muscle, gill and hepatopancreas. At a concentration of 5×10^{-3} M, both Na-PCP and DNP caused an inhibition of oxygen consumption of isolated tissues.

Introduction

Pentachlorophenol (PCP) and alkyldinitrophenols such as 2,4-dinitrophenol (DNP) are commonly used as uncoupling agents (Garbus and Weinbach, 1963; Nicholls *et al.*, 1967; Hanstein, 1976). PCP and its salts are used as pesticides, herbicides, molluscicides and bactericides in aquatic and terrestrial environments. Weinbach (1956) postulated that the uncoupling of oxidative phosphorylation is responsible for the biocidal actions of PCP. However, other studies indicate that PCP elicits much broader metabolic effects on organisms such as fish and molluscs (Holmberg *et al.*, 1972; Boström and Johansson, 1972; Ishak *et al.*, 1972).

Although the effect of PCP on freshwater organisms is well documented, very little is known about its effect on estuarine and marine organisms (Bevenue and Beckman, 1967; Zitko *et al.*, 1976). Measurement of oxygen consumption not only indicates metabolic rate but also provides an index of stress conditions in organisms. This study was designed to determine the effects of Na-PCP and DNP on the *in vivo* and *in vitro* oxygen consumption in crustaceans.

Experimental Animals and Measurement of Oxygen Consumption

The grass shrimp, *Palaemonetes pugio,* was used to determine the effect of Na-PCP on oxygen consumption *in vivo.* Identification of the stages in the molt cycle of grass shrimp was accomplished as described by Conklin and Rao (this symposium). The blue crab, *Callinectes sapidus,* was used in the *in vitro* experiments. The effects of Na-PCP on respiration in the hepatopancreas, gill and muscle tissues from the blue crab were determined. For comparative purposes, the effects of DNP on oxygen consumption was also determined.

Measurements of oxygen consumption were made using a Gilson Differential Respirometer. Each flask contained one shrimp in 5 ml of seawater with or without the test chemicals. Readings were made every 15 minutes and the respirometers were re-equilibrated every four to five hours. The animals were maintained at the same temperature and light conditions as in the laboratory and were not fed during experimentation. Because of the variability associated with the movement of shrimp in the flasks, a basal rate was determined for 18-24 hour periods. Since the flasks contained only 5 ml of water, concentrations of Na-PCP used here to cause alterations in oxygen consumption are somewhat higher than used in standard 96-hour bioassays.

Effects of Na-PCP and DNP on Oxygen Consumption in Intermolt (Stage C) Grass Shrimp

The oxygen consumption of 24 grass shrimp (Group I) was measured for 24 hours to determine a basal rate. At the end of this period 14 animals (Group IA) were placed in a medium containing 5 ppm Na-PCP. Ten shrimp remained in

TABLE 1. Effect of Sodium Pentachlorophenate (Na-PCP) and 2,4-Dinitrophenol (DNP) on Oxygen Consumption in the Grass Shrimp, *Palaemonetes pugio*

Group	N	Treatment	$\mu l\ O_2/g$ wet wt/hr	Level of signif. (P)
I	24	Untreated	326.6 ± 15.37[a]	
I A	14	5 ppm Na-PCP	292.3 ± 13.8	N.S.[b]
I A-1	8	10 ppm Na-PCP	346.9 ± 28.7[c]	N.S.
I A-2	6	20 ppm Na-PCP	466.5 ± 20.1[c]	< 0.001
I B	10	seawater	337.0 ± 16.2	N.S.[d]
II	13	Untreated	278.9 ± 18.9[a]	
II A	8	10 ppm Na-PCP	345.7 ± 22.5[c]	< 0.025
II B	5	seawater	301.5 ± 23.3	N.S.[d]
III	16	Untreated	325.8 ± 25.2[a]	
III A	11	20 ppm Na-PCP	406.8 ± 22.8[c]	< 0.025
III B	5	seawater	317.2 ± 20.4	N.S.[d]
IV	15	Untreated	297.2 ± 14.6[a]	
IV A	10	20 ppm DNP	364.7 ± 20.7	< 0.025
IV B	5	seawater with ethanol	290.8 ± 18.6	N.S.[d]

[a]Mean basal rate of O_2 consumption calculated from measurements over a 24 hour period preceding treatments.
[b]$P > 0.05$
[c]Based on O_2 consumption up to 2 hours preceding death of shrimp exposed to Na-PCP.
[d]Compared with the basal rate preceding treatments.

seawater (Group IB) and served as controls. The oxygen consumption of shrimp in 5 ppm Na-PCP was not significantly altered in 24 hours (Table 1). After this exposure to 5 ppm Na-PCP, 8 shrimp (Group 1A-1) were exposed to 10 ppm Na-PCP and 6 shrimp (Group 1A-2) were exposed to 20 ppm Na-PCP. Animals exposed to 10 ppm Na-PCP exhibited no changes in oxygen consumption. However, the oxygen consumption of shrimp exposed to 20 ppm Na-PCP increased significantly. This increase persisted for a few hours and was followed by a decline in the oxygen consumption and death of the shrimp. The data on the initial increase in oxygen consumption are shown in Table 1, while the overall pattern including the decline in oxygen consumption is shown in Figures 1 and 2.

The aims of the second and third experiments were to determine the effects of direct exposure to either 10 or 20 ppm Na-PCP without prior exposure to 5 ppm Na-PCP. In the second experiment the basal rate of 13 shrimp (Group II) was determined for 24 hours and then 8 shrimp (Group II-A) were exposed to 10 ppm Na-PCP while 5 shrimp were maintained in seawater (Group II-B). In the third experiment the basal rate of 16 shrimp (Group III) was determined for 24 hours and then 11 shrimp (Group IIIA) were exposed to 20 ppm Na-PCP while 5 shrimp were maintained in seawater (Group IIIB, controls). Transfer of shrimp to either 10 or 20 ppm Na-PCP resulted in a significant elevation of oxygen consumption for a short duration (Table 1) followed by a rapid decline and death of shrimp. The results in

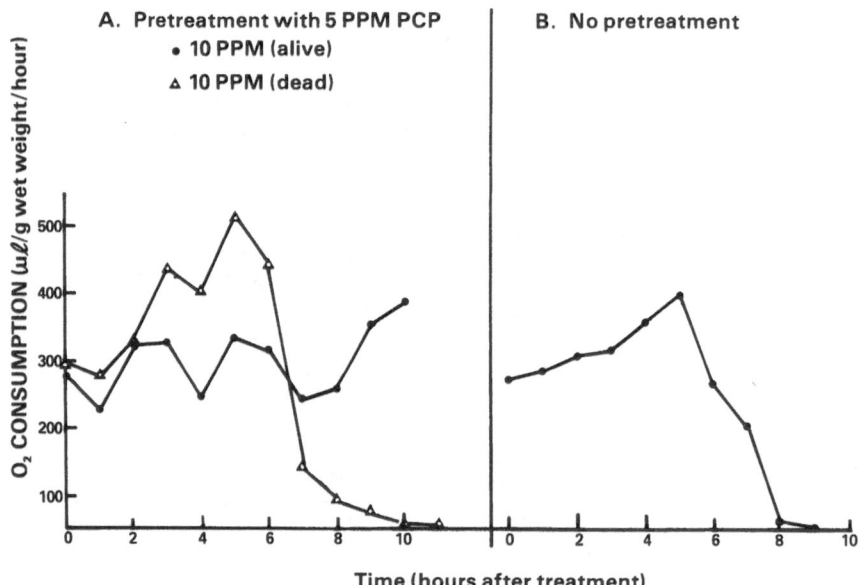

Figure 1. Effect of exposure to 10 ppm sodium pentachlorophenate (Na-PCP) on the oxygen consumption of the grass shrimp, *Palaemonetes pugio*. The basal rate of oxygen consumption was established over a 24 hour period. A: Shrimp exposed to 5 ppm Na-PCP for 24 hours prior to exposure to 10 ppm Na-PCP. B: Shrimp exposed directly to 10 ppm Na-PCP.

Table 1 reflect only the data related to the observed increases in oxygen consumption whereas the representative patterns of oxygen consumption including the decline are shown in Figures 1 and 2.

Pretreatment with 5 ppm Na-PCP for 24 hours increased the survival time for shrimp subsequently exposed to 10 and 20 ppm Na-PCP when compared to those shrimp exposed directly to the latter concentrations. Among the shrimp pretreated with 5 ppm Na-PCP and exposed to 10 ppm Na-PCP, 50% survived the 18-hour exposure while the remainder died within 8-10 hours. The oxygen consumption of the surviving shrimp showed a relatively stable pattern while the oxygen consumption of the animals that died showed an initial increase and then a rapid decline until death (Fig. 1A). In contrast, all of the shrimp transferred from seawater directly to 10 ppm Na-PCP died within 8 hours of exposure. The oxygen consumption of these shrimp also exhibited an initial increase followed by a dramatic decline (Fig. 1B). The shrimp pretreated with 5 ppm Na-PCP survived on an average for 8 hours after exposure to 20 ppm Na-PCP whereas those shrimp exposed directly to 20 ppm Na-PCP died within 4-5 hours of exposure. The oxygen consumption of these shrimp also increased initially and then declined rapidly until the shrimp died (Figs. 2A, 2B). Both groups of animals transferred directly to 10 or 20 ppm Na-PCP exhibited the same percent increase in oxygen consumption although the concentration of Na-PCP was doubled (Table 1).

The finding that pretreatment with 5 ppm Na-PCP increases the survival time for shrimp exposed subsequently to higher concentrations of Na-PCP indicates an

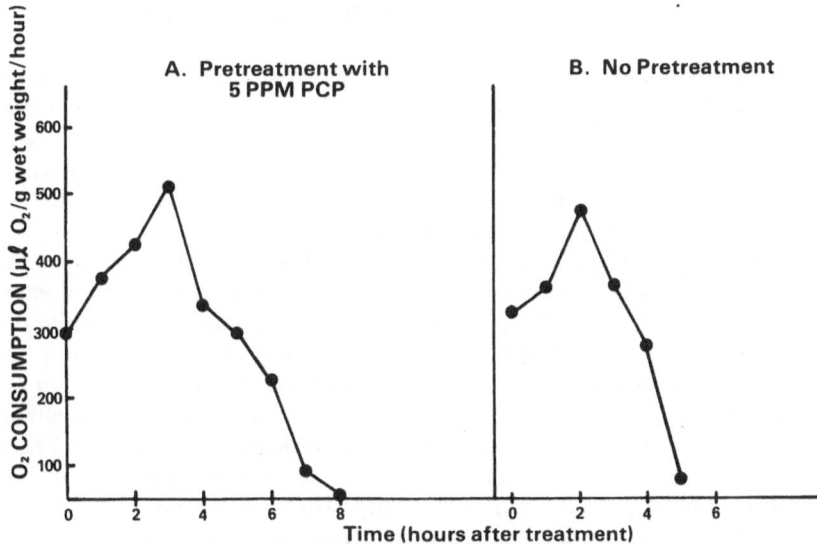

Figure 2. Effect of exposure to 20 ppm Na-PCP on the oxygen consumption of the grass shrimp, *Palaemonetes pugio*. The basal rate of oxygen consumption was established over a 24 hour period. A: Shrimp exposed to 5 ppm Na-PCP for 24 hours prior to exposure to 20 ppm Na-PCP. B: Shrimp exposed directly to 20 ppm Na-PCP.

increase in resistance resulting from an exposure to a lower concentration of the chemical. Norup (1972) has shown the same pattern of resistance to PCP in guppies.

Weinbach (1954) has shown an increase in oxygen consumption of snails exposed to Na-PCP. Crandall and Goodnight (1962) have also shown that fish fry increase their oxygen consumption when exposed to low concentrations of Na-PCP. Stimulation of oxygen consumption is consistent with the known action of PCP as an uncoupler of oxidative phosphorylation. The initial increase in oxygen consumption noted in shrimp exposed to Na-PCP may have been due to the uncoupling effects of this compound. However, the rapid decline in oxygen consumption noted in shrimp as the time of exposure to Na-PCP increased, may have been due to the accumulation of high concentrations of the chemical by the shrimp. Higher concentrations of PCP are known to inhibit respiration (Ishak *et al.*, 1972).

The last experiment on intermolt shrimp was aimed at examining the effects of 2,4-dinitrophenol on the oxygen consumption of grass shrimp. After establishing a 24-hour basal rate, 10 shrimp (Group IVA) were exposed to 20 ppm DNP. These animals exhibited oxygen consumption values significantly higher than the basal rates in normal seawater (Table 1). The relative increase in oxygen consumption was similar to that observed for animals exposed to 10 and 20 ppm Na-PCP. However, all of the shrimp exposed to 20 ppm DNP survived for the duration of the experiment and no decreases in oxygen consumption occurred. This indicates that DNP is relatively less toxic than Na-PCP to grass shrimp.

Effects of Na-PCP on Grass Shrimp in Relation to Ecdysis

It is well known that the period surrounding ecdysis is a vulnerable time when mortality rate is high and changes in the environment could lead to unsuccessful molting and death. Conklin and Rao have reported in this symposium that newly molted shrimp are more sensitive to Na-PCP than intermolt shrimp. Therefore, the next set of experiments was designed to examine the effects of Na-PCP on the oxygen consumption of shrimp in relation to ecdysis.

A basal rate of oxygen consumption was established for three groups of shrimp in late proecdysial (D_3-D_4) stages of the molt cycle. One group was then exposed to 1.5 ppm Na-PCP; the second group was exposed to 5.0 ppm Na-PCP; while the third group served as controls. Oxygen consumption was measured throughout the proecdysial period, during ecdysis (molting) and up to 8-10 hours after ecdysis. Changes in the rate of oxygen consumption of a representative control shrimp and shrimp exposed to 1.5 and 5.0 ppm Na-PCP are presented in Figure 3. The rate of oxygen consumption of shrimp 24 hours prior to ecdysis was not significantly higher than that of intermolt (Stage C) shrimp. The oxygen consumption increased by about 25% beginning 2-3 hours preceding ecdysis. The control animals maintained this elevated rate of oxygen consumption through ecdysis and up to 10 hours after ecdysis. The shrimp that completed ecdysis in seawater containing 1.5 ppm Na-PCP exhibited no deviations from the pattern of oxygen consumption in controls. The shrimp exposed to 5.0 ppm Na-PCP had oxygen consumption rates that were not significantly different from the controls during the proecdysial period and the hour during which ecdysis occurred. However, all of the shrimp that underwent ecdysis in 5.0 ppm Na-PCP died within 3 hours after ecdysis and exhibited a concurrent decline in oxygen consumption (Fig. 3). This decrease was similar to that noted when intermolt shrimp were exposed to higher concentrations of Na-PCP. The sensitivity of the shrimp to Na-PCP was greater at ecdysis as evidenced by the 100% mortality shortly after ecdysis in shrimp exposed to 5.0 ppm Na-PCP. This concentration neither caused any mortalities nor affected the respiration of intermolt shrimp.

The observed increase in sensitivity of newly molted shrimp to Na-PCP is most likely related to changes in permeability associated with molting. Many authors (Lockwood and Inman, 1973; Lockwood and Andrews, 1969; Dandifosse, 1966) have demonstrated an increase in permeability of crustacean tissues to salts and water at the time surrounding ecdysis. Experiments in this laboratory indicated that postecdysial shrimp take up significantly more ^{14}C-PCP than intermolt (Stage C) and proecdysial (Stage D) shrimp. These differences in the rate of uptake of PCP may explain why the same concentration of PCP (5 ppm Na-PCP) causes 0% mortality in intermolt animals and 100% mortality in early postecdysial shrimp.

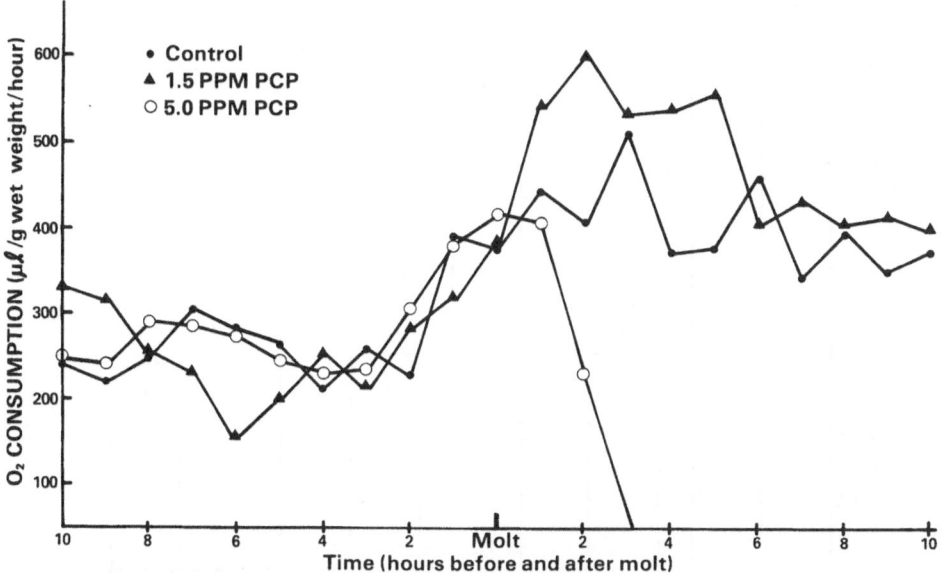

Figure 3. Effect of Na-PCP on the oxygen consumption in *Palaemonetes pugio* preceding, during and after ecdysis. Late proecdysial shrimp (Stages D_3-D_4) were used and a basal rate was determined prior to the addition of Na-PCP to the medium. Each curve is based on data from a representative shrimp from each group. The oxygen consumption during a period 10 hours prior to ecdysis and 10 hours after ecdysis in control shrimp (closed circles), shrimp exposed to 1.5 ppm Na-PCP (triangles) and 5 ppm Na-PCP (open circles) is shown.

These changes in permeability may also explain the pattern of oxygen consumption observed in animals undergoing ecdysis compared to that in intermolt shrimp exposed to Na-PCP. The newly molted animals did not exhibit any initial increase in oxygen consumption before death as did the intermolt animals. The rapid influx of PCP into the tissues after ecdysis could rapidly increase the concentration of PCP to the level necessary to inhibit oxygen consumption. Therefore, no increase in oxygen consumption would be observed; rather just a rapid decline in oxygen consumption followed by death.

Effects of Na-PCP on Isolated Tissues

The results described suggest that Na-PCP in low concentrations acts as an uncoupler of oxidative phosphorylation. However, once a critical concentration of Na-PCP is reached inside the shrimp a disruption of the metabolic activity occurs that leads to the death of the animal. In order to further evaluate the observed disruption of metabolism, the effects of Na-PCP and DNP on oxygen consumption of isolated tissues were determined. Tissues from individual *Palaemonetes pugio*

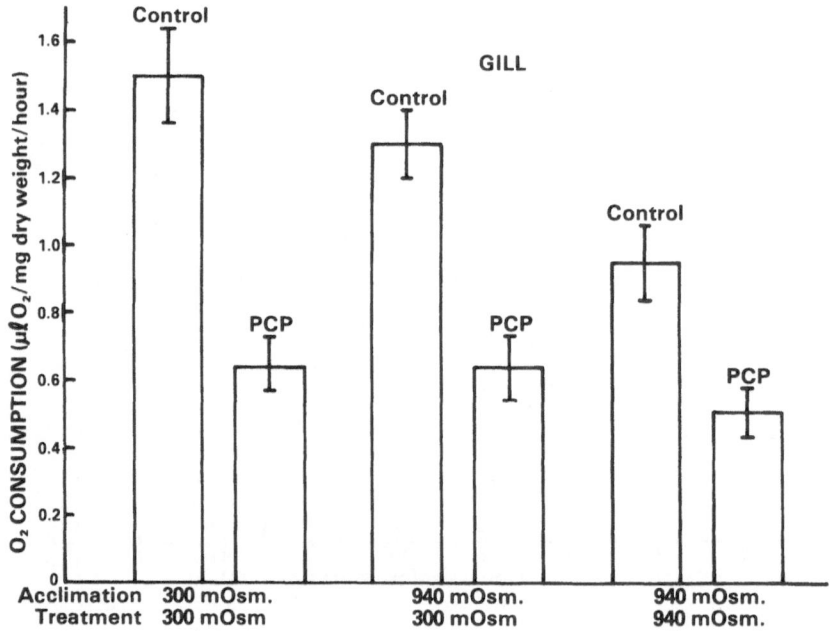

Figure 4. Effect of Na-PCP *in vitro* on the oxygen consumption of the gill tissue from the blue crab, *Callinectes sapidus*. All experimental flasks contained 5×10^{-3} M Na-PCP. The gills were removed from crabs acclimated to either 300 or 940 milliosmole seawater, and placed in 300 or 940 milliosmole seawater. The vertical line on each bar represents the Standard Error of the Mean.

were too small to accurately measure oxygen consumption. Thus, tissues from the blue crab, *Callinectes sapidus,* were chosen for this phase of study.

Estuarine organisms such as *Callinectes sapidus* and *Palaemonetes pugio* inhabit an environment that exhibits wide fluctuations in abiotic factors such as salinity. In order to live in this environment, these animals have evolved mechanisms of osmoregulation. However, a rapid alteration of the salinity of their environment still presents an osmotic stress to the organism. The regulatory responses to an osmotic stress require energy expenditure. We examined the effects of Na-PCP and DNP on crab tissues subjected to normal and acutely stressful osmotic conditions.

As in the studies with grass shrimp, respiration was measured using a Gilson Differential Respirometer. Specimens of *Callinectes sapidus* were acclimated to either 940 milliosmole (30°/oo) seawater or 300 milliosmole (10°/oo) seawater for two weeks prior to experimentation. Crabs were sacrificed and the gills, muscle and hepatopancreas were removed. Na-PCP and DNP were tested on tissues from crabs acclimated to 300 and 940 milliosmole sea water and on tissues from crabs

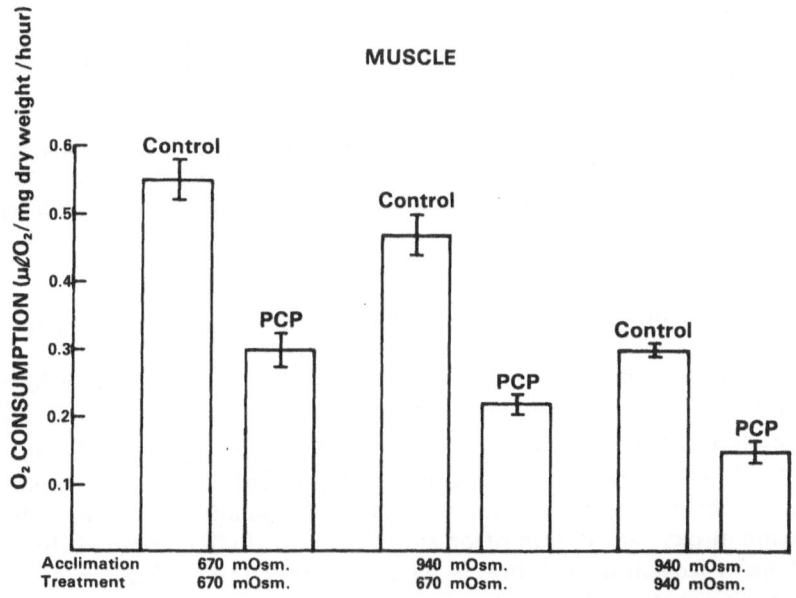

Figure 5. Effect of Na-PCP *in vitro* on the oxygen consumption of muscle tissue from the blue crab, *Callinectes sapidus.* All experimental flasks contained 5 × 10⁻³ M Na-PCP. The muscle was removed from crabs acclimated to 300 or 940 milliosmole seawater, and placed in 670 or 940 milliosmole seawater. The osmotic concentration of hemolymph in crabs acclimated to 300 milliosmole medium was 670 milliosmoles. The latter concentration was used in *in vitro* tests for tissues for crabs in 300 milliosmole medium and to cause an *in vitro* acute osmotic stress to tissues from crab in 940 milliosmole medium. The crabs in 940 milliosmole medium had a hemolymph osmotic concentration of 940 milliosmoles and this concentration was used in *in vitro* tests. The vertical line on each bar represents the Standard Error of the Mean.

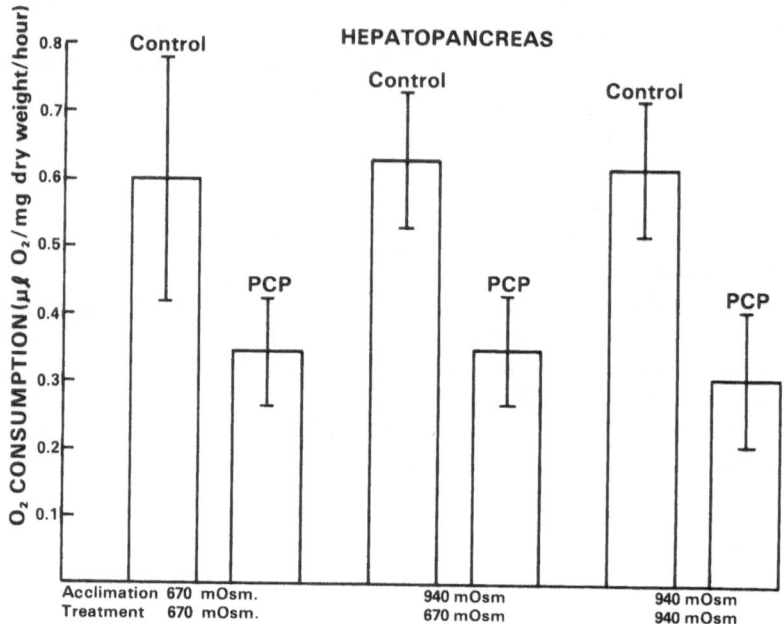

Figure 6. Effect of Na-PCP *in vitro* on the oxygen consumption of hepatopancreas from the blue crab, *Callinectes sapidus*. All experimental flasks contained 5×10^{-3} M Na-PCP. Other experimental details are similar to those described for muscle tissues (see legend to Figure 5).

acclimated to 940 milliosmole seawater but subjected to an acute hypoosmotic stress *in vitro*.

The results are summarized in Figures 4, 5, and 6. At a concentration of 5×10^{-3} M, Na-PCP decreased the oxygen consumption of all tissues significantly. The osmotic condition of the animal and the tissue did not appear to alter the inhibitory effect of Na-PCP on oxygen consumption. Even though the oxygen consumption was higher in the gill and muscle tissues under hypoosmotic conditions, the extent of inhibition of oxygen consumption exerted by Na-PCP remained the same as observed under isoosmotic conditions. Thurberg *et al.* (1973) found that although the oxygen consumption of the gill tissues of *Cancer irroratus* and *Carcinus maenas* varied inversely with salinity, the amount of inhibition of oxygen consumption by cadmium was independent of the test salinity.

DNP is less effective as an uncoupler of oxidative phosphorylation compared to PCP because the latter is known to bind more tightly with the mitochondrial proteins (Weinbach and Garbus, 1965). However, under most conditions tested in the present *in vitro* study, DNP caused a reduction in tissue oxygen consumption at levels comparable to that induced by Na-PCP (Fig. 7). However, the results from the *in vivo* study with shrimp indicated that DNP was less effective than Na-PCP in inhibiting oxygen consumption. Whether the observed differences between *in vivo*

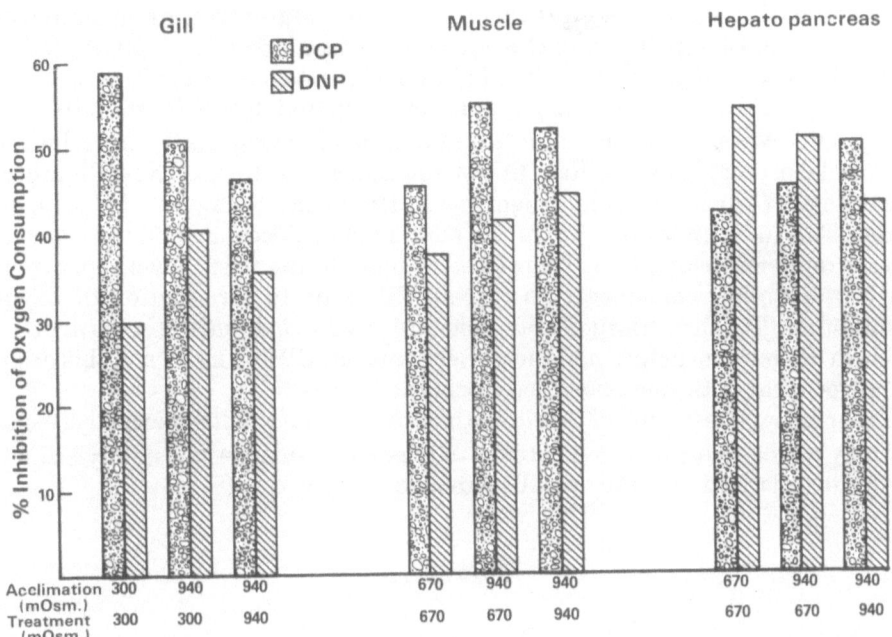

Figure 7. Comparative effects of Na-PCP and DNP on the *in vitro* oxygen consumption of muscle, gill and hepatopancreatic tissues from the blue crab, *Callinectes sapidus*. The concentration of Na-PCP and DNP was 5×10^{-3} M. Details of osmotic concentrations used are given in the legends for Figures 4, 5, and 6.

and *in vitro* studies are due to relative permeabilities of crustaceans to Na-PCP and DNP is not clear. However it is pertinent to note that DNP caused significantly less inhibition of oxygen consumption in gills from crabs acclimated to 300 milliosmole seawater (Fig. 7). It has been demonstrated that estuarine decapods are capable of reducing their permeability on exposure to hypoosmotic media (Smith, 1967, 1970; Capen, 1972; Cantelmo, 1977). It would be interesting to determine if these changes in permeability could affect the final concentration of DNP available to the cells.

The oxygen consumption of the gill, muscle, and hepatopancreas of *Callinectes sapidus* was reduced to an average of 50% of the control after treatment with Na-PCP. Thus each tissue was inhibited to the same extent by Na-PCP. Weinbach (1956) studied the effects of PCP on the oxygen consumption in mitochondria from the albumen gland of the aquatic snail *Lymnaea stagnalis*. He found that at concentrations of PCP below 2.5×10^{-5} M, oxygen consumption was mildly stimulated while phosphorylation was completely suppressed. At higher concentrations (5×10^{-4} M) phosphorylation was still completely suppressed and in addition, oxygen consumption was inhibited. Similar effects were noted in rat liver mitochondria (Weinbach, 1954). Ishak *et al.* (1972) examined the oxygen consumption in tissue homogenates of the snail *Biomphalaria alexandrina*. Homogenates exposed to 5×10^{-5} and 5×10^{-4} M PCP showed an inhibition of oxygen consumption, the degree of inhibition being comparable to that observed in this study with tissues from *Callinectes sapidus.* Many workers (Chance and Williams, 1956; Chappell, 1964; Crestea and Gurban, 1964) have attributed the decline in oxygen consumption to the buildup of oxaloacetate. Ishak *et al.* (1972) showed that the inhibition of oxygen consumption by PCP in snail tissue homogenates was mainly due to the accumulation of oxaloacetate. Investigations from our laboratory (Fox and Rao, this symposium) indicate that Na-PCP inhibits the activity of succinate dehydrogenase in the hepatopancreas of *Callinectes sapidus.* This inhibition would favor the buildup of oxaloacetate and the reduction of tissue oxygen consumption. Therefore, evidence at the tissue level supports the idea that a higher concentration of Na-PCP leads to a reduction of oxygen consumption. The three tissues examined appeared to exhibit the same degree of decline in oxygen consumption in the presence of Na-PCP and DNP, and this decline was independent of the metabolic activity of the tissues.

In summary, the results of this investigation indicate that the biocidal effects of PCP may not be solely due to its ability to uncouple oxidative phosphorylation but also due to a disruption of the overall metabolic activity.

References

Bevenue, A., and H. Beckman, 1967. Pentachlorophenol: A discussion of its properties and its occurrence as a residue in human and animal tissues. *Residue Rev.,* **19:** 83-134.

Boström, S.L., and R.G. Johansson, 1972. Effects of pentachlorophenol on enzymes involved in energy metabolism in the liver of the eel. *Comp. Biochem. Physiol.,* **41B:** 359-369.

Cantelmo, A., 1977. Water permeability of isolated tissues from decapod crustaceans. 1. Effect of osmotic conditions. *Comp. Biochem. Physiol.,* in press.

Capen, R.L., 1972. Studies of water uptake in the euryhaline crab, *Rhithropanopeus harrisi. J. Exp. Zool.,* **182:** 307-320.

Chance, B., and C.R. Williams, 1956. The respiratory chain and oxidative phosphorylation. *Adv. Enzymol.,* **17:** 65-134.

Chappell, J.B., 1964. Effects of 2, 4-dinitrophenol on mitochondrial oxidations. *Biochem. J.,* **90:** 237-248.

Crandall, C.A., and C.J. Goodnight, 1962. Effects of sublethal concentrations of several toxicants on growth of the common guppy, *Lebistes reticulatus. Limnol. Oceanogr.,* **7:** 232-239.

Dandifosse, G., 1966. Absorption d'eau au moment de la mue chez un crustace decapode: *Maja squinado* Herbst. *Arch. Intern. Physiol. Biochem.,* **74:** 329-331.

Garbus, J., and E.C. Weinbach, 1963. Restoration of oxidative phosphorylation and related reactions in uncoupled mitochondria by albumin. *Fed. Proc.,* **22:** 405.

Hanstein, W.G., 1976. Uncoupling of oxidative phosphorylation. *Biochim. Biophys. Acta,* **456:** 129-148.

Holmberg, B., J. Jensen, A. Larsson, K. Lewander, and M. Olsson, 1972. Metabolic effects of technical pentachlorophenol (PCP) on the eel *Anguilla anguilla* L. *Comp. Biochem. Physiol.,* **43B:** 171-183.

Ishak, M.M., A.A. Sharaf, and A.H. Mohamed, 1972. Studies on the mode of action of some molluscicides on the snail, *Biomphalaria alexandrina. Comp. Gen. Pharmacol.,* **3:** 385-390.

Lockwood, A.P.M., and W.R.H. Andrews, 1969. Active transport and sodium fluxes at molt in the amphipod *Gammarus duebeni*. *J. Exp. Biol.*, **51**: 591-605.

Lockwood, A.P.M., and C.B.E. Inman, 1973. Changes in the apparent permeability to water at molt in the amphipod, *Gammarus duebeni* and the isopod *Idotea linearis*. *Comp. Biochem. Physiol.*, **44A**: 943-952.

Nicholls, D.G., D. Shepherd, and P.B. Garland, 1967. A continuous recording technique for the measurement of carbon dioxide, and its application to mitochondrial oxidation and decarboxylation reactions. *Biochem. J.*, **103**: 677-691.

Norup, B., 1972. Toxicity of chemicals in paper factory effluents. *Water Res.*, **6**: 1585-1588.

Smith, R.I., 1967. Osmotic regulation and adaptive reduction of water permeability in the brackish water crab, *Rhithropanopeus harrisi* (Brachyura, Xanthidae). *Biol. Bull.*, **133**: 643-658.

Smith, R.I., 1970. The apparent water permeability of *Carcinus maenas* (Crustacea, Brachyura, portunidae) as a function of salinity. *Biol. Bull.*, **139**: 351-362.

Thurberg, F.P., M.A. Dawson, and R.S. Collier, 1973. Effects of copper and cadmium on osmoregulation and oxygen consumption in two species of estuarine crabs. *Marine Biol.*, **23**: 171-175.

Weinbach, E.C., 1954. The effect of pentachlorophenol on oxidative phosphorylation. *J. Biol. Chem.*, **210**: 545.

Weinbach, E.C., 1956. The influence of pentachlorophenol on oxidative and glycolytic phosphorylation in snail tissue. *Arch. Biochem. Biophys.*, **64**: 129-143.

Weinbach, E.C., and J. Garbus, 1965. The interaction of uncoupling phenols with mitochondria and mitochondrial protein. *J. Biol. Chem.*, **240**: 1811-1819.

Zitko, V., D.W. McLeese, W.G. Carson, and H.E. Welch, 1976. Toxicity of alkyldinitrophenols to some aquatic organisms. *Bull. Environ. Contam. Toxicol.*, **16**: 508-515.

Effects of Sodium Pentachlorophenate and 2,4-Dinitrophenol on Hepatopancreatic Enzymes in the Blue Crab, *Callinectes sapidus*

FERRIS R. FOX and K. RANGA RAO

Abstract—In view of the lack of information on the mechanism of PCP-induced toxicity in crustaceans, this investigation was undertaken to evaluate the effects of sodium pentachlorophenate (Na-PCP) *in vivo* and *in vitro* on certain hepatopancreatic enzymes in the blue crab, *Callinectes sapidus*. Fumarase, malate dehydrogenase and succinate dehydrogenase were inhibited by Na-PCP and DNP *in vivo,* whereas isocitrate dehydrogenase was stimulated. Of those tested, lactic dehydrogenase was the least affected cytoplasmic (soluble) enzyme *in vivo* while pyruvate kinase and glucose-6-phosphate dehydrogenase were inhibited at least 50% by Na-PCP. Glutamate-pyruvate transaminase was also inhibited. Na-PCP and DNP had an inhibitory effect on the various enzymes tested *in vitro* at concentrations of 10^{-4} M or higher. In general, the mitochondrial enzymes were more susceptible than cytoplasmic enzymes to DNP and Na-PCP. The calcium activated ATPase from the microsomal fraction of the crab hepatopancreas was inhibited by Na-PCP and DNP *in vitro* and *in vivo*. Na-PCP was more potent than DNP in inhibiting the ATPase activity. The effects of PCP on the blue crab enzymes have been compared to the results of previous investigations on other organisms.

Introduction

The uncoupling of oxidative phosphorylation is thought to be the biochemical basis for the toxicity of pentachlorophenol (Weinbach, 1957). Subsequent investigations of the effects of pentachlorophenol (PCP) on fish (Boström and

FERRIS R. FOX and K. RANGA RAO • Faculty of Biology, University of West Florida, Pensacola, Florida 32504, U.S.A.

Johansson, 1972; Holmberg *et al.,* 1972) and snail tissues (Ishak *et al.,* 1972) revealed that the compound has variable effects on several enzymes leading to alterations in carbohydrate and lipid metabolism. Uncouplers cause either no alterations in oxygen consumption or lead to an increase in oxygen consumption. Cantelmo *et al.* (this symposium) reported that Na-PCP causes an initial increase in oxygen consumption of the grass shrimp, *Palaemonetes pugio,* followed by a decrease in oxygen consumption. Incubation of gills, hepatopancreas and muscle from the blue crab, *Callinectes sapidus,* with Na-PCP, or 2,4-dinitrophenol caused an inhibition of oxygen consumption. In view of the lack of information on the mechanisms of PCP-induced toxicity in crustaceans, this investigation was undertaken to evaluate the effects of Na-PCP *in vivo* and *in vitro* on certain hepatopancreatic enzymes in the blue crab, *Callinectes sapidus.* For comparative purposes the effects of 2,4-dinitrophenol (DNP) were also determined.

Test Animals

Mature blue crabs, *Callinectes sapidus,* were purchased from the Rollo Seafood Company, Milton, Florida. They were maintained in the laboratory in 300 milliosmole seawater at 20 °C under controlled light conditions (12 hours light-12 hours dark). The crabs were kept in individual containers for two weeks prior to use. The water was changed three times a week and within eight hours after each feeding. The crabs were fed once a week. Crabs were used in the experiment only after a minimum period of five days after feeding.

Enzyme Preparation

The hepatopancreas was removed from intermolt crabs and washed in 0.25 M sucrose. The tissue was weighed and homogenized in 0.25 M sucrose using a Potter-Elvejehm homogenizer. The homogenate was fractionated as previously described (Fox and Rao, 1978). The mitochondrial and microsomal pellets were resuspended in 0.25 M sucrose, divided into aliquots and kept frozen (-40 °C) until needed. The soluble fraction was also divided into aliquots and frozen. The microsomal fraction was used within one week of preparation and the other fractions were kept no longer than 3 weeks. The enzymes were assayed as described in the legends to the figures.

Effect of Na-PCP on Tricarboxylic Acid (TCA) Cycle Enzymes

Weinbach and Garbus (1965) have examined the effect of PCP on rat liver mitochondria. They found that both DNP and PCP are bound to mitochondrial membrane proteins. Mitochondria can accumulate these phenols greatly in excess of concentrations required for uncoupling of oxidative phosphorylation. Although both phenols bind to mitochondria, only PCP is tightly bound after repeated

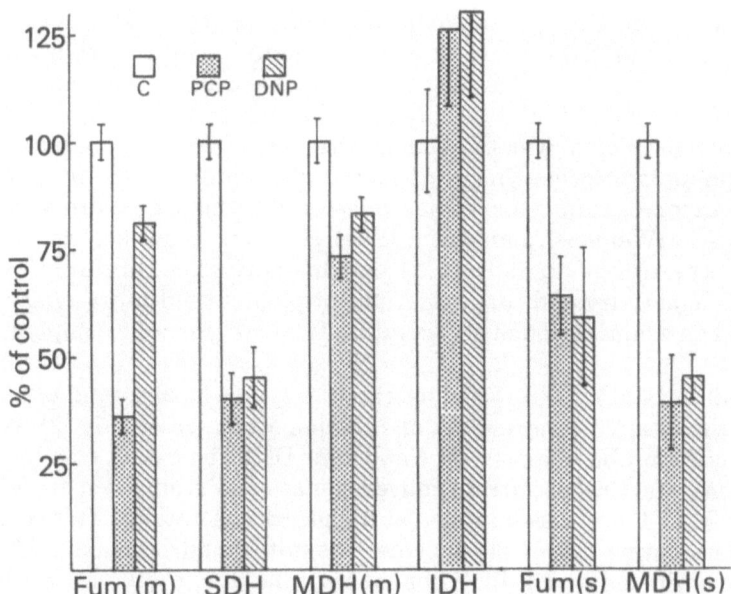

Figure 1. *In vivo* effect of Na-PCP and DNP on Tricarboxylic Acid Cycle (TCA) enzymes of the blue crab. The crabs were injected with 6 μg Na-PCP or DNP per gram body weight. After three hours the animals were dissected and the hepatopancreas was homogenized and differentially centrifuged (Fox and Rao, 1978) to yield the mitochondrial (m) and cytoplasmic (s) fractions.
Fumarase was assayed according to Sigma Chemical Company bulletin. Add 0.1 ml of enzyme solution to 2.9 ml of 0.05 M L-Malic acid buffered to pH 7.6 with 0.1 M phosphate. Record at 240 nm. Isocitrate dehydrogenase assay: 0.1 ml MnCl$_2$ (0.01 M), 0.1 ml NADH (3 mg/ml), 2.7 ml of isocitrate substrate (Sigma reagent No. 150-1A) and 0.1 ml enzyme solution. Malate dehydrogenase (Worthington Biochemical): 0.1 ml NADH (3.75 mM), 0.1 ml oxaloacetate (6 mM), 2.7 ml of 0.1 phosphate buffer, pH 7.4 and 0.1 ml enzyme solution. Succinate dehydrogenase was assayed according to Muller *et al.* (1968): 66 mM phosphate buffer pH 7.4 containing 33 μM 2,6-dichlorophenolindophenol, 1 mM phenazine methosulfate, 2 mM KCN, 20 mM sodium succinate and 0.1 ml enzyme solution. Record at 600 nm. All assays were performed at 25 °C using a Gilford 2400-2 recording spectrophotometer. The values are expressed as Mean ± S.D. of 5 experiments.

washings. The *in vivo* experiments revealed that Na-PCP has a stronger inhibitory effect than DNP on mitochondrial fumarase from the blue crab hepatopancreas (Fig. 1). However, the fumarase in the cytoplasmic (soluble) fraction was inhibited to a lesser extent in Na-PCP than by DNP. Boström and Johansson (1972) have found the fumarase in the homogenate of liver from eels to be stimulated by very low concentrations of Na-PCP.

Isocitrate dehydrogenase, mitochondrial malate dehydrogenase and succinate dehydrogenase are linked to oxidative phosphorylation by the electron transport system. Isocitrate dehydrogenase in crab hepatopancreas was stimulated by both

DNP and Na-PCP *in vivo*. Krueger *et al.* (1966) have found this enzyme to be stimulated in the fish *Cichlasoma bimaculatum* by PCP treatment. Malate dehydrogenase (MDH) is another enzyme found in the tricarboxylic acid cycle within the mitochondrion and in the cytoplasm as part of the malate shuttle. The cytoplasmic enzyme is inhibited more than the mitochondrial MDH in the blue crab hepatopancreas *in vivo* by both phenols. Succinate dehydrogenase is also inhibited by both phenols *in vivo*. Snails *(Australorbis glabratus)* kept in 10^{-5} M PCP exhibited an increase in cytochrome oxidase and succinic dehydrogenase activities in their tissues (Weinbach and Nolan, 1956). However, snails kept in 10^{-3} M PCP for 24 hours were found to have these enzyme activities inhibited by 50%. The variations in activity and concentrations of phenol used for *in vivo* experiments warranted an examination of the *in vitro* effects of phenolic compounds on these enzymes.

The inhibition or stimulation of these enzymes as observed *in vivo* was also found in the *in vitro* experiments on crab hepatopancreas (Fig. 2). While all the enzymes were inhibited at 10^{-3} M Na-PCP or DNP the extent of inhibition varied with the enzyme. Crab isocitrate dehydrogenase was stimulated at concentrations up to 10^{-5} M PCP and was inhibited at higher concentrations. This could be due to the allosteric properties of the enzyme. The mitochondrial malate dehydrogenase was only inhibited at Na-PCP concentrations above 10^{-5} M. Both succinate

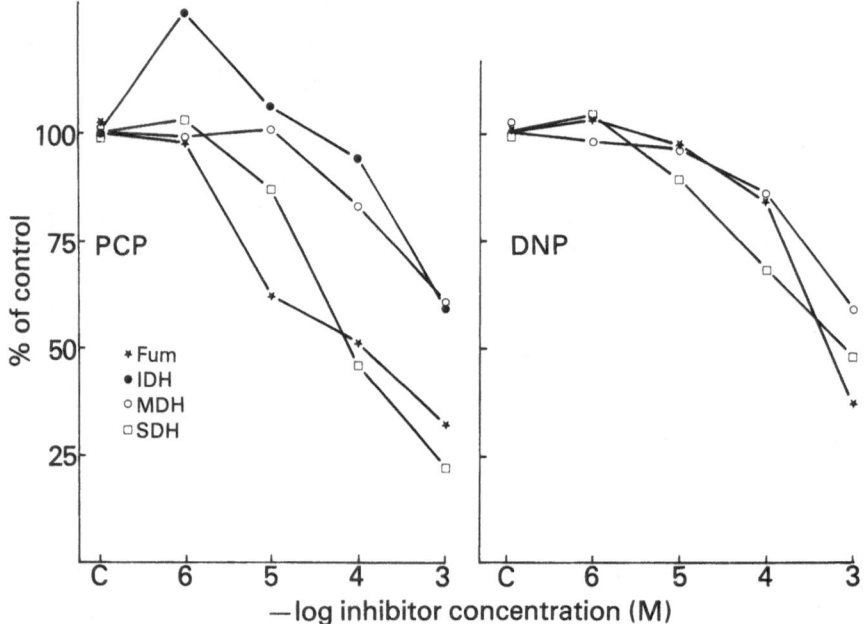

Figure 2. *In vitro* effects of Na-PCP and DNP on mitochondrial Tricarboxylic Acid Cycle enzymes of the blue crab. The assay conditions were the same as described in Figure 1 except the enzyme fraction was preincubated with the phenols for 5 minutes. The points are the Means of 5 experiments.

dehydrogenase and fumarase were very susceptible to Na-PCP with inhibition occurring at low concentrations. Dinitrophenol was much less effective as an inhibitor of the enzymes with inhibition occurring above 10^{-5} M. Boström and Johansson (1972) found that eel liver fumarase tolerated higher levels of Na-PCP with activity still at 100% in 10^{-4} M Na-PCP. Weinbach (1956b) has suggested that oxidative enzymes are inhibited by PCP. Ishak $et\ al.$ (1972) suggested that succinate dehydrogenase was inhibited by Na-PCP $in\ vitro$. They showed that 5×10^{-5} M Na-PCP inhibited succinate oxidation considerably.

While our investigations with crab tissues revealed that Na-PCP inhibits succinate dehydrogenase directly, Ishak $et\ al.$ (1972) have suggested that inhibition of this enzyme in snail tissue homogenates in the presence of Na-PCP is due to the accumulation of oxaloacetate. The fumarase of the crab was inhibited by Na-PCP $in\ vivo$ and $in\ vitro$ while Boström and Johansson (1972) found this enzyme to be stimulated by Na-PCP $in\ vivo$ and inhibited by Na-PCP $in\ vitro$ in the eel liver. The stimulation of fumarase $in\ vivo$ and inhibition $in\ vitro$ by Na-PCP suggests that the phenol may not be directly affecting the enzyme (Boström and Johansson, 1972). However, Weinbach and Garbus (1965) have shown that phenols do interact with mitochondrial protein. They suggest that this protein-phenol interaction may be an

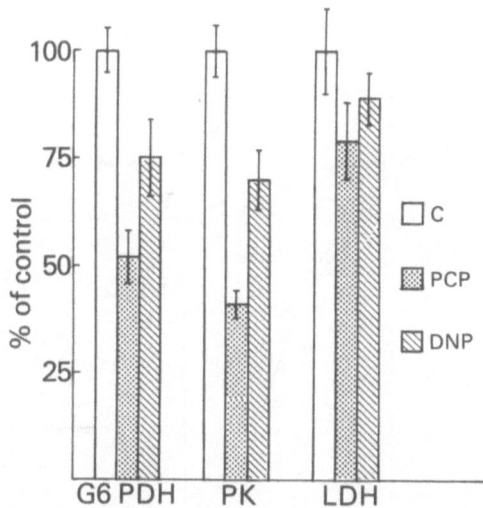

Figure 3. $In\ vivo$ effects of Na-PCP and DNP on carbohydrate metabolism in the blue crab. The enzymes were prepared as described in the legend for Figure 1. Glucose-6-phosphate dehydrogenase (Worthington) assay: 2.7 ml of 0.05 M Tris-HCl, pH 7.6 with 3.3 mM $MgCl_2$, 0.1 ml NADP (6 mM), 0.1 ml G-6-P (0.1 M) and 0.1 ml of enzyme solution. Lactic dehydrogenase (Worthington) assay: 2.7 ml of 0.03 M phosphate, pH 7.4, 0.1 ml pyruvate (0.01 M), 0.1 ml NADH (2 mM) and 0.1 ml of enzyme solution. Pyruvate kinase (Worthington) assay: 2.6 ml of 0.5 M imidazole, pH 7.6 containing KCl and $MgSO_4$, 0.1 ml ADP (20 mg/ml), 0.1 ml NADH (5 mg/ml), 0.1 ml PEP (5 mg/ml), 0.01 ml LDH (5 units) and 0.1 ml enzyme solution. Record at 340 nm and 25 °C. Each value represents the Mean ± S.D. of 5 experiments.

important factor in the uncoupling phenomenon as well as inhibition of enzymes associated with oxidative phosphorylation.

Effect of Na-PCP on Glucose Metabolism

The blue crab hepatopancreatic pyruvate kinase and lactic dehydrogenase were both inhibited by Na-PCP and DNP (Fig. 3) *in vivo*. Both the enzymes were inhibited more by Na-PCP than by DNP. Pyruvate kinase was relatively more affected than lactic dehydrogenase by Na-PCP. Glucose-6-phosphate dehydrogenase, a pentose phosphate enzyme, was also inhibited more in the presence of Na-PCP than DNP. While the two glycolytic enzymes, pyruvate kinase and lactic dehydrogenase, were inhibited by Na-PCP *in vivo* in the eel liver, the glucose-6-phosphate dehydrogenase activity was stimulated (Boström and Johansson, 1972). These effects were seen in eels exposed to a medium containing 0.1 ppm Na-PCP. Liu (1969) reported that in cichlid fish the pentose phosphate pathway was inhibited in the presence of K-PCP. The carbohydrate metabolism shifted almost entirely to glycolysis when fish were kept in 1.5 ppm K-PCP. However, aldolase and lactic dehydrogenase of cichlid fish were inhibited *in vivo* at 0.1 ppm Na-PCP (Cheng, 1965). When these animals were exposed to media

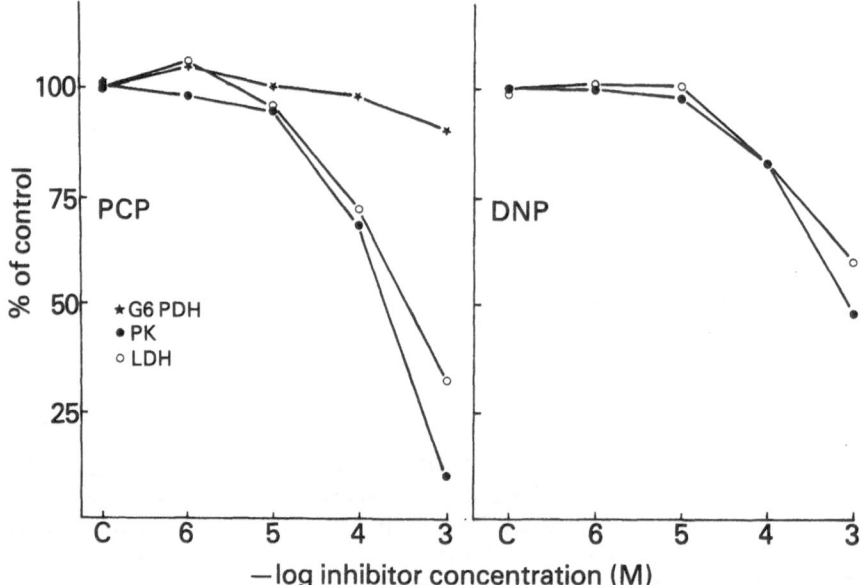

Figure 4. *In vitro* effects of Na-PCP and DNP on carbohydrate metabolism in the blue crab. The enzymes were assayed under the same conditions as given in the legend for Figure 3 except the enzyme fractions were preincubated with the phenols for 5 minutes prior to addition to the assay medium. Each value represents the Mean of five experiments.

containing PCP it appeared that glycolysis was the pathway of catabolism but even this pathway was inhibited to some extent. The blue crab hepatopancreatic pyruvate kinase and lactic dehydrogenase were inhibited above 10^{-5} M Na-PCP or 10^{-5} M DNP *in vitro*, with extensive inhibition at 10^{-3} M Na-PCP (Fig. 4). DNP was again a less effective inhibitor of glycolysis than Na-PCP. Weinbach (1956b) found no significant effect on glycolytic phosphorylation *in vitro* in the snail tissues at 5 × 10^{-4} M PCP while higher levels inhibited glycolysis completely. More recently Boström and Johansson (1972) showed eel liver pyruvate kinase to be inhibited at 3 × 10^{-5} M Na-PCP and completely inhibited at 10^{-3} M. Eel lactic dehydrogenase was more resistant to inhibition showing 59% activity at 10^{-3} M Na-PCP. Glucose-6-phosphate dehydrogenase seemed to be the most stable enzyme in the presence of Na-PCP. The *in vitro* results for both the crab and the eel showed little inhibition even at 10^{-3} M Na-PCP.

The differences between the results of *in vivo* and *in vitro* experiments with the crabs indicate that *in vivo* Na-PCP might be affecting glucose-6-phosphate dehydrogenase indirectly. Boström and Johansson (1972) suggested that the effect of Na-PCP on glycolysis in eels may be a cause of the increased aerobic activity. They also suggested that the *in vivo* activity of these enzymes in eels may be a consequence of hormonal mediation. The question of whether Na-PCP elicits a direct or indirect effect on the crab hepatopancreatic enzymes requires further elucidation.

Effect of Na-PCP on Other Metabolic Enzymes

Na-PCP and DNP showed inhibitory activity toward glutamate-pyruvate transaminase from the blue crab hepatopancreas (Table 1). The enzyme, which is involved in amino acid metabolism, is inhibited extensively by Na-PCP *in vivo* and

TABLE 1. Effect of Phenols on Glutamate-Pyruvate Transaminase Activity

		Phenol	
Experiment	Phenol concn. (M)	Na-PCP	DNP
in vitro	10^{-6}	106[a]	115
	10^{-5}	99	112
	10^{-4}	86	94
	10^{-3}	24	65
in vivo	6 µg/g body wt.	28	67

[a]Results are given as a percentage of normal enzyme activity. The *in vivo* concentration is the level of phenol injected into the animals.

TABLE 2. Effect of Na-PCP and DNP on Calcium Activated
ATPase *in vivo*

	Control	Experimental	
Sample	seawater	DNP	Na-PCP
Conc.	——	6 μg/g	6 μg/g
Activity	0.339 ± 0.012	0.322 ± 0.007	0.279 ± 0.011
% Activity	100%	95%	82%

ASSAY CONDITIONS: 6 μg Na-PCP or DNP per gram body weight were injected into crabs. After three hours the animals were dissected and the hepatopancreas was homogenized and differentially centrifuged (Fox and Rao, 1978) to obtain the various fractions. The microsomal preparation was incubated in a medium of 50 mM Tris-HCl (pH 7.1), 10 mM $CaCl_2$, 2.5 mM Tris-ATP for 30 minutes at 25 °C in 1.5 ml. The reaction was stopped by the addition of 1.5 ml of 10% trichloroacetic acid and centrifuged to precipitate the protein. Inorganic phosphate was determined by the Fiske-Subba Row method (1925). Activity is measured in μmoles Pi/mg protein/min. and is the Mean ± S.D. of five experiments. The specific activity noted in the sample treated with Na-PCP is significantly different from the control ($P < 0.01$).

in vitro. DNP inhibited this enzyme only half as much as Na-PCP. Holmberg *et al.* (1972) have also shown this transaminase in eel *(Anguilla anguilla)* liver and muscle to be less active in the presence of Na-PCP. They exposed the eels to low levels of Na-PCP (0.1 ppm) in both fresh water and sea water. Cheng (1965) showed slight stimulation of glutamate-oxaloacetate transaminase and glutamate-pyruvate transaminase in fish exposed to 0.1 ppm K-PCP while exposure to 0.2 ppm K-PCP caused 30% inhibition of these enzymes.

Another pathway exposed to PCP and not examined in the blue crab is lipid metabolism. Holmberg *et al.* (1972) examined some aspects of lipid metabolism and found that Na-PCP increased the breakdown of lipids. Hanes *et al.* (1968) also studied fatty acid metabolism and found coho salmon *(Oncorhynchus kisutch)* catabolized twice as much total lipid when exposed to K-PCP than the controls. The limited work on lipid metabolism in relation to PCP-toxicity provides the general trend of increased catabolism to free fatty acids yet shows no effect on the enzymes involved.

Effect of Na-PCP on ATPases

Adenosine triphosphatases are vital for muscle function, ionic transport and oxidative phosphorylation. Their requirements for stimulation are dependent on the tissue as well as their function. We studied the effects of PCP and DNP on a calcium activated ATPase of the hepatopancreatic microsomes from the blue crab. The enzyme was stimulated by calcium and to some degree by other ions such as barium and strontium (Fox and Rao, 1978). The results shown in Table 2 indicate that Na-PCP was more effective than DNP in inhibiting the ATPase *in vivo*. Similar results were obtained in *in vitro* tests, as shown in Figure 5. The activity of the enzyme decreased with an increase of phenol concentration. The greatest inhibition occurred at 10^{-3} M where Na-PCP inhibited 37% and DNP inhibited 29% of enzyme activity. The concentrations of the phenols were in the range most investigators used for either stimulation or inhibition of ATPase activity. Weinbach (1956b) reported that the mitochondrial magnesium-ATPase from snail tissues was not stimulated by low concentrations of PCP yet was inhibited at higher concentrations. The albumen gland from the snail *Lymnaea stagnalis* yielded a mitochondrial ATPase that had properties associated with oxidative phosphorylation. This enzyme was stimulated considerably at 5×10^{-5} M PCP. A biphasic response of PCP was seen in rat liver ATPase; PCP stimulated the enzyme at low concentrations and inhibited at high concentrations (Weinbach, 1954).

An interesting property of the rat liver ATPase was the reversibility of inhibition by PCP through washing (Weinbach, 1956a). Unlike PCP, dinitrophenol seemed to stimulate mitochondrial ATPase at higher concentrations (Lardy and Wellman, 1953). Kameyama *et al.* (1974) found a muscle calcium ATPase to be stimulated by low concentrations of PCP while an EDTA- and potassium-ATPase were inhibited at the same concentration. PCP also had inhibitory effects on ATPase from the bacterium, *Bacillus megaterium* (Ishida and Mizushima, 1969). This bacterial enzyme was stimulated by either magnesium or calcium ions. PCP

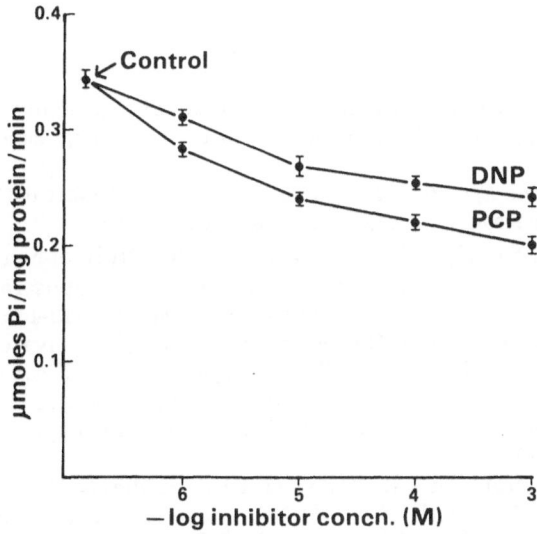

Figure 5. *In vitro* effect of Na-PCP and DNP on a microsomal calcium activated ATPase of the hepatopancreas from the blue crab. The assay conditions were the same as described in Table 1 except the microsomal fraction was preincubated with the phenols for 5 minutes. The values are expressed as Mean ± S.D. for 5 experiments.

showed no effect on the magnesium activity when low concentrations are used, however, 2 mM PCP almost eliminated ATPase activity. Calcium-stimulated activity decreased at concentrations below 10 μM PCP. ATPases were not the only enzymes to be inhibited by PCP. Dall-Larsen *et al.* (1976) have shown an ATP phosphoribosyltransferase to be inhibited by PCP and DNP at concentrations of 300 μM and 150 μM, respectively.

Although considerable work has been done on the relationship between PCP and ATPases, no definite mechanism of action can be postulated. Mitochondrial ATPases appear to follow the general trend of stimulation at low concentrations of PCP and inhibition at higher concentrations. However, the ATPases from invertebrate tissues did not appear to be stimulated by PCP but are inhibited at high concentrations of PCP. This phenol appears to affect carbohydrate metabolism, energy production, lipid metabolism, ion transport and possibly protein metabolism. Is the effect a general one where the phenol is bound non-specifically to membrane proteins and changing their configuration? Or, is it more specific in its interaction with each enzyme involved? Although further investigations are needed to answer these questions, it appears clear that uncoupling of oxidative phosphorylation is not the sole basis for the toxicity of PCP.

ACKNOWLEDGMENTS

This investigation was supported by Grant R-804541-01 from the United States Environmental Protection Agency.

References

Boström, S.-L., and R.G. Johansson, 1972. Effects of pentachlorophenol on enzymes involved in energy metabolism in the liver of the eel. *Comp. Biochem. Physiol.,* **41B:** 359-369.

Cheng, J.T., 1965. *The Effects of Potassium Pentachlorophenate on Selected Enzymes in Fish.* Master's thesis, Oregon State University, Corvallis.

Dall-Larsen, T., H. Kryvi, and L. Klungsoyr, 1976. Dinitrophenol, dicoumarol and pentachlorophenol as inhibitors and parasite substrates in the ATP phosphoribosyltransferase reaction. *Eur. J. Biochem.,* **66:** 443-446.

Fox, F.R., and K.R. Rao, 1978. Characteristics of a Ca^{2+}-activated ATPase from the hepatopancreas of the blue crab, *Callinectes sapidus. Comp. Biochem. Physiol.,* in press.

Holmberg, B., S. Jensen, A. Larsson, K. Lewander, and M. Olsson, 1972. Metabolic effects of technical pentachlorophenol (PCP) on the eel *Anguilla anguilla* L. *Comp. Biochem. Physiol.,* **43B:** 171-183.

Ishak, M.M., A.A. Sharaf, and A.H. Mohamed, 1972. Studies on the mode of action of some molluscicides on the snail, *Biomphalaria alexandrina.* II. Inhibition of succinate oxidation by Bayluscide, sodium pentachlorophenate and copper sulfate. *Comp. Gen. Pharmacol.,* **3:** 385-390.

Ishida, M., and S. Mizushima, 1969. Membrane ATPase of *Bacillus megaterium.* I. Properties of membrane ATPase and its solubilized form. *J. Biochem.,* **66:** 33-43.

Kameyama, T., S. Hayakawa, and T. Sekine, 1974. Effect of phenol derivatives and chemical modification on the adenosine triphosphatase activities of heavy meromyosin and subfragment 1. *J. Biochem.*, **75:** 381-387.

Krueger, H., S.D. Lu, G. Chapman, and J.T. Cheng, 1966. Effects of pentachlorophenol on the fish, *Cichlasoma bimaculatum.* Abstracts from the 3rd Int. Pharm. Cong. S. Paulo, Brazil, 24-30 July 1966.

Lardy, H.A., and H. Wellmann, 1953. The catalytic effect of 2,4-dinitrophenol on adenosinetriphosphate hydrolysis by cell particles and soluble enzymes. *J. Biol. Chem.*, **201:** 357-370.

Liu, D.H.W., 1969. *Alterations of Carbohydrate Metabolism by Pentachlorophenol in Cichlid Fish.* PhD. dissertation, Oregon State University, Corvallis.

Muller, F., J.F. Hogg, and C. DeDuve, 1968. Distribution of tricarboxylic acid cycle enzymes and glyoxylate cycle enzymes between mitochondria and peroxisomes in *Tetrahymena pyriformis. J. Biol. Chem.*, **243:** 5385-5395.

Weinbach, E.C., 1954. The effect of pentachlorophenol on oxidative phosphorylation. *J. Biol. Chem.*, **210:** 545-550.

Weinbach, E.C., 1956a. Pentachlorophenol and mitochondrial adenosinetriphosphatase, *J. Biol. Chem.*, **221:** 609-618.

Weinbach, E.C., 1956b. The influence of pentachlorophenol on oxidative and glycolytic phosphorylation in snail tissue. *Arch. Biochem. Biophys.*, **64:** 129-143.

Weinbach, E.C., 1957. Biochemical basis for the toxicity of pentachlorophenol. *Proc. National Acad. Sci.*, **43:** 393-397.

Weinbach, E.C., and J. Garbus, 1965. The interaction of uncoupling phenols with mitochondria and with mitochondrial protein. *J. Biol. Chem.*, **240:** 1811-1819.

Weinbach, E.C., and M.O. Nolan, 1956. The effect of pentachlorophenol on the metabolism of the snail *Australorbis glabratus. Exp. Parasit.*, **5:** 276-284.

Effect of Pentachlorophenol on the ATPases in Rat Tissues

D. DESAIAH

Abstract—The effect of pentachlorophenol (PCP) on the ATPase system in the rat hepatic, brain and kidney fractions was investigated using *in vitro* techniques. PCP showed a biphasic effect on oligomycin-sensitive (mitochondrial) Mg^{++} ATPase activities in liver and kidney tissues: at low concentrations (10^{-5} M) it produced stimulatory effect; at higher concentrations (10^{-4} M) inhibition was observed. However, in brain, only stimulation was observed at all the concentrations used. The oligomycin-insensitive Mg^{++} ATPase activity in liver was not altered by the addition of PCP but the brain and kidney enzyme activities were significantly inhibited. Na^+, K^+-ATPase in brain and kidney showed a contrasting response to PCP. A biphasic effect was seen with brain enzyme, whereas the kidney ATPase was significantly inhibited in a dose dependent manner. In addition, the K^+-dependent *p*-nitrophenyl phosphatase (K^+-PNPPase) activity in rat brain microsomes was inhibited by PCP. It appears from the results that PCP may be acting as an uncoupler of oxidative phosphorylation at low concentrations and inhibiting the same at high concentrations. The results also indicate that the Na^+, K^+-ATPase (cation transporting enzyme) may be a locus of action of PCP.

Introduction

The biocidal effects of pentachlorophenol (PCP) and its salts are thought to be due to their ability to uncouple oxidative phosphorylation (Weinbach, 1954; 1956).

D. DESAIAH • Department of Pharmacology and Toxicology, University of Mississippi Medical Center, Jackson, Mississippi 39216, U.S.A.

PCP and 2,4-dinitrophenol (DNP) stimulate the ATPase (Adenosine triphosphate phosphohydrolase) activity in the mitochondrial preparations from rat liver and fish brain (Weinbach, 1956; Yap *et al.*, 1975). The effects of PCP and its salts on other ATPases such as Na^+, K^+-ATPase have not been previously described. This report deals with the effects of PCP *in vitro* on the mitochondrial Mg^{++} ATPase activities in rat liver, brain, kidney and on the Na^+, K^+-ATPase activities in brain and kidney tissues.

Preparation of Tissue Fractions

Male Sprague-Dawley rats (weighing about 250 grams) obtained from Charles River Co., and maintained in animal facilities away from any known inducers were used in this study. The tissues, brain, kidney and liver were removed and homogenized in sucrose solution (containing 0.32 M sucrose, 1 mM EDTA and 10 mM imidazole. pH 7.5) using a ground glass homogenizer. All three homogenates were centrifuged at 750 × g for 10 min to remove nuclei and cell debri. The resulting supernatant of liver was centrifuged at 9000 × g for 20 min to obtain heavy mitochondria, whereas the brain and kidney supernatants were centrifuged at 13,000 × g for 20 min to obtain nerve endings in addition to the mitochondria. The pellets were resuspended in sucrose solution, quick frozen in liquid N_2 and stored at -85° C. The sample may be stored at this temperature for a period of 2-3 months without appreciable loss of enzyme activity. Rat brain microsomes were obtained by centrifuging the 13,000 × g supernatant at 100,000 × g for 1 hr. The pellet was resuspended in sucrose solution and used for K^+-PNPPase (K^+-dependent p-nitrophenyl phosphatase) assay.

Enzyme Assays

ATPase activity was measured essentially according to the method of Fritz and Hamrick (1966). A 3 ml reaction mixture (unless otherwise mentioned) contained: 4.5 mM ATP, 5 mM Mg^{++}, 100 mM Na^+, 20 mM K^+, 135 mM imidazole-Cl buffer, pH 7.5, 0.2 mM NADH, 0.5 mM phosphoenol pyruvate, 0.02% bovine serum albumin, approximately 9 units of pyruvate kinase and 12 units of lactic acid dehydrogenase. A 50 μl tissue fraction was used with a protein content of 30-50 μg. Absorbance changes in the reaction mixture were measured at 340 nm using a Beckman Acta III recording spectrophotometer with temperature controlled at 37° C. The change in optical density at 340 nm over a period of 10 min was used in calculating the specific activity. Enzyme activities were expressed as μmole Pi/mg protein/hr. Protein was determined by the method of Lowry *et al.* (1951) using bovine serum albumin as standard.

Total ATPase activity in brain and kidney fractions was measured with Mg^{++}, Na^+ and K^+ present in the reaction mixture. Mg^{++} ATPase activity was measured in the presence of 1 mM ouabain, which is a specific inhibitor of Na^+, K^+-ATPase. Na^+, K^+-ATPase activity (ouabain-sensitive) was obtained by the difference between total ATPase activity and Mg^{++} ATPase activity. Na^+, K^+-ATPase activity in liver fractions was very low, hence it was not determined. Mg^{++} ATPase in brain, kidney and liver fractions was further delineated into oligomycin-sensitive (mitochondrial) and -insensitive Mg^{++} ATPase activities by adding 1 μl of oligomycin (5 × 10^{-6} M) in ethanol (Lardy *et al.*, 1958 and Desaiah *et al.*, 1972).

K^{+}-dependent *p*-nitrophenyl phosphatase (K^{+}-PNPPase) was assayed essentially as described by Ahmed and Judah (1964) using rat brain microsomal preparation. *p*-nitrophenylphosphate was used as substrate. The enzyme activity was calculated as the difference between the activity measured in the presence of Mg^{++} (5 mM) and K^{+} (20 mM) and the activity detected with Mg^{++} alone. The Mg^{++} dependent PNPPase was less than 5% of the total activity and was not altered by PCP at the concentrations tested.

PCP (analytical grade sodium pentachlorophenate, supplied by Dr. K. Ranga Rao, University of West Florida) was dissolved in H$_2$O and different concentrations as indicated in the tables, were added to the reaction mixture. The enzyme was preincubated with PCP for 3 min before the reaction was started with the addition of substrate. The enzyme activities were determined in the presence and absence of different concentrations of PCP. Each value represents the mean of three separate homogenate preparations (n = 3). Each homogenate was assayed two times and the average was taken. Significance of differences between control and treatments was checked by Student's *t* test.

Effect of PCP on Oligomycin-sensitive (Mitochondrial) Mg^{++} ATPase

The data in Table 1 show that the mitochondrial Mg^{++} ATPase activity in liver and kidney was significantly stimulated up to 20 μM concentration of PCP.

TABLE 1. Effect of Pentachlorophenol on Oligomycin-sensitive (Mitochondrial) Mg^{++} ATPase Activity in Rat Tissues

PCP conc. (μM)	Specific activity[a]		
	Liver	Brain	Kidney
0	20.7 ± 0.42	7.8 ± 0.14	30.2 ± 2.14
10	24.0 ± 0.79[b]	14.5 ± 0.60[c]	41.2 ± 3.99
20	25.3 ± 0.22[c]	15.2 ± 0.38[c]	45.4 ± 4.77[b]
40	23.3 ± 1.83	13.3 ± 0.71[c]	39.0 ± 5.51
80	15.4 ± 1.67[b]	11.2 ± 0.45[c]	27.6 ± 1.59
120	10.9 ± 0.79[c]	9.4 ± 0.26[b]	21.5 ± 0.38[c]

[a]Mean ± standard error; μmole Pi/mg protein/hr
[b]Significantly different from control (P < 0.05)
[c]Significantly different from control (P < 0.01)

TABLE 2. Effect of Pentachlorophenol on Oligomycin-
insensitive Mg⁺⁺ ATPase Activity in Rat Tissues

PCP conc. (μM)	Specific activity[a]		
	Liver	Brain	Kidney
0	4.19 ± 0.37	12.7 ± 0.60	23.0 ± 1.54
10	4.36 ± 0.26	14.5 ± 0.56	21.8 ± 0.90
20	4.12 ± 0.26	13.3 ± 0.35	17.5 ± 1.46[b]
40	4.27 ± 0.30	12.4 ± 0.50	15.8 ± 1.48[b]
80	3.67 ± 0.52	10.1 ± 0.48[b]	14.8 ± 2.02[b]
120	3.61 ± 0.64	9.7 ± 0.36[b]	13.2 ± 2.16[b]

[a]Mean ± standard error; μmole Pi/mg protein/hr
[b]Significantly different from control (P < 0.05)

However, when the concentration of PCP increased to 80 μM and above the enzyme activities in these tissues were significantly inhibited. PCP was more effective on brain mitochondrial Mg⁺⁺ ATPase as compared to liver and kidney. The enzyme activity in brain was significantly stimulated by PCP at all the concentrations tested as compared to a biphasic effect seen in liver and kidney. Interestingly, maximal (90%) stimulation was observed at 20 μM PCP while the extent of stimulation gradually decreased with an increase in concentration of PCP.

Racker et al. (1975) have shown that the oligomycin-sensitive Mg⁺⁺ ATPase in the mitochondria is involved in oxidative phosphorylation. The biphasic effects of PCP on Mg⁺⁺ ATPase suggest that PCP may be acting as an uncoupler and an inhibitor of oxidative phosphorylation.

Effect of PCP on Oligomycin-insensitive (OI) Mg⁺⁺ ATPase

The results in Table 2 demonstrate that PCP has no significant effect on liver OI Mg⁺⁺ ATPase activity. However, a 14% inhibition was observed at 120 μM PCP but it was not statistically significant. The brain OI Mg⁺⁺ ATPase activity showed a biphasic response to PCP, in that, the lower concentrations of PCP caused a stimulation while the higher concentrations caused inhibition. On the other hand, OI Mg⁺⁺ ATPase activity in kidney was inhibited at all concentrations of PCP tested, reaching a 42% level of inhibition at 120 μM PCP. The implications of these findings are not clear since the functional role of this enzyme is not fully understood. However, previous studies indicate that the OI Mg⁺⁺ ATPase is present in a variety of tissues.

**TABLE 3. Effect of Pentachlorophenol on
Na⁺, K⁺-ATPase Activity in Rat Tissues**

PCP conc. (μM)	Specific activity[a]	
	Brain	Kidney
0	26.5 ± 1.80	25.3 ± 1.87
10	31.9 ± 1.52[b]	23.8 ± 2.05
20	35.6 ± 1.66[b]	19.9 ± 1.83
40	25.6 ± 3.16	19.0 ± 1.71[b]
80	19.1 ± 1.01[c]	17.9 ± 1.27[b]
120	16.2 ± 1.40[c]	16.6 ± 1.59[b]

[a]Mean ± standard error; μmole Pi/mg protein/hr
[b]Significantly different from control (P < 0.05)
[c]Significantly different from control (P < 0.01)

Effect of PCP on Na⁺, K⁺-ATPase Activity

The Na⁺, K⁺-ATPase activity in liver fraction (9,000 × g pellet) was too low to be correctly measured, hence the effect of PCP was not determined. The data presented in Table 3 reveal that PCP produced a biphasic effect on brain Na⁺, K⁺-ATPase. A significant stimulation of 30% was observed at 20 μM PCP and by increasing the concentration of PCP to 120 μM a 39% inhibition was observed. In contrast, the kidney Na⁺, K⁺-ATPase was inhibited by PCP at all concentrations used. However, the inhibition at low concentrations was not statistically significant.

Na⁺, K⁺-stimulated ATPase activity was shown to be the biochemical manifestation of the Na⁺ pump (Skou, 1965 and Schwartz *et al.,* 1972). The functional role of Na⁺, K⁺-ATPase is the active ion transport across the cell membrane (Skou, 1957). The effects of PCP on this important enzyme, as noted in the present investigation, could lead to marked alterations in ion transport across the cell membrane.

The mechanism of operation of the Na⁺, K⁺-ATPase is believed to involve a series of sequential steps as illustrated below:

$$\text{Step 1:} \quad E_1 \underset{\text{ATP, ADP}}{\overset{\text{Na}^+,\text{Mg}^{++}}{\rightleftharpoons}} E_1P$$

$$\text{Step 2:} \quad E_1P \overset{\text{Mg}^{++}}{\rightleftharpoons} E_2P$$

$$\text{Step 3:} \quad E_2P \overset{\text{K}^+}{\rightleftharpoons} E_1 + Pi$$

**TABLE 4. Effect of Pentachlorophenol
on Rat Brain Microsomal K⁺-PNPPase**

PCP conc. (μM)	Specific activity[a]
0	3.81 ± 0.16
10	2.98 ± 0.17
20	2.94 ± 0.01
40	2.87 ± 0.14
80	2.17 ± 0.08[b]
120	1.12 ± 0.12[b]

[a]Mean ± standard error; μmole pNPP/mg protein/hr
[b]Significantly different from control ($P < 0.01$)

In the presence of free enzyme, Na⁺, ATP and Mg⁺⁺ an E_1-phosphoenzyme complex is formed (step 1). This complex undergoes a change in the enzyme conformation to form an E_2-phosphoenzyme complex (step 2) which is sensitive to hydrolysis by K⁺ (Fahn *et al.*, 1966). The step 3 is represented by the K⁺-dependent p-nitrophenyl phosphatase reaction (Ahmed and Judah, 1964). In order to elucidate the mechanism of Na⁺, K⁺-ATPase inhibition by PCP, its effect was determined on K⁺-PNPPase using rat brain microsomal preparation. The results obtained (Table 4) suggest that PCP may be inhibiting the Na⁺, K⁺-ATPase by inhibiting the terminal step of the reaction represented by K-PNPPase. However, further work is needed to evaluate the effects of PCP on the remaining reactions involved in the operation of Na⁺, K⁺-ATPase.

The present results on the effects of PCP on ATPase system may support the theory that chlorinated pesticides exert their toxic action by inhibiting the ATPase activity (Desaiah *et al.*, 1974). However, the action of PCP on ATPases is unique in that it produced a biphasic effect suggesting that PCP may be interfering with more than one active site on the enzyme molecule.

ACKNOWLEDGMENTS

I thank Dr. I.K. Ho for providing support and facilities for carrying out this research in his laboratory.

References

Ahmed, K., and J.D. Judah, 1964. Preparation of lipoproteins containing cation-dependent ATPase. *Biochim. Biophys. Acta,* **93:** 603-613.
Desaiah, D., L.K. Cutkomp, and R.B. Koch, 1974. A comparison of DDT and its

biodegradable analogues tested on ATPase enzymes in cockroach. *Pest. Biochem. Physiol.,* **4:** 232-238.

Desaiah, D., L.K. Cutkomp, H.H. Yap, and R.B. Koch, 1972. Inhibition of oligomycin-sensitive and -insensitive Mg++ adenosine triphosphatase activity in fish by polychlorinated biphenyls. *Biochem. Pharmacol.,* **21:** 857-865.

Fahn, S., G.J. Koval, and R.W. Albers, 1966. Sodium-potassium-activated adenosine triphosphatase of *Electrophorus* electric organ. *J. Biol. Chem.,* **241:** 1882-1889.

Fritz, P.J., and M.G. Hamrick, 1966. Enzymatic analysis of adenosine triphosphatase. *Enzymol. Acta Biocat.,* **30:** 57-64.

Lardy, H.A., P. Witonsky, and W.C. McMurray, 1958. Antibiotics as tools for metabolic studies. 1. A survey of toxic antibiotics in respiratory phosphorylation and glycolytic systems. *Arch. Biochem. Biophys.,* **78:** 587-597.

Lowry, O.H., N.J. Rosebrough, A.L. Farr, and R.L. Randall, 1951. Protein measurement with the folin phenol reagent. *J. Biol. Chem.,* **193:** 265-275.

Racker, E., A.F. Knowles, and E. Eytan, 1975. Resolution and reconstitution of ion transport systems. *Annals N.Y. Acad. Sci.,* **264:** 17-33.

Schwartz, A., G.E. Lindenmayer, and J.C. Allen, 1972. The Na+-K+ ATPase membrane transport system: importance in cellular function. In, *Current Topics in Membranes and Transport,* (F. Bronner and A. Kleinzeller, ed.), Vol. **3,** pp. 1-82, Academic Press, Inc., New York.

Skou, J.C. 1957. The influence of some cations on an adenosine triphosphatase from peripheral nerves. *Biochim. Biophys. Acta,* **23:** 394-401.

Skou, J.C. 1965. Enzymatic basis for active transport of Na+ and K+ through cell membrane. *Physiol. Rev.,* **45:** 596-617.

Weinbach, E.C., 1954. The effect of pentachlorophenol on oxidative phosphorylation. *J. Biol. Chem.,* **210:** 545-550.

Weinbach, E.C., 1956. Pentachlorophenol and mitochondrial adenosine triphosphatase. *J. Biol. Chem.,* **221:** 609-618.

Yap, H.H., D. Desaiah, L.K. Cutkomp, and R.B. Koch, 1975. *In vitro* inhibition of fish brain ATPase activity by cyclodiene insecticides and related compounds. *Bull. Environ. Contam. Toxicol.,* **14:** 163-167.

Effects of Sodium Pentachlorophenate on Survival and Energy Metabolism of Embryonic and Larval Steelhead Trout

GARY A. CHAPMAN and DEAN L. SHUMWAY

Abstract—A study was conducted to determine the effects of technical grade sodium pentachlorophenate (Na-PCP) on the early developmental stages of the steelhead trout *(Salmo gairdneri)*. In an experiment where embryos were exposed to Na-PCP from fertilization to hatching, 100% mortality occurred within one week after fertilization at concentrations down to 300 ppb (μg/l); within 24 hours post-hatch, 100% mortality occurred down to 50 ppb of Na-PCP. Alevin dry weight at hatch was decreased by exposure to Na-PCP and hatching was delayed. In 5-day tests, alevins usually died within 24 hours at concentrations down to 200 ppb, but little mortality occurred at lower concentrations.

Continuous exposure to Na-PCP from fertilization to complete yolk absorption produced 100% mortality at 40 ppb Na-PCP but little mortality at 20 or 10 ppb. However, in water containing 5 mg O_2/l, 20 ppb Na-PCP was 100% lethal and at 3 mg O_2/l, 10 ppb was 100% lethal. Little mortality occurred at these oxygen levels in the absence of Na-PCP. Oxygen consumption rates of alevins in 40 ppb Na-PCP were higher than those of control alevins. Exposure to Na-PCP reduced yolk utilization efficiency and growth. The bioenergetic data obtained in the study are consistent with the concept that PCP disrupts energy metabolism.

GARY A. CHAPMAN • Western Fish Toxicology Station, Corvallis Environmental Research Laboratory, U.S. Environmental Protection Agency, Corvallis, Oregon 97330, U.S.A.

DEAN L. SHUMWAY • Bureau of Power, Federal Power Commission, Washington, D.C. 20426, U.S.A.

The views expressed herein are those of the authors and do not necessarily represent the official views and policies of the Environmental Protection Agency or the Federal Power Commission.

Introduction

The experiments described in this paper were conducted at Oregon State University to follow survival, growth, development, oxygen consumption, and bioenergetics of steelhead trout *(Salmo gairdneri)* embryos and alevins in media containing sodium pentachlorophenate (Na-PCP). The terminology applied to life history stages is as follows:

embryo — developmental stages to the moment of hatching

alevin — developmental stages from hatching until all yolk has been absorbed

The embryonic and alevin stages each comprise roughly one-half of the developmental period. Approximately halfway through the alevin period, the alevin attains a stage called the "swim-up." Once the swim-up stage is attained, the alevins swim freely about in the water.

The total reliance upon yolk for nutrition lasts throughout the embryonic stages and through the alevin period up to the swim-up stage. The swim-up alevin can utilize a dual source of nutrition, yolk and exogenous food.

A number of laboratory studies on the growth and bioenergetics of embryos and alevins of salmonids have been conducted. Most notable are the studies by Smith (1947, 1952, 1958), Gray (1926, 1928), and the series of experiments summarized by Hayes (1949). Almost all investigations of this type have taken advantage of the closed nutritive system afforded by the yolk supply and have routinely denied the swim-up alevins a supply of exogenous food. Such an approach simplifies the analysis of growth data by eliminating the need to determine food consumption and assimilation efficiencies.

The growth of salmonid alevins denied exogenous food is limited by the amount of yolk present and the relative proportions of the yolk which are utilized for tissue elaboration, maintenance, and activity. Thousands of eggs can be obtained from a single fish and fertilized at the same time, giving a population of embryos with essentially uniform size, age, and genetic composition. With such a population, the amount of yolk is rather constant between eggs, and therefore, the growth potential of each embryo is the same. Growth and bioenergetic parameters of embryos and alevins from such a population can be measured under a variety of conditions with relatively little variation within a given sample, thus permitting the detection of rather small changes in these parameters.

Experimental Animals

Eggs of the steelhead trout, a sea-run form of the rainbow trout, were obtained from trout hatcheries in Oregon and transported to the laboratory within 2 hours of fertilization. Some batches of eggs were immediately used in experiments while others were held in running water in 2-liter separatory funnels. Just prior to hatching, the eggs held in the funnels were transferred to shallow troughs. The newly hatched alevins were held in the troughs until used in the experiments. Aerated water at 10 °C was continuously supplied to the troughs and separatory funnels at rates of about 500 ml/min.

Sodium Pentachlorophenate

A single lot of Santobrite® (Monsanto Chemical Company) containing a minimum of 90% sodium pentachlorophenate (Na-PCP) was used in all experiments. All concentrations of Na-PCP were calculated on the basis that Santobrite had a purity of 100% Na-PCP.

Static Tests

Static toxicity tests were conducted in shallow glass vessels, 25 × 25 × 5 cm. The embryo and alevin tests were performed at 10 and 15°C in constant temperature rooms, respectively.

An embryo test, lasting about 1 month, was begun in the field 15 minutes after the eggs were fertilized; the eggs were immediately returned to the laboratory and the test was terminated shortly after hatching occurred. The test solutions were constantly aerated and were renewed every other day. Unfertilized eggs and eggs containing dead embryos were removed when the solutions were changed.

A series of 5-day toxicity tests was conducted using alevins hatched from the eggs of a single female. Tests were started when alevins were 3, 8, 14, and 22 days old (age from hatch). A second series was similarly performed using 4-, 10-, and 18-day old alevins from another female. Alevins were transferred to fresh solutions once or twice daily, depending on the experiment; this procedure was adopted to minimize the effects of detoxification and deoxygenation of the solutions by the alevins. Dead alevins were removed when the solutions were renewed.

Flow-Through Tests

In all experiments, filtered stream water (pH about 7.8) was brought into a constant temperature room through polyethylene pipes and into a head box via a float valve. The water in the head box was aerated and heated.

The apparatus used in experiments with Na-PCP consisted of a series of components integrated into a system to deliver water with desired dissolved oxygen and Na-PCP concentrations to respirometers designed for rearing salmonid embryos and alevins (Chapman, 1969).

At the beginning of an experiment, approximately equal numbers of alevins (or eggs) were placed into each respirometer. In addition, a representative sample of alevins or eggs was taken to determine the mean weight of the alevins and the mean weight of yolk present at the start of each experiment. Dead alevins were usually removed and counted daily. Thirty to 250 alevins, depending on the number remaining in each respirometer, their weight, and degree of development, were removed from each respirometer at intervals spaced so that changes in alevin and yolk weights could be adequately defined.

The alevins in each sample were divided into two sub-samples and killed in a solution of MS-222. The alevins in one sub-sample were blotted on paper towels, counted, and weighed in aggregate. Alevins in the other sub-sample were handled in the same manner except that their yolk was removed prior to weighing. The removal of yolk from young alevins was easily accomplished by excising the entire yolk sac. However, as the amount of yolk diminished and significant dermal overgrowth of the yolk sac occurred, it became necessary to make a mid-ventral slit along the body cavity in order to ensure complete removal of yolk and prevent the excision of significant quantities of non-yolk tissue such as skin and viscera. After being weighed, the samples were dried at 75 to 80°C for 5 days, cooled in a desiccator, and weighed again. The amount of yolk present at a given time was determined by taking the difference between the weight of alevins with yolk and the weight of alevins from which the yolk had been excised.

In several experiments the caloric content of alevin and yolk tisssue was routinely determined using bomb calorimetry. Dried samples were finely ground and 100 to 200 mg portions combusted in a Parr Model 1411 bomb calorimeter. At least two determinations were made on each sample, and duplicate samples usually yielded results within 1%. Caloric values for yolk were calculated from results obtained on samples of alevins with yolk and of alevins from which the yolk had been excised.

Oxygen determinations were made by using the azide modification of the Winkler method. Determinations of oxygen consumption were normally made twice a day, once in the morning and once in the late afternoon. Oxygen consumption rates were calculated and expressed as mg O_2/g alevin dry weight/hr.

Results of Static Tests

Embryos

All eggs held in Na-PCP concentrations of 300 ppb (μg/l) or greater died during the first week of development. Na-PCP concentrations of 50, 80 and 180 ppb caused 100% mortality, with most of the mortality occurring after hatching at 50 and 80 ppb and before hatching at 180 ppb. The dry weight at hatch of alevins reared in Na-PCP was markedly less than that of control alevins.

Alevins

During the 5-day static toxicity tests with alevins, most mortality occurred within 24 hours. Concentrations of Na-PCP higher than 300 ppb always caused 100% mortality within 12 hours. The lowest concentration which produced 50% mortality in any of the toxicity tests was 150 ppb; the highest concentration which failed to produce 50% mortality was 250 ppb.

The relationship between Na-PCP concentration in the static toxicity tests and the time required to produce 50% mortality of steelhead trout alevins is shown in Figure 1. Data from similar tests with chinook salmon *(Oncorhynchus tshawytscha)*

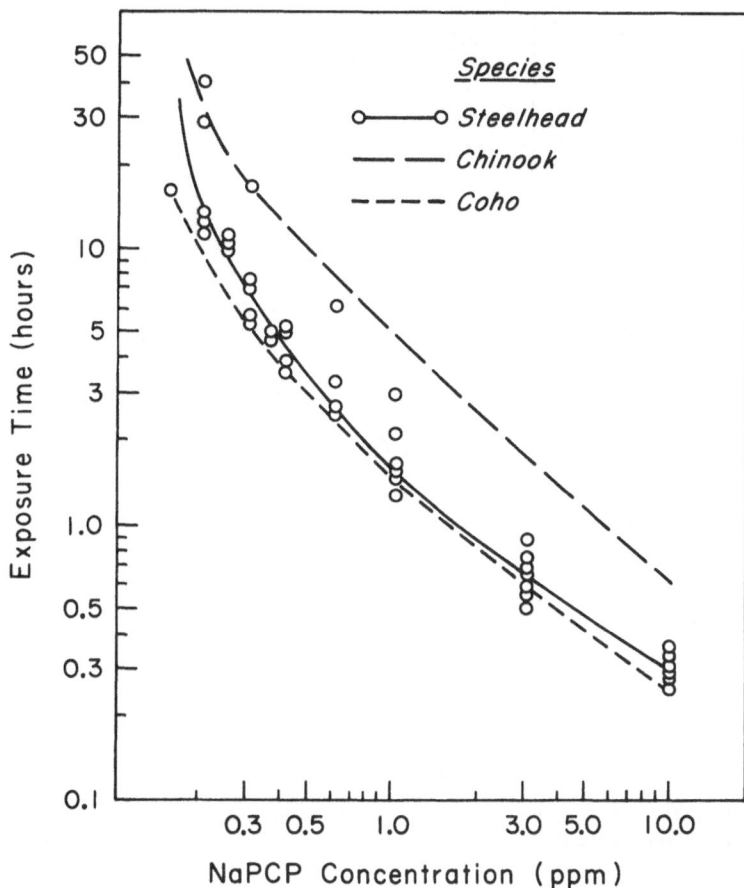

Figure 1. The time required to produce 50% mortality of alevins of three species of salmonids at the indicated concentrations of Na-PCP. All data points are for steelhead trout. Chinook and coho salmon curves were fitted by eye to data from 5 tests with chinook salmon and one test with coho salmon.

alevins and coho salmon *(O. kisutch)* alevins are included for comparison. Alevins of steelhead trout, coho salmon, and chinook salmon appeared to be similar in their response to Na-PCP, although chinook alevins survived somewhat longer at most concentrations.

Results of Flow-Through Tests with Trout Alevins

Dissolved Oxygen and Na-PCP Toxicity

The mortality rate of alevins exposed to 40 ppb Na-PCP was highest at a dissolved oxygen concentration of 3 mg/l (Fig. 2). Although no mortality was seen

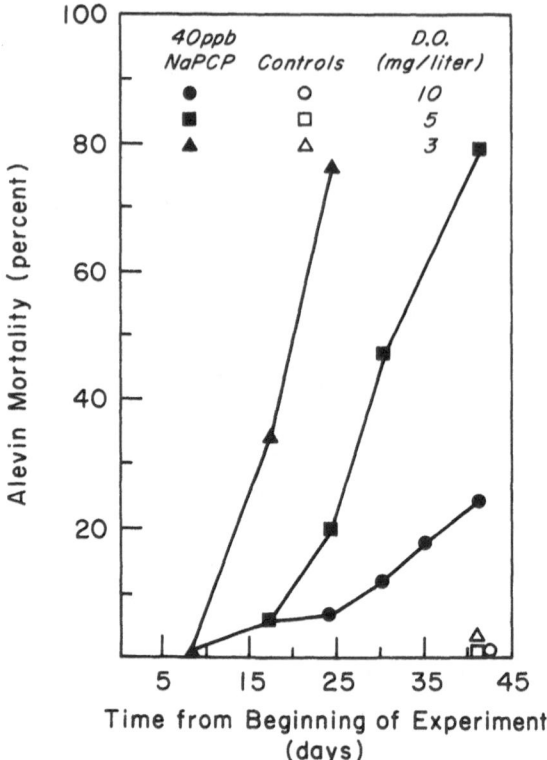

Figure 2. The effect of dissolved oxygen concentration on the mortality of steelhead trout alevins reared in 40 ppb Na-PCP.

within the first 8 days of exposure to 40 ppb Na-PCP, after 24 days of exposure the alevins reared at 3 mg O_2/l suffered 76% mortality compared to only 20% at 5 mg O_2/l and 7% at 10 mg O_2/l. After 41 days, mortality in 40 ppb Na-PCP was 79% at 5 mg O_2/l and 24% at 10 mg O_2/l. The dissolved oxygen concentrations were of themselves nonlethal, as observed in control alevins.

Low levels of dissolved oxygen had a slight effect on the growth of control alevins, while Na-PCP treated alevins exhibited less growth (Fig. 3). The time required to attain maximum weight was prolonged at 5 mg O_2/l and 3 mg O_2/l. The maximum dry weight gain of alevins reared at 40 ppb Na-PCP at the high dissolved oxygen concentration was 22.5% lower than that attained by control alevins. Alevins reared in 40 ppb Na-PCP at 5 and 3 mg O_2/l never attained maximum weight before the last surviving alevins were removed; however, the nature of the growth curve of the surviving alevins reared at 5 mg O_2/l suggests that the final weight noted was near the maximum.

The effects of dissolved oxygen concentration and 40 ppb Na-PCP, both singly and in combination, on the efficiency of yolk utilization were studied. The efficiency

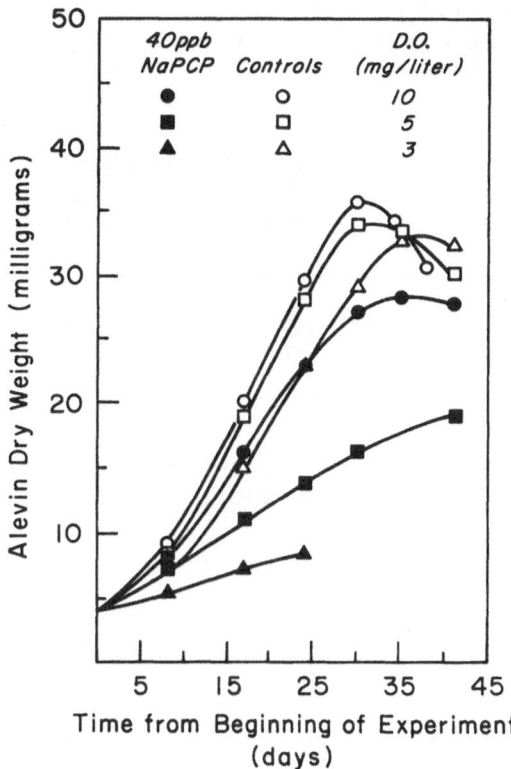

Figure 3. The effect of dissolved oxygen concentration on the growth (dry weight) of steelhead trout alevins reared under control conditions or at 40 ppb Na-PCP.

reduction at 40 ppb Na-PCP was greater at low O_2 concentrations, while low dissolved oxygen concentrations alone only slightly reduced the efficiency of yolk utilization.

The relative contributions of increased yolk catabolism rates and decreased yolk uptake rates to the reduced efficiencies of yolk utilization seen under conditions of exposure to 40 ppb Na-PCP at various dissolved oxygen concentrations are indicated in Table 1. In all cases, an increased rate of yolk catabolism was the primary factor in decreasing the yolk utilization efficiency in alevins exposed to Na-PCP. In the absence of Na-PCP, dissolved oxygen concentrations of 5 and 3 mg/l had no clear-cut effect on the rates of yolk uptake or yolk catabolism.

Exposure of alevins to water of reduced oxygen concentration in the presence or absence of 40 ppb Na-PCP caused the caloric increments of the alevins to be lower than that of control alevins reared at 10 mg O_2/l. The magnitudes of the reductions in caloric increment were similar to those produced in dry weight increment with slight reductions occurring in control alevins at oxygen concentrations of 5 and 3 mg/l and greater reductions in alevins exposed to 40 ppb Na-PCP, especially at lower oxygen concentrations (Table 2).

TABLE 1. The Effect of Dissolved Oxygen
Concentration and Na-PCP on Yolk Uptake and
Yolk Catabolism of Steelhead Trout Alevins[a]

	Mean yolk uptake rate (mg yolk/g alevin/day)	
O_2 (mg/l)	Control	40 ppb Na-PCP
10	85	82
5	89	71
3	84	73
	Mean yolk catabolism rate (mg yolk/g alevin/day)	
	Control	40 ppb Na-PCP
10	16	20
5	20	26
3	15	45

[a]Alevins exposed to 40 ppb Na-PCP for 24 days.

TABLE 2. The Effect of Dissolved Oxygen Concentration on the Growth
and Growth Efficiency (Dry Weight and Caloric) of Steelhead Trout Alevins
Reared Under Control Conditions or at 40 ppb Na-PCP

	O_2 (mg/l)	Maximum gain		Percent reduction		Efficiency		
		wt (mg)	cal (cal)	wt	cal	wt (%)	cal (%)	cal/ wt[a]
Controls	10	31.5	168.9	—	—	82.9	75.2	0.91
	5	29.7	155.2	5.5	8.0	80.3	69.8	0.87
	3	28.6	154.5	9.1	8.5	78.6	70.0	0.89
40 ppb Na-PCP	10	24.0	125.6	23.8	25.6	65.0	56.8	0.87
	5	14.9	77.2	52.7	54.3	55.2	48.6	0.88
	3	4.2	21.7	85.3	87.2	38.5	30.1	0.78

[a]The ratio between caloric efficiency and weight efficiency.

The efficiency of yolk utilization for growth was lower on a caloric basis than on a dry weight basis. The ratios between caloric efficiencies and dry weight efficiencies were 0.89 ± 0.02 under all rearing conditions except 40 ppb Na-PCP at 3 mg O_2/l where the ratio was 0.78.

Oxygen Consumption

The oxygen consumption of alevins was highest during the first week after hatching, declined steadily throughout the period of alevin growth, and reached a minimum when the alevins attained maximum dry weight (Fig. 4). Once alevins started to lose weight, the oxygen consumption rate increased sharply. The oxygen consumption rates of most groups of alevins showed a positive correlation with growth rate (mg growth/gram dry weight/day). The oxygen consumption rates of alevins exposed to 40 ppb Na-PCP were higher than the oxygen consumption rates of control alevins (Fig. 5). Oxygen consumption rates of Na-PCP exposed alevins were also higher than controls when related to age and to a lesser degree when related to weight (Fig. 6).

The effects of 40 ppb Na-PCP and various concentrations of dissolved oxygen on the relationship between alevin growth and the cumulative quantity of oxygen consumed per alevin are shown in Figure 7. Control alevins reared at 10, 5, and 3 mg O_2/l utilized nearly identical quantities of oxygen in the attainment of a given

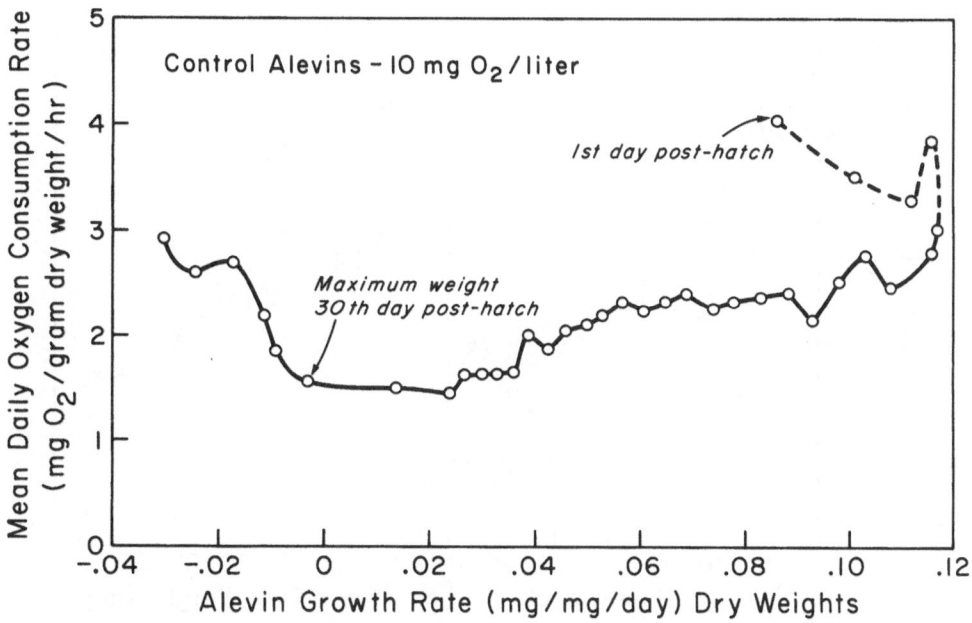

Figure 4. Relationship between growth rate and oxygen consumption rate of steelhead trout alevins reared under control conditions.

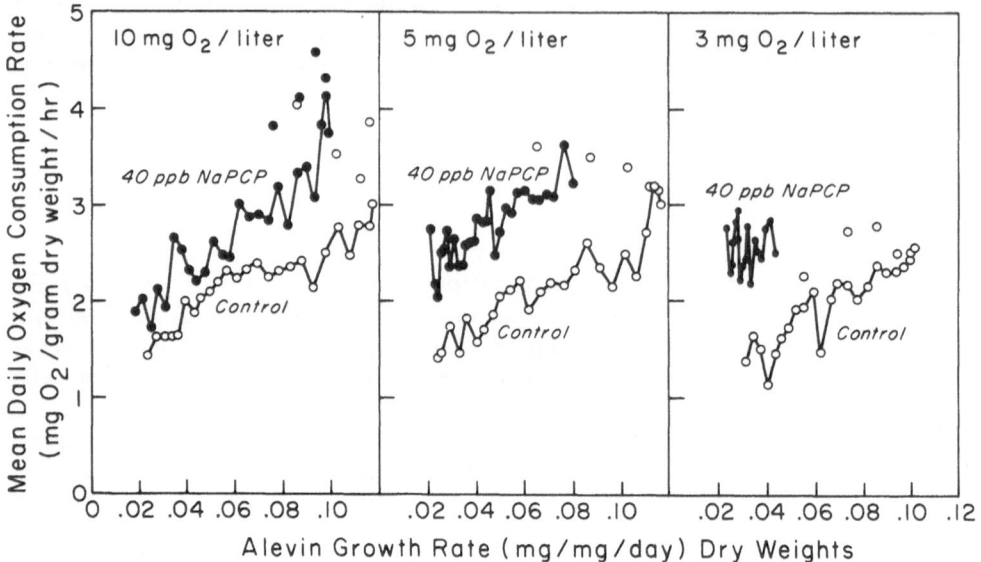

Figure 5. Effects of 40 ppb Na-PCP on the relationship between growth rate and oxygen consumption rate of steelhead trout alevins reared at different oxygen levels.

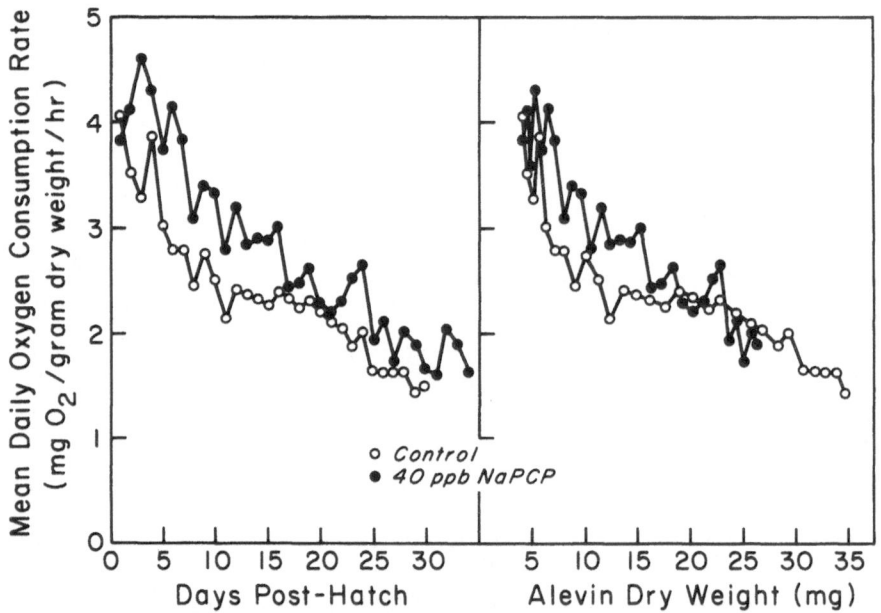

Figure 6. Effects of 40 ppb Na-PCP on the relationships between oxygen consumption rates and age (days post-hatch) and alevin dry weight.

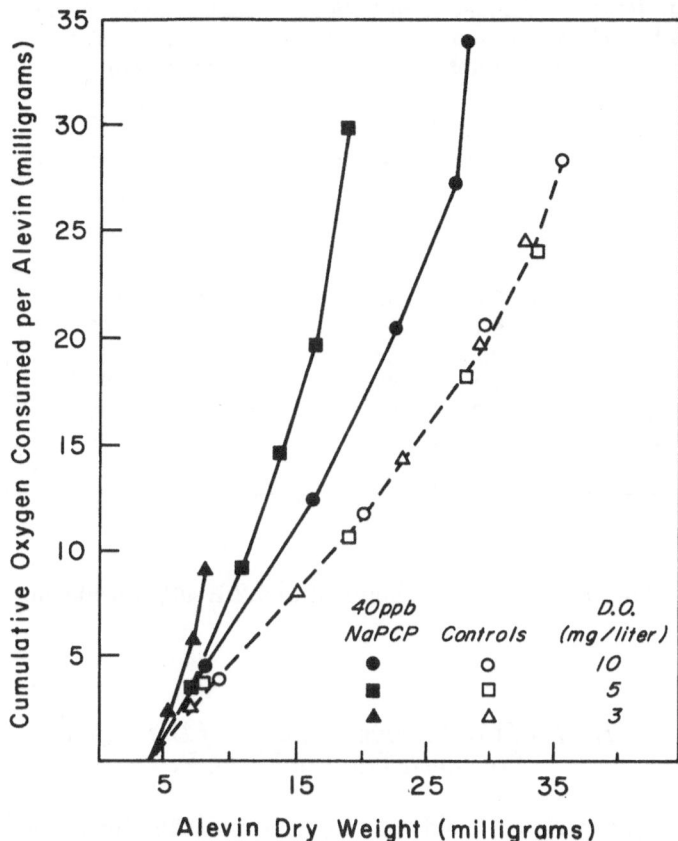

Figure 7. The effect of 40 ppb Na-PCP and various dissolved oxygen levels on the relationship between the amount of oxygen consumed and the dry weight attained by steelhead trout alevins.

weight. Alevins reared at 40 ppb Na-PCP consumed greater quantities of oxygen in growing to a given weight than did control alevins. The lower the dissolved oxygen concentration, the greater the amount of oxygen consumed by Na-PCP-reared alevins in attaining a given weight.

The exposure of alevins to Na-PCP and the resultant increase in metabolic demands disrupted the normal bioenergetic pattern of tissue elaboration and yolk catabolism (Table 3). Typically, control alevins utilized primarily lipids for energy demands, reserving protein for tissue elaboration. Alevins exposed to Na-PCP apparently were forced to use a considerable portion of yolk protein for energy production. Yolk lipid (ether extractable) had a heat of combustion of about 9.1 Kcal/g while that for lipid-free yolk was approximately 5.2 Kcal/g. The yolk material used for energy in control alevins had a heat of combustion of 8.5 to 9.1 Kcal/g while that used by Na-PCP exposed alevins was 6.7 to 7.4 Kcal/g. Exposure to Na-PCP slightly reduced the caloric content of alevin tissue elaborated, with

TABLE 3. Heat of Combustion of Yolk Material Utilized for Energy and of
Alevin Tissue Elaborated During the Alevin Growth Phase
(from Hatch to Indicated Age Post-Hatch)

Treatment		Yolk catabolized		Tissue elaborated		Age
Na-PCP (ppb)	O$_2$ (mg/l)	Kcal/g	cal/alevin	Kcal/g	cal/alevin	days
0	10	8.6	56	5.4	169	30
0	5	9.1	67	5.2	155	30
0	3	8.5	66	5.4	155	35
40	10	7.4	96	5.2	126	35
40	5	6.7	82	5.2	77	41
40	3	7.3	50	5.2	22	24

controls having 5.2 to 5.4 Kcal/g as opposed to 5.2 Kcal/g for alevins reared in Na-PCP.

Effects of Continuous Exposure of Developing Steelhead Trout to Na-PCP

In a separate experiment conducted to determine the effect of exposure to Na-PCP from fertilization to the time of complete yolk utilization, eggs and alevins were continuously exposed to 10, 20 or 40 ppb Na-PCP, each in water with dissolved oxygen concentrations of 10, 5, and 3 mg/l (Table 4).

No mortality of eggs was noted until after the eggs were eyed and had been shocked by transferring them to another vessel via a siphon; this action caused sufficient disturbance that dead and nonviable eggs took up water and became opaque. Shocking, which is a routine fish culture operation, was carried out after the embryos were eyed, and hence past the fragile stage. After the dead eggs were removed, the remaining eggs were placed back in the respirometers.

Control embryos at all oxygen concentrations and embryos reared in Na-PCP (10, 20 and 40 ppb) at oxygen levels of 10 and 5 mg/l had about 90% survival at the eyed stage. Approximately one-third of the embryos in each group exposed to Na-PCP at 3 mg O$_2$/l were dead at the eyed stage. Mortality prior to hatching exceeded 50% only at 20 and 40 ppb Na-PCP at 3 mg O$_2$/l (Table 4). Hatching of control eggs was delayed by one day at 5 mg O$_2$/l and by three days at 3 mg O$_2$/l. Alevins exposed to Na-PCP were delayed in hatching by two days at 5 mg O$_2$/l and by five days at 3 mg O$_2$/l.

Control alevins were capable of bursts of swimming activity shortly after hatching, and when at rest, continuous rhythmic movements of mouth, opercles,

TABLE 4. Mortality (%) of Steelhead Trout Exposed to Na-PCP and
Low Dissolved Oxygen Concentrations

Na-PCP (ppb)	10 mg O$_2$/l		5 mg O$_2$/l		3 mg O$_2$/l	
	embryo[a]	total[b]	embryo	total	embryo	total
0	9.3	9.3	14.0	14.0	12.8	13.3
10	6.5	9.2	14.5	27.4	36.0	100.0
20	13.8	23.3	15.6	90.8	60.4	100.0
40	11.5	100.0	24.4	100.0	99.6	100.0

[a]Mortality of embryos.
[b]Total mortality occurring from fertilization through the end of the alevin stage.

and pectoral fins could be seen. The alevins which hatched at 40 ppb Na-PCP and 5 mg O$_2$/l and at 20 ppb Na-PCP and 3 mg O$_2$/l were moribund; many had only a heartbeat as a vital sign. One-hundred percent mortality of alevins eventually occurred at all concentrations of dissolved oxygen at 40 ppb of Na-PCP and at O$_2$ concentrations of 3 mg/l at 20 and 10 ppb of Na-PCP (Table 4). Mortality of 90% occurred at 20 ppb Na-PCP and 5 mg O$_2$/l. About 25% mortality was observed at 20 ppb Na-PCP and 10 mg O$_2$/l, and at 10 ppb Na-PCP and 5 mg O$_2$/l. Little or no alevin mortality was noted in any control groups or in 10 ppb Na-PCP and 10 mg O$_2$/l.

Dissolved oxygen concentrations appeared to have more marked affects than Na-PCP on the dry weights of newly hatched alevins. However, the alevins

TABLE 5. Maximum Dry Weights (mg) Attained by Alevins of
Steelhead Trout at Various Concentrations of
Dissolved Oxygen and Na-PCP

O$_2$ (mg/l)	Na-PCP		
	Control	10 ppb	20 ppb
10	41.5 (66)[a]	40.5 (66)	37.7 (71)
5	41.8 (76)	37.4 (86)	26.6 (92)
3	37.9 (81)	————	————

[a]Figures in parentheses indicate the number of days from fertilization
required to attain the maximum weight.

surviving a long-term exposure to 20 ppb Na-PCP at 5 mg O_2/l appear to achieve relatively lower maximum weights compared to corresponding controls (Table 5).

Discussion and Conclusions

Embryos and alevins reared in Na-PCP at low O_2 concentrations were more severely affected than those at high O_2 levels. This intensification by low O_2 levels of the effects of toxic materials has been reported before with ammonia, prussic acid, cyanide, and p-cresol (Wuhrmann, 1952). This phenomenon is usually ascribed to one of two causes. The first is that at low levels of dissolved O_2 the fish is required to pass more water over the gills in order to obtain a sufficient quantity of O_2, and as a result, absorbs more toxicant via the gills than at higher O_2 levels and lower ventilation volumes. A second explanation is that some chemicals (e.g., prussic acid) have a mechanism of action which produces lesions in gill tissue, and this interferes with normal gas exchange. At low O_2 levels, a partial loss of effective gill area would be more severe than at high O_2 levels.

The mechanism of action of PCP is generally regarded to be one of uncoupling oxidative phosphorylation. The response to partial uncoupling of oxidative phosphorylation or to a more rapid decomposition of ATP would seem to require an increase in O_2 consumption correlated with an increase in substrate oxidation (yolk catabolism) and electron transport. Thus, to obtain a given quantity of ATP, if oxidative phosphorylation were uncoupled to the extent that P:O ratios fell to 2 (normal 3), would require the consumption of 50% more O_2 and the catabolism of 50% more substrate. In fish exposed to PCP, the synergistic effect of low dissolved O_2 concentrations becomes obvious where increased O_2 consumption rates are required to maintain necessary quantities of ATP. Furthermore, the seemingly requisite increase in ventilation volume necessary for a fish to obtain greater amounts of O_2 would also produce the previously mentioned effect of increasing toxicant uptake.

The concept of increased ventilation rates causing the apparently synergistic effect of Na-PCP and low dissolved O_2 is difficult to extend to the case of embryos, since the O_2 uptake of embryos would seem to be passive and dependent on the O_2 concentration gradients between the water, perivitelline fluid and embryo. With the development of the embryonic vascular system and hemoglobin, compensations for living in an environment with low levels of O_2 might conceivably occur. Most compensations which would actually or effectively increase contact between the embryo and the perivitelline fluid would presumably increase the uptake of PCP. Regardless of the mechanism of synergism, there is no doubt that embryos as well as alevins are less able to cope with the stress of Na-PCP when the dissolved O_2 concentration is low.

While determinations of O_2 consumption, growth, and yolk utilization efficiency are obviously too general to produce definitive information as to the mechanism of action of PCP, the bioenergetic data obtained in this study are consistent with the concept that PCP disrupts energy metabolism. Steelhead trout alevins exposed to Na-PCP grew less rapidly than controls due to higher rates of yolk catabolism and resultant lower efficiencies of yolk utilization, while also exhibiting higher rates of O_2 consumption. Similarly, Webb and Brett (1973) have found that sublethal concentrations of Na-PCP cause reduced growth rate and food conversion efficiency in underyearling sockeye salmon *(Oncorhynchus nerka)*. Crandall and Goodnight (1962) found that exposure of mature guppies *(Lebistes reticulatus)* to Na-PCP resulted in higher oxygen consumption, and that long-term (90 days) exposure of newborn guppies to the toxicant resulted in decreased growth, increased mortality, and delayed sexual maturity.

References

Chapman, G., 1969. *Toxicity of Pentachlorophenol to Trout Alevins.* Ph.D. dissertation. Oregon State University, Corvallis, Oregon.

Crandall, C.A., and C.J. Goodnight, 1962. Effects of sublethal concentrations of several toxicants on growth of the common guppy, *Lebistes reticulatus. Limnol. Oceanogr.,* 7: 233-239.

Gray, J., 1926. The growth of fish. I. The relationship between embryo and yolk in *Salmo fario. J. Exp. Biol.,* 4: 215-225.

Gray, J., 1928. The growth of fish. II. The growth-rate of the embryo of *Salmo fario. J. Exp. Biol.,* 6: 110-124.

Hayes, F.R., 1949. The growth, general chemistry, and temperature relations of salmonid eggs. *Quart. Rev. Biol.,* 24: 281-308.

Smith, S., 1947. Studies in the development of the rainbow trout *(Salmo irideus).* I. The heat production and nitrogenous excretion. *J. Exp. Biol.,* 23: 357-378.

Smith, S., 1952. Studies in the development of the rainbow trout *(Salmo irideus).* II. The metabolism of carbohydrates and fats. *J. Exp. Biol.,* 28: 650-666.

Smith, S., 1958. Yolk utilization in fishes. In *Embryonic nutrition,* (D. Rudnick, Ed.), pp. 33-53, Chicago University Press, Chicago.

Webb, P.W., and J.R. Brett, 1973. Effects of sublethal concentrations of sodium pentachlorophenate on growth rate, food conversion efficiency, and swimming performance in underyearling sockeye salmon *(Oncorhynchus nerka). J. Fish. Res. Board Can.,* 30: 499-507.

Wuhrmann, K., 1952. Concerning some principles of the toxicology of fishes. Montreal. 11 p. Fisheries Research Board of Canada. Translation Series No. 243. Translated from Bulletin du Centre Belge d'Etude et de Documentation des Eaux. 15: 49.

Results of Two-Year Toxicity and Reproduction Studies on Pentachlorophenol in Rats

B.A. SCHWETZ, J.F. QUAST, P.A. KEELER, C.G. HUMISTON, and R.J. KOCIBA

Abstract—Male and female Sprague-Dawley rats were maintained on diets containing pentachlorophenol (PCP), characterized by a low content of nonphenolic impurities, for up to 24 months. Pentachlorophenol (PCP) was not found to be carcinogenic when administered to rats in their diet on a chronic basis at dose levels sufficiently high to cause mild signs of toxicity (1, 3, 10, or 30 mg/kg/day). Effects at the high dose level included decreased body weight gain (females), increased serum glutamic pyruvic transaminase activity (males and females), and increased urine specific gravity (females). An accumulation of pigment was observed in the livers and kidneys of females at 30 and 10 mg PCP/kg/day and males at 30 mg/kg/day. Ingestion of 3 mg PCP/kg/day or less by females and 10 mg/kg/day or less by males was not associated with significant toxicologic effects. To evaluate the effects on reproduction, rats were fed 3 or 30 mg PCP/kg/day for 62 days prior to mating, during 15 days of mating, and subsequently throughout gestation and lactation. A reduction in the mean body weight was observed among the adult rats at the highest dose level. Except for a significant decrease in neonatal survival and growth among litters of females ingesting 30 mg PCP/kg/day, measures of reproductive capacity were unaffected at both dose levels of pentachlorophenol. Ingestion of 3 mg PCP/kg/day had no effect on reproduction or neonatal growth, survival, or development.

B.A. SCHWETZ, J.F. QUAST, P.A. KEELER, C.G. HUMISTON and R.J. KOCIBA
 • Toxicology Research Laboratory, Health and Environmental Research, Dow Chemical U.S.A., Midland, Michigan 48640, U.S.A.

Introduction

Pentachlorophenol is a registered antimicrobial agent used widely for the preservation of wood. The acute oral LD_{50} of the pentachlorophenol (PCP) used in these studies was approximately 150 mg/kg in rats (unpublished data; Dow Chemical Company, Midland, Michigan). In subchronic toxicity studies conducted on this same pentachlorophenol, no gross pathological or histopathological changes were noted in rats maintained for 90 days on diets providing 3, 10, or 30 mg PCP/kg/day. Increased kidney weights were observed at 30 mg PCP/kg/day and increased liver weights were observed at 10 and 30 mg PCP/kg/day (Johnson *et al.*, 1973). PCP has been found to be toxic to the developing embryo and fetus of rats but was not teratogenic (Schwetz *et al.*, 1974). The purpose of the studies reported herein was to determine the potential of PCP to interfere with the reproductive capacity of rats as well as to evaluate the toxic effects, including carcinogenicity, that might be associated with the chronic ingestion of pentachlorophenol.

Experimental Procedures

Pentachlorophenol was mixed in the diet of the animals and was provided free choice at all times. To evaluate the effects on reproduction, Sprague-Dawley rats (Spartan substrain, Spartan Research Animals, Haslett, Michigan) were fed 0, 3, or 30 mg PCP/kg/day. Each group consisted of 10 males and 20 females. The rats were fed PCP for 62 days prior to mating as well as during 15 days of mating. The males were then separated from the females during gestation and lactation but all rats continued on the test diets. At the time of weaning, all adult and weanling rats were necropsied. The skeletons of the weanlings were prepared and stained for examination.

In the chronic toxicity study, rats were fed 0, 1, 3, 10, and 30 mg PCP/kg/day. Female rats were maintained on the test diets for 24 months but the male rats were terminated after 22 months because of high mortality among the control and the experimental rats. At each dose level, 25 rats of each sex were utilized. An additional two rats of each sex were included at each dose level to provide tissues for chemical analysis. Hematology and urinalysis data were collected at the end of 12 months. Observations at the termination of the study included hematology, clinical chemistry, urinalysis as well as measurement of organ weights, and a gross and microscopic examination of tissues. Observations during the study included measurements of body weights and food consumption at monthly intervals as well as observations for overt alterations in behavior.

The results of chemical analysis of the sample of PCP used in these studies are given in Table 1. This sample was representative of commercially available pentachlorophenol identified as DOWICIDE⁽ᵐ⁾ EC-7 antimicrobial (Dow Chemical Company). This is a purified grade of pentachlorophenol characterized by a lower content of nonphenolic contaminants than is found in technical grade pentachlorophenol. The test diets were prepared by mixing PCP dissolved in anisole with ground Purina laboratory chow to make a 1% premix. The test diets were then

**TABLE 1. Analytical Description of the
Pentachlorophenol (PCP) Sample
Used in these Studies**

Phenolics	Percent[a]		
Pentachlorophenol	90.4	±	1.0
Tetrachlorophenol	10.4	±	0.2
Trichlorophenol	<0.1[c]		
Nonphenolics	ppm[b]		
Dibenzo-*p*-dioxins			
2,3,7,8-tetrachloro-	<0.05[c]		
hexachloro-	1.0	±	0.1
heptachloro-	6.5	±	1.0
octachloro-	15.0	±	3.0
Dibenzofurans			
Hexachloro-	3.4	±	0.4
heptachloro-	1.8	±	0.3
octachloro-	<1[c]		
Hexachlorobenzene	400	±	40

[a]Average of 4 analyses by gas-liquid chromatography.

[b]Average of 4 analyses by gas chromatography-mass
spectrometry.

[c]Concentration in the sample was less than the
indicated level of sensitivity for the analytical
method.

prepared weekly from the 1% premix. The concentrations of PCP in the diet were adjusted on a monthly basis to maintain the designated dose levels on a mg/kg body weight/day basis according to the food consumption and body weights of the rats. Groups of control rats of each sex received the laboratory chow to which anisole without PCP was added.

In both studies, statistical evaluation of all continuous data consisted of an analysis of variance together with Dunnett's test for comparison of means of experimental and control groups (Steel and Torrie, 1960). Incidence data were analyzed by the Fisher exact probability test (Siegel, 1956). P values less than 0.05 were considered significant.

304   B.A. Schwetz *et al.*

TABLE 2. Body Weights of Adults and Survival of Neonates of Rats Fed
Pentachlorophenol (PCP) in a Reproduction Study

	Dose level (mg PCP/kg/day)		
	0	3	30
	Body weight, g, Mean ± S.D.		
MALES			
Number of rats	10	10	10
Day 0	220 ± 7	225 ± 11	213 ± 8
29	405 ± 28	394 ± 32	374 ± 26
62	501 ± 46	480 ± 44	463 ± 36
169	619 ± 62	590 ± 67	570 ± 42
FEMALES			
Number of pregnant rats	16	17	19
Day 0	187 ± 9	184 ± 11	187 ± 10
29	259 ± 11	259 ± 15	256 ± 16
62	298 ± 13	293 ± 18	287 ± 16
21 postpartum (~ 110 days)	344 ± 21	334 ± 21	308 ± 25[a]
	Reproduction indices (percent)		
Pregnancy	80 (16/20)	85 (17/20)	95 (19/20)
Liveborn pups	97 (174/180)	96 (195/204)	91 (188/206)[b]
24-Hour survival	98 (171/174)	98 (191/195)	96 (180/188)
7-Day survival	96 (167/174)	95 (186/195)	81 (153/188)[b]
14-Day survival	95 (166/174)	94 (183/195)	79 (148/188)[b]
21-Day survival	95 (165/174)	94 (183/195)	76 (142/188)[b]

[a]Significantly different from control by Dunnett's test, $p < 0.05$.
[b]Significantly different from control by the Fisher Exact Probability test, $p < 0.05$.

Effects of PCP on Reproduction in Rats

Toward the end of the study, the mean body weight of the female rats at the high dose level (30 mg/kg/day) was significantly less than that of control rats (Table 2). Daily observation of the male and female adult rats and their litters revealed no treatment-related effect on demeanor or physical appearance. Ingestion of 3 mg PCP/kg/day had no effect on the percent of pregnancy among females or on

**TABLE 3. Litter Size and Weight of Rats Fed
Pentachlorophenol (PCP) in a Reproduction Study**

	Dose level (mg PCP/kg/day)		
	0	3	30
	Litter Size, Mean ± S.D.		
Birth	11.6 ± 3.0	11.5 ± 3.2	9.9 ± 2.1
Day 1	11.4 ± 3.1	11.2 ± 3.3	9.5 ± 1.8
Day 7	11.1 ± 2.9	10.9 ± 3.4	8.1 ± 2.9[a]
Day 14	11.1 ± 2.8	10.8 ± 3.3	7.8 ± 2.8[a]
Day 21	11.0 ± 2.8	10.8 ± 3.3	7.7 ± 2.7[a]
	Neonatal body weight, g, Mean ± S.D.		
Day 1	7.1 ± 0.6	7.0 ± 0.4	6.1 ± 0.7[a]
Day 7	14.1 ± 2.3	13.8 ± 1.4	11.2 ± 2.0[a]
Day 14	26.3 ± 4.1	25.5 ± 3.4	19.9 ± 4.1[a]
Day 21, male	41 ± 9	38 ± 7	30 ± 5[a]
Day 21, female	40 ± 8	38 ± 7	30 ± 5[a]

[a]Significantly different from control by Dunnett's test, $p < 0.05$.

the survival indices of neonates from the time of birth to the end of lactation, 21 days of age. Ingestion of 30 mg PCP/kg/day had no effect on the percent of pregnancy but did cause a significant decrease in the percent of pups that were born alive as well as significantly decreased survival to days 7, 14, and 21 of lactation. Every male in each treatment group was fertile, that is, at least one of the females with each male became pregnant.

The average litter size throughout lactation was not affected by the ingestion of 3 mg PCP/kg/day in the diet (Table 3). The average litter size among rats receiving 30 mg/kg/day was significantly lower than among controls on days 7, 14, and 21 of lactation. The mean neonatal body weight was significantly less than the control value at the high dose level, 30 mg/kg/day, but not at the low dose level, 3 mg/kg/day.

Necropsy of the weanling rats at the time of sacrifice revealed no changes that were visible externally or upon visceral examination that were related to ingestion of PCP. There was a significantly increased number of litters at the high dose level of PCP which showed variations in the development of skeletal structures — lumbar spurs and vertebra with unfused centra. These skeletal variants probably reflect perinatal toxicity rather than an effect on the neonate alone.

As in the previous study on rat embryos (Schwetz *et al.*, 1974) the results of this reproduction study indicate that developing or neonatal rats are more sensitive to the toxicologic effects of PCP than are adult rats. This is further substantiated by an age-related difference in the LD_{50} of PCP in rats — being about 150 mg/kg in adult rats and about 65 mg/kg in 3-4 day old rats (unpublished data, Dow Chemical Company, Midland, Michigan).

The Two-Year Toxicity Study of PCP in Rats

Ingestion of PCP did not affect the demeanor of animals, the mean food consumption or their survival. Hematology measures including the hemoglobin levels, packed cell volume, total erythrocyte counts, and total and differential leukocyte counts were not affected by ingestion of PCP. The mean body weight of the females at the high dose level was significantly less than that of control rats throughout much of the study. The mean of the monthly body weights of the rats during the last twelve months of the study was 428 grams for the female control rats compared to 378 grams for the females at the high dose level, 30 mg/kg/day. There was a significant increase in the serum glutamic pyruvic transaminase activity among the male and female rats at the high dose level at the termination of the study (males: control - 38 ± 16, high dose - 63 ± 13; females: control - 36 ± 8, high dose - 48 ± 15). There was an increase in the specific gravity of urine among the female rats at the high dose level after one year but not at the termination of the study. Among the rats killed at the termination of the study, the weights of the brain, heart, liver, kidneys, and testes of the rats in the experimental groups were comparable to those of the controls.

The incidence of non-tumorous pathologic changes in rats fed PCP are

TABLE 4. Non-tumorous Pathologic Changes in Rats Fed Pentachlorophenol (PCP) for 22 Months (Males) and 24 Months (Females)

	Males					Females				
Dose: mg PCP/kg/day	0	1	3	10	30	0	1	3	10	30
Number of rats examined	27	26	27	27	27	27	27	27	27	27
GROSS OBSERVATIONS				*Number of rats affected*						
Liver — dark, discolored	0	0	0	0	0	0	0	0	2	14
Kidney — dark, discolored	0	0	0	0	0	0	1	0	3	12
MICROSCOPIC OBSERVATIONS										
Liver — brown granular pigment within hepatocytes and reticulo-endothelial cells surrounding the central vein	0	0	0	0	1	0	0	0	8	16
Kidney — brown granular pigment within the tubular epithelial cells	0	0	0	0	0	0	0	0	7	19

summarized in Table 4. An accumulation of pigment was observed in the liver and kidneys of the experimental rats. This was grossly recognizable among the female but not male rats as a dark coloration of these organs. Microscopic examination of hematoxylin and eosin stained sections revealed the presence of brown granular pigmented material in these organs - in the liver, primarily in the hepatocytes surrounding the central veins; in the kidneys, primarily in epithelial cells of the proximal convoluted tubules. The material did not stain with any of the following stains: oil red-O, nile blue sulfate, acid fast, Gomori's stain for iron, McManus' Periodic Acid Schiff, Hall's method for bilirubin and Stein's method for bile pigments. Approximately half of the females at the high dose level and a few at 10 mg/kg/day showed the dark colored liver and kidney. This was not observed macroscopically in any of the male rats. Microscopic examination of sections of the liver and kidney revealed the brown pigment in a larger number of the female rats at 10 and 30 mg/kg and in the liver of a single male rat at the high dose level. No other non-tumorous pathologic changes were observed which were related to ingestion of pentachlorophenol.

The results of examinations of tumorous formations in the rats which died spontaneously or were necropsied in a moribund state or at the terminal necropsy are summarized in Table 5. As expected in this strain of rat, the highest incidence of tumors among male rats involved endocrine organs such as pituitary, adrenal, and thyroid glands, testes and pancreas. The frequently occurring tumors among female rats were located in the mammary gland region. The pituitary, thyroid glands, and uterus were the other organs with the most frequently occurring tumorous changes.

TABLE 5. Incidence of Primary Tumors (Based on Histopathological Diagnosis) in Rats Fed Pentachlorophenol (PCP) for 22 Months (Males) and 24 Months (Females)

Dose: mg PCP/kg/day	Males					Females				
	0	1	3	10	30	0	1	3	10	30
Number of rats examined	27	26	27	27	27	27	27	27	27	27
Number of rats with tumors	11	13	13	12	11	27	26	25	25	25
Number of tumors	17	14	17	15	16	62	67	42	63	63
Number of tumors/rats with tumors	1.6	1.1	1.3	1.3	1.4	2.3	2.6	1.7	2.5	2.5
Number of morphologic malignant tumors	1	3	2	1	0	2	7	2	3	2

The highest incidence of tumors among the experimental groups was observed at the lowest dose level of PCP (1 mg/kg/day); however, no tumors were observed in any group of experimental male or female rats at an incidence significantly different from that in the control rats. Rats ingesting diets containing PCP were not submitted for necropsy examination earlier than control rats.

The findings in this chronic toxicity study are consistent with those reported by other workers. Goldstein *et al.*, (1976) have reported on the effects associated with the ingestion of either pure or technical grade PCP in the diet for 8 months by female Sherman rats. At the time of sacrifice, the livers of some of the rats given PCP in the diet were very dark in color. At least some but not all of the pigment appeared to be due to porphyrins since some of the dark livers contained levels of porphyrins no higher than that found in livers of some control rats. Goldstein *et al.*, (1976) concluded that the presence of chlorodioxins in technical grade pentachlorophenol was almost totally responsible for the hepatic effects of PCP. Kimbrough and Linder (1975) observed a prominent brown pigment in the macrophages and Kupffer cells in the livers of rats fed technical grade PCP for three months. The livers of rats fed technical PCP showed more ultrastructural changes than the livers of rats fed pure pentachlorophenol. Previous studies in our own laboratory have also associated toxicological effects in animals with the nonphenolic contaminants in pentachlorophenol (Johnson *et al.*, 1973).

Conclusions

In summary, the following conclusions have been drawn from these studies. In the single generation reproduction study in which PCP was given in the diet at dose levels of 3 and 30 mg/kg/day, there was no significant effect on the fertility of rats but there was a significant decrease in neonatal survival and growth at the high dose level, 30 mg/kg/day. The no-adverse effect dose level of PCP in the reproduction study was 3 mg/kg/day. In the chronic toxicity study in which PCP was given in the diet at dose levels ranging from 1-30 mg/kg/day, there was no carcinogenicity when rats were given dose levels of PCP which were significantly high to cause signs of mild toxicity. The no-adverse-effect dose level of pentachlorophenol in the chronic toxicity study was 10 mg/kg/day among the males and 3 mg/kg/day among the females.

References

Goldstein, J.A., R.E. Linder, P. Hickman, and H. Bergman, (1976). Effects of pentachlorophenol on hepatic drug metabolism and porphyria related to contamination with chlorinated dibenzo-*p*-dioxins. Fifteenth annual meeting of the Society of Toxicology, Atlanta, Georgia, March 14-18, 1976.

Johnson, R.L., P.J. Gehring, R.J. Kociba, and B.A. Schwetz, (1973). Chlorinated dibenzodioxins and pentachlorophenol. *Environ. Hlth. Persp.*, **5**: 171-175.

Kimbrough, R.D., and R.E. Linder, (1975). The effect of technical and 99% pure pentachlorophenol on the rat liver. Light microscopy and ultrastructure. *Toxicol. Appl. Pharmacol.*, **33**: 131-132.

Schwetz, B.A., P.A. Keeler, and P.J. Gehring, (1974). The effect of purified and commercial grade pentachlorophenol on rat embryonal and fetal development. *Toxicol. Appl. Pharmacol.*, **28**: 151-161.

Siegel, S. (1956). *Non-parametric Statistics for the Behavioral Sciences.* McGraw-Hill Book Company, Inc., New York.

Steel, R.G.D., and H.H. Torrie, (1960). *Principles and Procedures of Statistics,* McGraw-Hill Book Company, Inc., New York, New York.

III

Chlorophenols and Their Contaminants

Chemistry, Toxicology, Environmental Residues and Environmental Impact

Impurities in Commercial Products Related to Pentachlorophenol

CARL-AXEL NILSSON, ÅKE NORSTRÖM, KURT ANDERSSON
and CHRISTOFFER RAPPE

Abstract—When discussing the effects of pentachlorophenol (PCP) on the environment one should have in mind that PCP itself is just a part of the problem. Most commercial chlorophenol formulations contain other chlorophenols, in PCP mainly tetrachlorophenol. In one PCP formulation sold in Sweden the PCP can be regarded as an impurity as the main constituent is tetrachlorophenol (80%). In addition to other chlorophenols, technical chlorophenols have been shown to contain a variety of dimeric impurities. Although present only in minor amounts most of the interest has been devoted to the chlorinated dioxins due to their exceptionally high toxicity. However, chlorinated phenoxyphenols, which can act as precursors of dioxins, are often present in 1-5%. Other dimeric impurities found are chlorinated diphenyl ethers, dibenzofurans and dihydroxybiphenyls. Considerable effort has been made in a number of laboratories to prepare some of these impurities for use as analytical standards and for evaluation of their biological effects. Different methods of synthesizing chlorinated dioxins, dibenzofurans, diphenyl ethers and phenoxyphenols are reviewed.

This paper is based on the Ph.D. thesis, 'Studies of chlorinated 2-phenoxyphenol precursors of chlorinated dibenzo-*p*-dioxins', by Carl-Axel Nilsson; and on the Ph.D. thesis, 'Studies of potential environmental hazardous halogenated aromatics', by Åke Norström, Umeå University, 1977.

CARL-AXEL NILSSON • National Board of Occupational Safety and Health, Department of Occupational Health in Umeå, Chemical Division, Umeå Hospital, S-901 85 Umeå, Sweden.
ÅKE NORSTRÖM, KURT ANDERSSON and CHRISTOFFER RAPPE • Department of organic chemistry, University of Umeå, S-901 87 Umeå, Sweden.

313

Introduction

Chlorinated phenols are manufactured in large amounts and have widely varying applications. They are used as insecticides, fungicides, mould inhibitors, antiseptics and disinfectants, their effectiveness generally increasing with the degree of chlorination. Other fields of application are as starting materials for certain pesticides and also as dyes and pigments (Doedens, 1964). The annual world production of chlorinated phenols and their derivatives is hard to estimate but a figure in the region of 150,000 tons seems reasonable.

The most widely used methods by which chlorinated phenols are prepared are direct chlorination and alkaline hydrolysis of the appropriate chlorobenzene, the particular method used depending on the isomer desired (Doedens, 1964; Melnikov, 1971).

The isomers manufactured via direct chlorination include 2,4-dichlorophenol, 2,4,6-trichlorophenol, 2,3,4,6-tetrachlorophenol and pentachlorophenol.

Isomers manufactured by direct chlorination

| 2,4 Dichlorophenol | 2,4,6 Trichloro-phenol | 2,3,4,6 Tetra-chlorophenol | Pentachloro-phenol |

2,4-dichlorophenol is most often used as the starting material in the production of 2,4-D.

The most important use of 2,4,6-trichlorophenol, 2,3,4,6-tetrachlorophenol and pentachlorophenol is as wood preservatives, both as phenols and the corresponding sodium or potassium salts. In addition, they are used as bactericides, pentachlorophenol also being used in slime control in pulp and paper mills.

The chlorophenols manufactured via hydrolysis of chlorobenzenes include 2,4,5-trichlorophenol and pentachlorophenol.

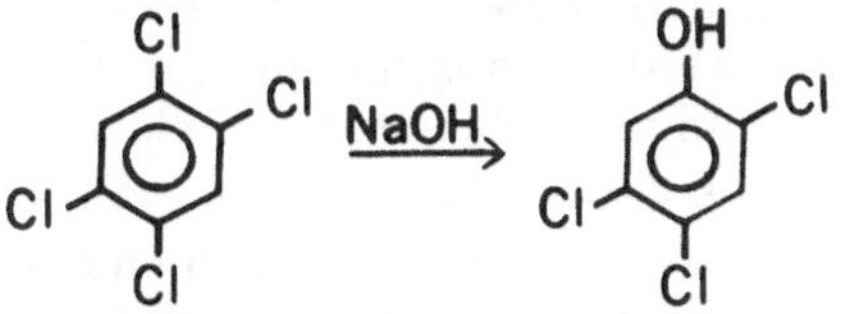

Isomers manufactured by
hydrolysis of chlorobenzenes:

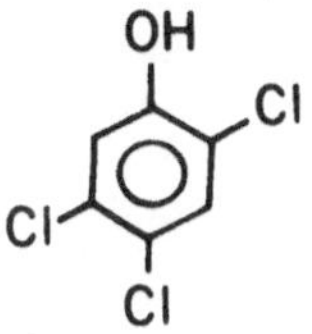

2,4,5,- trichlorophenol

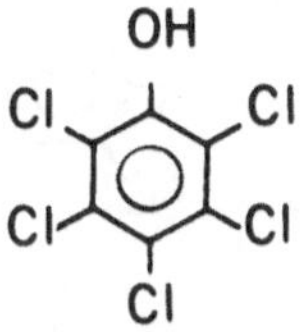

pentachloro-
phenol

The hydrolysis of chlorobenzenes can give rise to the unwanted formation of chlorinated dibenzo-*p*-dioxins as by-products (Milnes, 1971). This side-reaction is favoured by high temperatures.

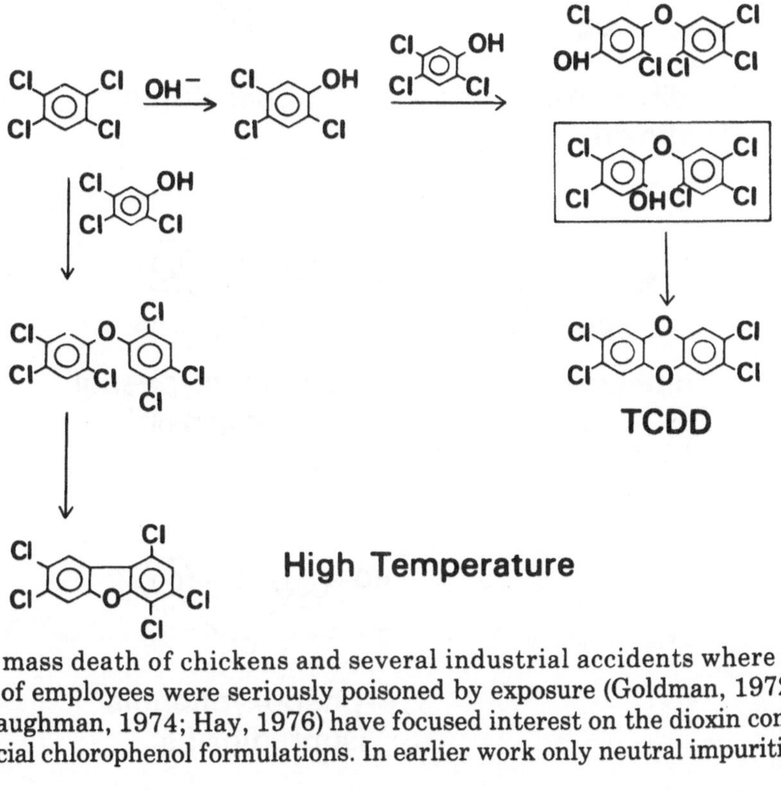

TCDD

High Temperature

The mass death of chickens and several industrial accidents where a large number of employees were seriously poisoned by exposure (Goldman, 1972; May, 1973; Baughman, 1974; Hay, 1976) have focused interest on the dioxin content of commercial chlorophenol formulations. In earlier work only neutral impurities such as the chlorinated dioxins, dibenzofurans and diphenyl ethers were taken into account. However, since the presence of chlorinated phenoxyphenols in the chlorophenols and their ability to undergo thermal and photochemical ring closure to dioxins were demonstrated (Jensen and Renberg, 1972; Rappe and Nilsson, 1972; Nilsson and Renberg, 1974; Nilsson *et al.*, 1974; Levin *et al.*, 1976), most chemists have taken the phenoxyphenols into account and have modified their analytical procedures to eliminate possible interference by phenoxyphenols in dioxin quantifications (Baughman, 1974; Crummet and Stehl, 1973; Buser and Bosshardt, 1976; Blaser *et al.*, 1976; Pfeiffer, 1976).

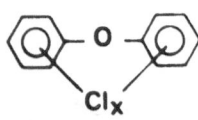

dibenzo-p-dioxin

dibenzofuran

diphenyl ether

2-phenoxyphenol "predioxin"

4-phenoxyphenols "isopredioxins"

Analysis of Impurities in Technical Chlorophenol Formulations

The analysis of chlorinated dioxins in technical chlorophenol formulations for the most part involves some extraction procedure for separating the neutral impurities (such as chlorinated dioxins, diphenyl ethers and dibenzofurans) from the acidic components. However, this extraction, as the only clean-up prior to the dioxin analysis, creates complications. As pointed out by several workers, (Crummet and Stehl, 1973; Plimmer, 1973; Pfeiffer, 1976), polychlorinated phenoxyphenols are often incompletely extracted and this can cause some misinterpretations in the quantification of dioxins. We demonstrated that thermal ring closure of nonachloro-2-phenoxyphenol to octachlorodioxin occurring in the injection port of the gas chromatograph can give rise to a wrong quantification of octachlorodioxin.

$$\text{(nonachloro-2-phenoxyphenol)} \xrightarrow{275^{\circ}} \text{(octachlorodioxin)}$$

As a method of eliminating that risk we suggested that methylation with diazomethane would provide a simple and satisfactory derivatization, and we found that the ring closure was thereby prevented.

Other methods of separating residual phenoxyphenols in the neutral fraction involve some type of column chromatography. Buser and Bosshardt (1974, 1976) performed the clean-up by alkaline extraction and subsequent fractionation of the neutral components on a basic alumina micro column which completely removed residual phenoxyphenols in the extract. Several workers have used an ion-exchange resin column in the analysis of neutral impurities (Crummet and Stehl, 1973; Blaser et al., 1976; Pfeiffer, 1976). This can be done without previous extraction.

In the determination of neutral as well as acidic components related to chlorophenols we used thin-layer chromatography on silica gel for fractionation of the samples (Nilsson and Renberg, 1974; Nilsson et al., 1974; Levin et al., 1976).

The identification and quantification of the various components are usually carried out by GC-EC or GC-MS. The most specific and sensitive tool for the identification of the various impurities is the GC-MS system. In the analysis of chlorophenol formulations, the quantities are generally adequate to allow a complete spectrum to be run, but if very low concentrations are present the mass fragmentography technique affords very sensitive and more specific detection than does GC-EC analysis (Buser and Bosshardt, 1974). The capillary column technique has recently been used in the dioxin analysis to increase further the specificity of the mass fragmentography technique (Buser, 1975, 1976). However, if enough substance is present, useful information about the isomeric structure can be obtained from a complete mass spectrum.

The amount of various impurities present varies considerably between different formulations but typical figures for formulations used as fungicides for wood treatment are:

Chlorinated dioxins	< 1 —	2000 ppm
Chlorinated diphenylethers	100 —	1000 ppm
Chlorinated dibenzofurans	50 —	200 ppm
Chlorinated phenoxyphenols	~ 1%	

Analytical method

Basic extraction

↓

TLC   GC/MS

Acidic extraction

↓

TLC   GC/MS

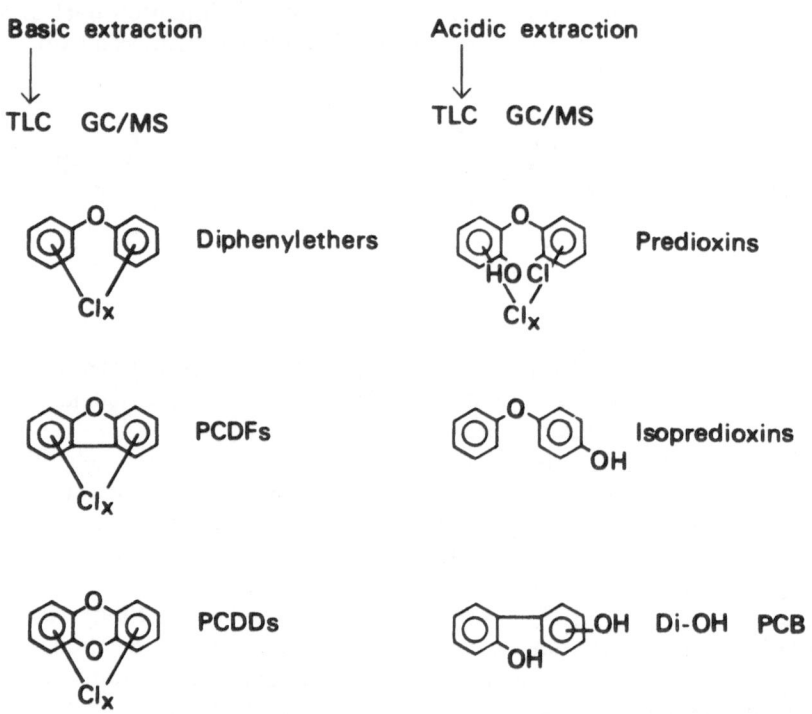

Diphenylethers

Predioxins

PCDFs

Isopredioxins

PCDDs

Di-OH   PCB

Formation of Chlorinated Dioxins and Dibenzofurans

Pyrolysis

The pyrolysis of chlorinated phenoxyphenols has been shown to yield chlorinated dioxins. This reaction has also been utilized in the synthesis of some dioxin isomers. Furthermore, this thermal ring closure is of interest since large amounts of sawdust and other sawmill waste contaminated with chlorophenols are burned in large combustion facilities. If the combustion is not controlled to ensure high enough temperatures and excess air, the possibility of dioxin formation cannot be excluded.

Photochemistry

In one paper (Nilsson *et al.,* 1974) we reported the photochemical ring closure of 4,5,6-trichloro-2-(2,4-dichlorophenoxy)phenol to a tetrachlorodioxin in a yield of 0.4%. The main pathway, however, was dechlorination.

main reaction
dechlorination

In two recent papers we have demonstrated the photochemical formation of chlorinated dibenzofurans from chlorinated diphenyl ethers (Norström *et al.,* 1976a, 1977). In one case this reaction has synthetic value since 2,8-dichlorodibenzofuran could be isolated from the photolysis of 2,4,4′-trichlorodiphenyl ether.

Further dechlorination

(from Norström *et al.,* 1976a)

Synthesis of Chlorinated Dioxins

The synthetic routes most frequently used are summarized below. Route 1 is preferred in the synthesis of 2,3,7,8-tetrachlorodioxin, since under controlled conditions this is the main product. This route has also been used in the synthesis of ^{37}Cl-labeled TCDD (Baughman, 1974). Route 2 has been used by Poland and Yang (1972) with high yields, and the cleanup of the reaction mixture was minimized to recrystallizations. Route 3 has been used by Buser (1975) to prepare a number of dioxin isomers by micro scale pyrolysis. In Route 4 the use of chlorinated 2-phenoxyphenols as starting material in the synthesis of dioxins offers a large number of isomers since the number of phenoxyphenols now available is large (Nilsson and Andersson, 1977).

Methods for the synthesis of chlorinated dioxins

1) Chlorination of dioxins

2,3,7,8,- TCDD

2) From chlorocatechols and chloronitrobenzenes

1,2,3,4-TCDD

3) Pyrolysis of chlorophenate

1,3,6,8-TCDD

4) From chlorinated 2-phenoxyphenols

1,2,3,8-TCDD

Synthesis of Chlorinated Dibenzofurans

The routes to chlorinated dibenzofurans are listed below. The most useful route is the palladium (II) acetate promoted cyclisation (Norström *et al.*, 1976b) of chlorinated diphenyl ethers developed in our laboratory. This is a very convenient route and a large number of chlorinated diphenyl ether isomers suitable as starting material are available.

Methods for the synthesis of chlorinated dibenzofurans

1) Chlorination of dibenzofuran

~2:1

2) Diazotisation of o-NH₂-chlorodiphenyl ether

3) From o,o'-diphenyldiols

4) From polychlorinated diphenylethers

Synthesis of Chlorinated Phenoxyphenols

The routes 1 and 3 proceed in high yields and offer convenient ways to most isomers. The diphenyliodonium salt coupling with chlorinated guaiacols has recently become even more useful, since some higher chlorinated diphenyliodonium salts not previously described have been prepared in our laboratory.

Principal routes for the synthesis of chlorinated phenoxyphenols

1) Chlorination of chlorinated phenoxyphenols

2) From nitrochlorobenzene and chlorophenol or chloroguaiacol

3) Diphenyliodonium salt coupling with chlorinated guaiacol

4) Ullman biaryl ether synthesis

$$CH_3O\text{-}\bigcirc\text{-}Br \ + \ \overset{\ominus}{O}\text{-}\bigcirc\text{-}Cl \ \xrightarrow[Cu]{\Delta} \ CH_3O\text{-}\bigcirc\text{-}O\text{-}\bigcirc\text{-}Cl$$

$$\xrightarrow{BBr_3} \ HO\text{-}\bigcirc\text{-}O\text{-}\bigcirc\text{-}Cl$$

Synthesis of Chlorinated Diphenyl Ethers

The most convenient route to chlorinated diphenyl ethers is the diphenyliodonium salt coupling with suitable chlorophenols. This route has been used by us to prepare a large number of isomers with 1-8 chlorines (Norström and Andersson, 1977; Nilsson *et al.*, 1977). Other routes used are direct chlorination and the Ullman biaryl ether coupling reaction.

References

Baughman, R.W., 1974. Tetrachlorodibenzo-*p*-dioxins in the Environment: High Resolution Mass Spectrometry at the Picogram Level, Thesis. Harvard University, Cambridge, Massachusetts.

Blaser, W.W., R.A. Bredgeweg, L.A. Shadoff, and R.H. Stehl, 1976. Determination of chlorinated dibenzo-*p*-dioxins in pentachlorophenol by gas chromatography-mass spectrometry. *Anal. Chem.*, **48**: 984.

Buser, H.-R., 1975. Polychlorinated dibenzo-*p*-dioxins, separation and identification of isomers by gas chromatography-mass spectrometry. *J. Chromatogr.*, **114**: 95.

Buser, H.-R., 1976. Analysis of TCDD's by gas chromatography-mass spectrometry using glass capillary columns. *Workshop on TCDD*, Milan, Italy, Oct. 23-24.

Buser, H.-R., and H.-P. Bosshardt, 1974. Determination of 2,3,7,8-tetrachlorodibenzo-1,4-dioxin at parts per billion levels in technical-grade 2,4,5-trichlorophenoxyacetic acid, in 2,4,5-T alkyl ester and 2,4,5-T amine salt herbicide formulations by quadrupole mass fragmentography. *J. Chromatogr.*, **90**: 71.

Buser, H.-R., and H.-P. Bosshardt, 1976. Determination of polychlorinated dibenzo-*p*-dioxins and dibenzofurans in commercial pentachlorophenols by combined gas chromatography-mass spectrometry. *J. Ass. Offic. Anal. Chem.*, **59**: 562.

Crummet, W.B., and R.H. Stehl, 1973. Determination of chlorinated dibenzo-*p*-dioxins and dibenzofurans in various materials. *Environ. Health Persp.*, **5**: 15.

Doedens, J.D., 1964. *Chlorophenols, Kirk-Othmer Encycl. Chem. Technol.*, 2nd Ed. **5**: 325.

Goldman, P.J., 1972. Schwerste akute Chlorakne durch Trichlorophenol-Zersetzungsprodukte. *Arbeitsmed. Sozialmed. Arbeitshyg.* **7**: 12.

Hay, A., 1976. Toxic cloud over Seveso. *Nature,* **262**: 636.

Jensen, S., and L. Renberg, 1972. Contaminants in pentachlorophenol: chlorinated dioxins and predioxins. *Ambio,* **1**: 62.

Levin, J.-O., C. Rappe, and C.-A., Nilsson, 1976. Use of chlorophenols as fungicides in sawmills. *Scandinavian J. Work Environ. Health,* **2**: 71-81.

May, G., 1973. Chloracne from the accidental production of tetrachlorodibenzodioxin. *British J. Ind. Medicine,* **30:** 276.

Melnikov, N.N., 1971. *Chemistry of Pesticides,* Springer-Verlag, New York.

Milnes, M.H., 1971. Formation of 2,3,7,8-tetrachlorodioxin by thermal decomposition of sodium 2,4,5-trichlorophenate. *Nature,* **232:** 395.

Nilsson, C.-A., and K. Andersson, 1977. Synthesis of chlorinated 2-phenoxyphenols. *Chemosphere,* **6:** 249.

Nilsson, C.-A., K. Andersson, C. Rappe, and S.-O., Westermark, 1974. Chromatographic evidence for the formation of chlorodioxins from chloro-2-phenoxyphenols. *J. Chromatogr.,* **96:** 137-147.

Nilsson, C.-A., Å. Norström, M. Hansson, and K. Andersson, 1977. The synthesis of halogenated diphenyliodonium salts and their coupling products with halogenated phenols. *Chemosphere,* **6:**599.

Nilsson, C.-A., and L. Renberg, 1974. Further studies on impurities in chlorophenols. *J. Chromatogr.,* **89:** 325-333.

Norström, Å., and K. Andersson, 1977. Preparation of chlorinated diphenyl ethers from biaryliodonium salts. *Chemosphere,* **6:** 237.

Norström, Å., K. Andersson and C. Rappe, 1976a. Formation of chlorodibenzofurans by irradiation of chlorinated diphenyl ethers. *Chemosphere,* **5:** 21.

Norström, Å., K. Andersson, and C. Rappe, 1976b. Palladium (II) acetate promoted cyclization of polychlorinated diphenyl ethers to the corresponding dibenzofurans. *Chemosphere,* **5:** 419.

Norström, Å., and K. Andersson, and C. Rappe, 1977. Studies on the formation of chlorodibenzofurans by irradiation or pyrolysis of chlorinated diphenyl ethers. *Chemosphere,* **6:** 241.

Pfeiffer, C.D., 1976. Determination of chlorinated dibenzo-*p*-dioxins in pentachlorophenol by liquid chromatography. *J. Chromatogr. Sci.,* **14:** 386.

Plimmer, J.R., 1973. Technical pentachlorophenol: origin and analysis of base-insoluble contaminants. *Environ. Health Persp.,* **5:** 41.

Poland, A.E., and G.C. Yang, 1972. Preparation and characterization of chlorinated dibenzo-*p*-dioxins. *J. Agr. Food Chem.,* **20:** 1093.

Rappe, C., and C.-A. Nilsson, 1972. An artifact in the gas chromatographic determination of impurities in pentachlorophenol. *J. Chromatogr.,* **67:** 247-253.

Villanueva, E.C., R.W. Jennings, V.W. Burse, and R.D. Kimbrough, 1975. A comparison of analytical methods for chlorodibenzo-*p*-dioxins in pentachlorophenol. *J. Agr. Food Chem.,* **23:** 1089.

Genetic Activity of Chlorophenols and Chlorophenol Impurities

RUDOLF FAHRIG, CARL-AXEL NILSSON and CHRISTOFFER RAPPE

Abstract—Studies were made of the genetic activity of 2,4,6-trichlorophenol, pentachlorophenol, and their impurities:

2,3,7,8-tetrachlorodibenzofuran
2,4,4',6-tetrachlorodiphenyl ether
2,4,4',5-tetrachlorodiphenyl ether
4,5,6-trichloro-2-(2,4-dichlorophenoxy)phenol, (Cl_5-predioxin)
4-chloro-2-(2,4-dichlorophenoxy)phenol, (Cl_3-predioxin)

The investigation was made in two steps. In the first step, all substances were tested *in vitro* with the multipurpose strain MP-1 of *Saccharomyces cerevisiae* suitable for simultaneous detection of forward mutations, intergenic recombinations (mitotic crossing over), and intragenic recombinations (mitotic gene conversion). Those substances which gave a positive response in at least one of the three systems were tested in the mammalian spot test, an *in vivo* method for the detection of genetic alterations — especially point mutations — in somatic cells of mice.

Four of the seven substances tested showed an effect in one or two of the genetic systems of the MP-1 strain, but no substance in all three. 2,4,6-Trichlorophenol showed a very weak but significant activity in the mutation system

RUDOLF FAHRIG • Zentrallaboratorium für Mutagenitätsprüfung der Deutschen Forschungsgemeinschaft, 78 Freiburg i. Br., West Germany

CARL-AXEL NILSSON • Umeå Hospital, Chemical Division, S-901 85 Umeå, Sweden
CHRISTOFFER RAPPE • Umeå University, Department of Organic Chemistry, S-901 87 Umeå, Sweden

at a concentration of 400 mg/l, while pentachlorophenol at the same concentration was active in the mutation system as well as intragenic recombination system.

The chlorophenol impurities 2,3,7,8-tetrachlorodibenzofuran, 2,4,4',6-tetrachlorodiphenyl ether and 2,4,4',5-tetrachlorodiphenyl ether showed no genetic activity at a concentration up to 1000 mg/l. The other chlorophenol impurities, Cl_3- and Cl_5-predioxin showed activity in the mutation and the intergenic recombination system: the Cl_5-predioxin at 50 mg/l, the Cl_3-predioxin at 200 mg/l. The four substances active *in vitro* were tested in a concentration of 50 or 100 mg/kg in the mammalian spot test. All four substances were effective in this test system, but the activity of the chlorophenols was apparently lower than that of the predioxins.

Introduction

Chlorinated phenols are widely used as fungicides, bactericides, herbicides, etc. as well as starting materials for a series of industrial chemicals and pesticides, such as chlorinated phenoxy acids (2,4-D, 2,4,5-T). The annual world production of chlorinated phenols is in the range of 100,000 to 200,000 tons.

The chlorophenols are produced either via direct chlorination of phenol or by hydrolysis of polychlorobenzenes. Technical formulations of tri-, tetra- and pentachlorophenol have previously been studied (Nilsson and Renberg, 1974; Buser, 1975; Buser and Bosshardt, 1976; Levin *et al.*, 1976) and were found to contain dimeric impurities, such as chlorinated phenoxyphenols, diphenyl ethers, dibenzodioxins, dibenzofurans and dihydroxybiphenyls. The main constituents are chlorophenoxyphenols with 4-9 chlorine atoms (I) and they are often present in the amount of 1-5% (Nilsson and Renberg, 1974).

I

x = 4 - 9

The use of chlorophenol formulations as fungicides in sawmills is known to cause occupational health problems (Levin *et al.*, 1976). It has been observed that the workers in trimming-grading plants, where the sawn timber is handled after treating with chlorophenol and drying, often complain about cutaneous irritations, respiratory difficulties, headache, etc. Moreover, several serious intoxications are reported in the literature (Truhaut *et al.*, 1952; Menon, 1958). The cases concern severe and sometimes fatal exposure to chlorophenol solution in connection with dipping of timber in this solution. It has been shown that the dimeric impurities are present in the work environment of the sawmills (Levin *et al.*, 1976).

The biological effects of the chlorophenols, some chlorodibenzodioxins and a few chlorodibenzofurans have been studied (Kimbrough, 1974), but very little is known about the toxicity and other effects of chlorophenoxyphenols (Lyman and Furia, 1969) and chlorodiphenyl ethers. McConnell (1975) and Stalling (personal communication) have found the acute toxicity of a pentachlorophenoxyphenol to be low or comparable to pentachlorophenol.

Since the workers handling chlorophenol formulations are exposed not only to the chlorophenols, but also to the various impurities in the formulations, we have studied not only the genetic activity of 2,4,6-trichlorophenol and pentachlorophenol, but also their impurities.

Purity of the Chemicals Tested

2,4,6-Trichlorophenol and pentachlorophenol were recrystallized to an isomeric purity of 99%. The content of various dimers, chlorinated dioxins, dibenzofurans, diphenyl ethers and phenoxyphenols were checked and found to be less than 0.1 ppm. 2,3,7,8-Tetrachlorodibenzofuran was 95% isomer pure and the remaining 5% was another tetrachloroisomer. 2,4,4′,6-Tetrachlorodiphenyl ether, and 2,4,4',5-tetrachlorodiphenyl ether were 99% pure. 4,5,6-Trichloro-2-(2,4-dichlorophenoxy)phenol, (Cl_5-predioxin) and 4-chloro-2-(2,4-dichlorophenoxy) phenol, (Cl_3-predioxin) were both 99% isomer pure and contained less than 0.1 ppm of dioxins.

In Vitro Test with Saccharomyces cerevisiae MP-1 (Fahrig, 1975)

Description of Yeast Strain

The diploid multipurpose strain MP-1 allows for screening of intergenic recombination (mitotic crossing over), intragenic recombination (mitotic gene conversion) and forward mutation. It has the following constitution:

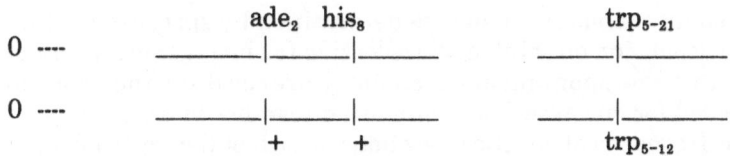

Induction of intergenic recombination (mitotic crossing over) results in homozygosity of the ade_2 marker and for a complete requirement for adenine. Such colonies are deeply red. In order to exclude mutation and intragenic recombination, simultaneous induction of homozygosity of the flanking marker his_8 must be demonstrated.

The system of intragenic recombination (mitotic gene conversion) is based on the induction of trp-prototrophic cells in a trp-auxotrophic population. In contrast to most reverse mutation systems, intragenic recombination could never be specifically caused by only one type of mutagen. A large variety of mutagens, such as ionizing radiation, UV irradiation, alkylating and arylating agents, acridines, nitrous acid and hydroxyurea have been shown to induce mitotic gene conversion (Zimmermann, 1971; Fahrig, 1976).

A forward mutation from actidione sensitivity to actidione resistance in yeast can be induced by mutations in at least eight different genes. Therefore, its spontaneous and induced mutation frequency is much higher than that of only one locus. Actidione resistance is due to an alteration of ribosomal protein. A nonsense or frameshift mutation in the respective structural gene(s) will certainly lead to the formation of a completely nonfunctional protein which will be lethal (Brusick, 1972). Consequently, frameshift inducing agents like aminoacridines are not able to induce actidione resistance (Fahrig, unpublished results).

Media

Synthetic complete medium: The medium is prepared by adding to Bacto yeast nitrogen base without amino acids (Difco Laboratories, Detroit, Mich., No. 0919-15) the following ingredients per liter of medium: L-Arginine-HCl (10 mg), L-Aspartic acid (10 mg), L-Glutamic acid (100 mg), L-Leucine (60 mg), L-Lysine-HCl (10 mg), L-Methionine (10 mg), L-Phenylalanine (50 mg), L-Serine (20 mg), L-Tyrosine (30 mg), L-Valine (30 mg), Uracil (10 mg).

According to the system used, the following ingredients are added or not added per liter of medium: Adenine sulfate (10 mg), L-Histidine-HCl (10 mg), L-Tryptophan (10 mg), Actidione (1 mg).

Liquid complete medium: The liquid medium YEP is composed of 2% Bacto peptone (Difco No. 0118-01), 1% yeast extract (Difco No. 0127-01) and 2% glucose.

Culturing of Yeast Cells

About 10^3 yeast cells are inoculated in a 300 ml Erlenmeyer flask containing 100 ml YEP, set on a shaker and allowed to grow for 3 days at 25 °C into the stationary phase. For the detection of the spontaneous frequency of intragenic recombination, the cultures of strains are stored at 4 °C for 3-4 days. During this time their spontaneous frequency is determined by spreading 0.1 ml from the YEP/yeast suspension on solid media selective for intragenic recombinants. Only cultures with a low spontaneous frequency are used for the experiments. The cultures needed for one experiment are mixed together in order to obtain a similar spontaneous frequency of genetic alterations in each of the experiments.

In Vitro Test

The cell cultures are washed twice with distilled water and the cell titers are adjusted to $3 \cdot 10^8$ cells per ml of 0.1 M phosphate buffer, pH 7.

These cell suspensions are incubated in a test tube on a shaker at 25° C with different concentrations of the test substances. At defined time-intervals treatments are stopped, and 4 x 0.1 ml of the suspensions containing $3 \cdot 10^7$ cells are spread on four plates of a solid nutrient-deficient media, selective for intragenic recombinants and mutants, respectively. Similarly, after 5 dilutions 1 : 10 in distilled water suspensions of $3 \cdot 10^2$ cells are plated out on ten plates of complete medium to attain the number of survivors (white colonies) and intergenic recombinants (red colonies or sectors).

These cultures are incubated at 25 °C, and the survivor and recombinant colonies are counted after 4 days. Actidione-resistant colonies are incubated 8 days.

Spot Test with Mice (Russell and Major, 1957; Fahrig, 1975)

Method

Embryos of the genotype $a/a;\ b/+;\ c^{ch}p/+\ +;\ d\ se/+\ +;\ s/+$ (black coat, dark eyes) are produced by mating about ten weeks old virgin females of the inbred C57BL/6JHan strain (a/a; otherwise wild type) to fertile males of the rotation bred T-stock (a/a = nonagouti; b/b = brown; $c^{ch}p/c^{ch}p$ = chinchilla and pink-eyed dilution; $d\ se/d\ se$ = dilute and short ear; s/s = piebald spotting).

These embryos which are heterozygous for four different recessive coat-color genes are treated in utero during the tenth day of fetal development by injection of the mutagen into the peritoneal cavity of the mother animal. If this treatment leads in a pigment precursor cell to an alteration of the wild type allele of one of the genes under study or to its loss, a color spot in the adult coat can be seen.

Application of the Compounds

The predioxins are diluted in Hank's balanced salt solution (HBSS), the chlorophenols in 10% dimethylformamide (DMF). The concentration of the compounds in the stock solutions allows the intraperitoneal injection of 0.1 ml of the solution per 10 g body weight.

Time of Observation

The first observation for color spots is made at the end of the second week of age, because at that time dark gray dorsal spots can be clearly distinguished from the black fur. At a later age, these spots can be missed. On the other hand, spots in ventral regions are sometimes visible only at a later age, probably owing to the hair not yet being full length at two weeks of age. It is, therefore, checked for color spots twice a week between two and five weeks of age.

Genetic Activity of Chlorophenols and Chlorophenol Impurities in the

In Vitro Tests with S. cerevisiae MP-1

Four of the seven substances tested showed an effect in one or two of the genetic systems of the MP-1 strain, but no substance was active in all three (Table 1).

TABLE 1: Induction of Forward Mutation, Intragenic and Intergenic Recombination in *S. cerevisiae* MP-1 *in Vitro* with Chlorophenols and Chlorophenol Impurities at a Treatment Time of 3,5 Hours

Substance	Genetic alteration	Experiment				Control		
		N^a	Concentration (mg/l)	Survival^b (%)	Colonies^c of genetically altered cells per survivors	N^a	Colonies^c of genetically altered cells per survivors	(P)
(trichlorophenol)	mutation	4	400	95 ± 16 (12487)	10.29 ± 2.75 (1249)	4	5.63 ± 0.91 (719)	< 0.02
	intergen. rec.				0.44 ± 0.07 (54)		0.44 ± 0.06 (58)	> 0.9
	intragen. rec.				10.63 ± 3.16 (1298)		7.59 ± 1.12 (974)	> 0.1
(pentachlorophenol)	mutation	4	400	59 ± 8 (10813)	2.00 ± 0.22 (216)	4	0.61 ± 0.07 (113)	< 0.001
	intergen. rec.				0.47 ± 0.14 (50)		0.49 ± 0.08 (91)	> 0.8
	intragen. rec.				5.64 ± 0.45 (609)		2.93 ± 0.10 (542)	< 0.001
(chlorinated dibenzofuran)	mutation	3	1000	91 ± 10 (4623)	2.88 ± 0.36 (134)	3	2.25 ± 0.72 (116)	> 0.2
	intergen. rec.				0.56 ± 0.06 (26)		0.57 ± 0.04 (29)	> 0.7
	intragen. rec.				5.15 ± 0.58 (236)		5.07 ± 0.58 (259)	> 0.8

Structure		a				
	mutation	3	1000	108 ± 8 (5578)	2.60 ± 0.37 (146)	2.25 ± 0.72 (116)	> 0.4
	intergen. rec.				0.58 ± 0.13 (33)	0.57 ± 0.04 (29)	> 0.8
	intragen. rec.				5.23 ± 0.90 (292)	5.07 ± 0.58 (259)	0.8
	mutation	3	1000	96 ± 5 (5862)	2.10 ± 0.42 (123)	2.17 ± 0.27 (132)	> 0.7
	intergen. rec.				0.31 ± 0.20 (18)	0.31 ± 0.12 (19)	> 0.9
	intragen. rec.				1.81 ± 0.28 (106)	1.39 ± 0.22 (84)	> 0.1
	mutation	3	50	106 ± 14 (30519)	15.64 ± 2.83 (4700)	4.28 ± 1.56 (1200)	< 0.005
	intergen. rec.				0.74 ± 0.13 (223)	0.39 ± 0.06 (109)	< 0.02
	intragen. rec.				5.12 ± 0.70 (1545)	5.90 ± 0.21 (1650)	> 0.1
	mutation	3	200	82 ± 7 (24441)	3.33 ± 0.16 (815)	1.86 ± 0.52 (555)	< 0.01
	intergen. rec.				0.41 ± 0.08 (100)	0.20 ± 0.03 (59)	< 0.02
	intragen. rec.				3.04 ± 0.44 (740)	2.57 ± 0.14 (765)	> 0.1

[a]Number of experiments
[b]Survival control = 100%
[c]Mutants, convertants (intragenic recombination)/10^7 survivors, recombinants (intergenic recombination)/10^2 survivors.
The numbers in parenthesis give the actual numbers of colonies counted.

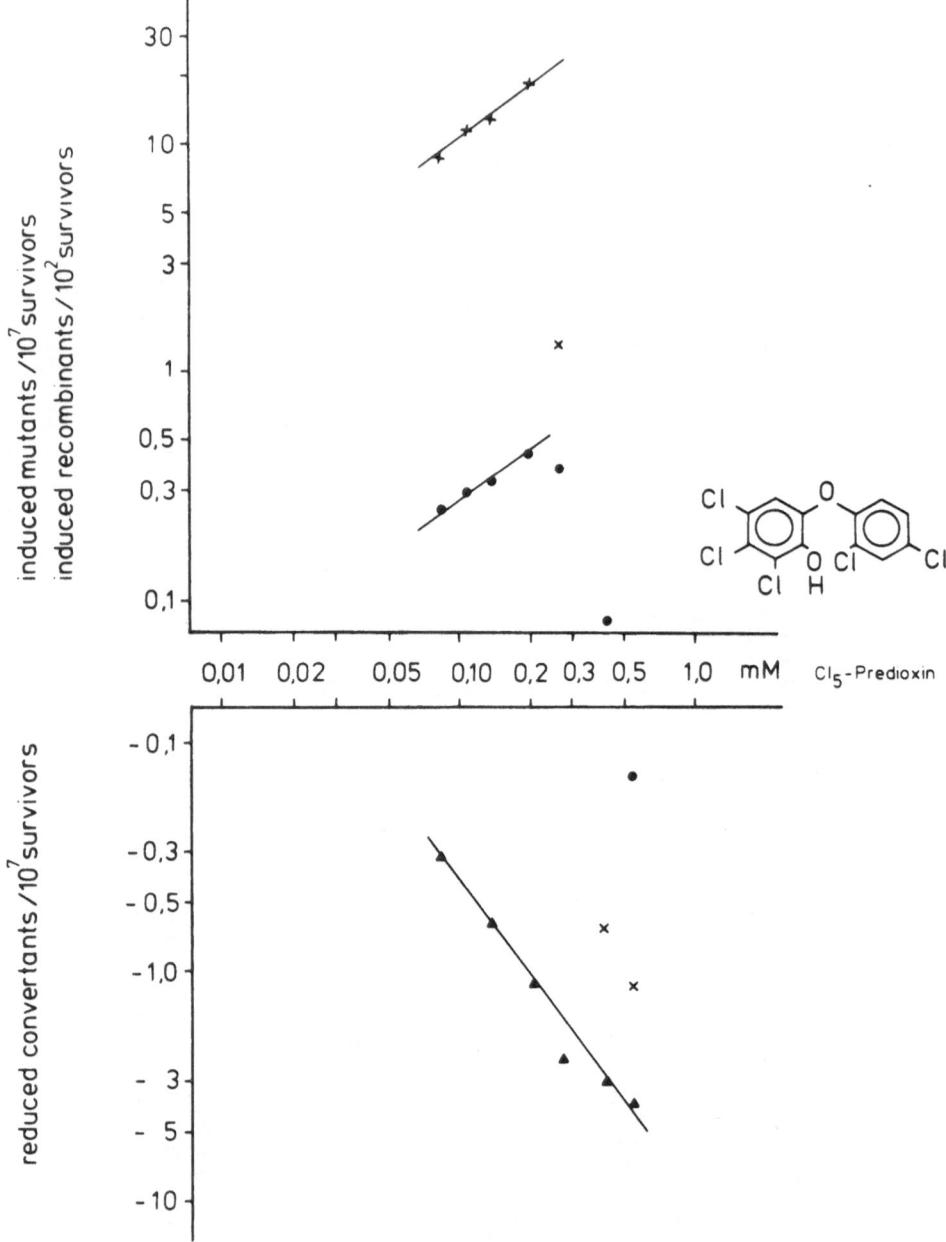

Figure 1: Induced mutants (x), intergenic recombinants (recombinants ●) and intragenic recombinants (convertants ▲) in *Saccharomyces cerevisiae* MP-1 after treatment with different concentrations of Cl_5-predioxin for 210 min at pH 7 and 25 °C in 0.10 M phosphate buffer. Each point is the mean value of the data of three single experiments.

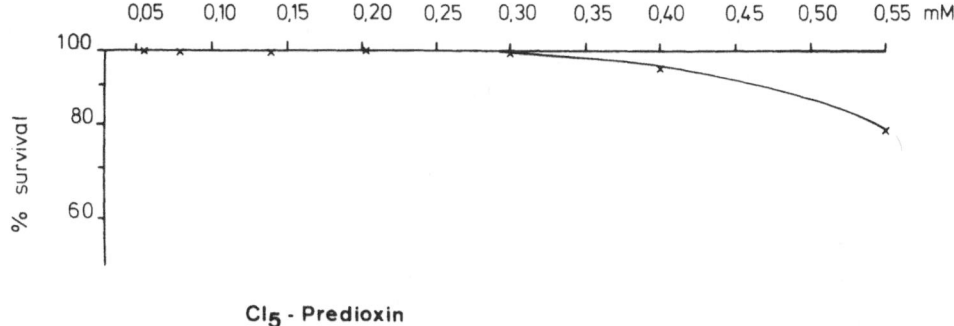

Figure 2: Survival of *Saccharomyces cerevisiae* MP-1 after treatment with different concentrations of Cl$_5$-predioxin for 210 min at pH 7 and 25 °C in 0.10 M phosphate buffer. Each point is the mean value of three single experiments.

Treatment time was 3,5 hours for all experiments. 2,4,6-trichlorophenol showed a very weak but significant activity in the mutation system at a concentration of 400 mg/l, while pentachlorophenol at the same concentration was active in the mutation system as well as in the intragenic recombination system.

The chlorophenol impurities 2,3,7,8-tetrachlorodibenzofuran, 2,4,4′,6-tetrachlorodiphenyl ether and 2,4,4′,5-tetrachlorodiphenyl ether showed no genetic activity at concentrations up to 1000 mg/l.

The other chlorophenol impurities, Cl$_3$- and Cl$_5$-predioxin showed activity in the mutation and the intergenic recombination system: the Cl$_5$-predioxin at 50 mg/l, the Cl$_3$-predioxin at 200 mg/l.

With the most effective compound, Cl$_5$-predioxin, determinations were made of the effect of different doses on the induction of genetic alterations (Fig. 1) and inactivation of yeast cells (Fig. 2). After subtracting the control frequency of genetic alterations and plotting the resulting induced frequency on a double logarithmic scale, the following dose/effect-curves result: In the concentration range of 0.08 to 0.2 mM the dose/effect-curves for mutants and intergenic recombinants are straight lines with a slope of about 1, i.e., the dose/effect-relationship is linear. Beginning at a concentration of about 0.3 mM a mutation and intergenic recombination frequency decrease can be observed. Beginning at a concentration of about 0.4 mM, this decrease leads to a reduction of the spontaneous frequency of mutants and intergenic recombinants. In the intragenic recombination system over the whole range of concentration a decrease can be observed leading to a linear reduction of the spontaneous frequency. The toxic effect of Cl$_5$-predioxin is not very strong. At 0.55 mM, the highest concentration used, the survival is about 80%. An inactivation of yeast cells can be observed at 0.3 mM, a point where the decrease in mutants and intergenic recombinants begins.

TABLE 2: The Activity of Chlorophenols and Predioxins in the Mammalian Spot Test

Mutagen	Number of			Animals with spots of questionable genetic relevance			Animals with spots of genetic relevance			
	Dose (mg/kg)	Females[a] treated	Offspring survived to observation	White	White-gray	Other	Light gray	Gray	Light brown	Brown
(trichlorophenol structure)	50	36	181	2	1	–	1	–	–	–
	50	38	159	–	3	–	1	–	–	–
	100	42	175	–	–	–	–	1	–	–
(chlorophenol structure)	50	42	169	1	–	2[c]	–	–	1	–
	50	44	147	–	–	–	–	1	–	–
	100	40	157	1	1	–	1	1	–	–
(predioxin structure)	50	46	132	1	1	–	–	3	–	–
	50	43	200	–	1	–	1	–	–	4
(predioxin structure)	50	46	177	1	–	–	–	1	1	–
	50	46	142	–	–	1[d]	–	–	–	2
Control	0	152	967	4	1	–	–	–	1	–

[a]Including females that had no litters; [b]Checking for color spots was made between 2 and 5 weeks of age; [c]Yellow-gray; [d]Yellow-white

Genetic Activity of Chlorophenols and Chlorophenol Impurities in the

Spot Test with Mice

The four substances active *in vitro* were tested in a concentration of 50 or 100 mg/kg in the mammalian spot test (Table 2). All four substances were effective in inducing color spots, but the activity of the chlorophenols was apparently lower than that of the predioxins. In total, among 1639 mutagen-treated animals there were 35 with a color spot:

6 white	: Hairs contained no pigment. Ventral position of spots.
7 white-gray	: Hairs contained at least traces of pigment. Ventral position of spots.
4 light gray	: Hairs contained at least some black pigment granules. Spots were randomly distributed. The intensity of pigmentation was lower ventrally.
7 gray	: In hairs reduction of black pigment granules. Spots were randomly distributed.
2 light brown	: In hairs reduced level of pigmentation, the pigment was greatly reduced distally.
6 brown	: In hairs brown pigment instead of black one. Spots were randomly distributed.
2 yellow-gray	: Similar to hairs found near the ears, mammae and genitalia (represents abnormal differentiation rather than genetic change).
1 yellow-white	: White hairs with yellow tips. This spot can perhaps be equated with a light brown one with unusual large reduction of pigment.

Among 967 untreated animals whose coats were examined, there was one with a light brown, one with a white-gray and four with midventral white spots. The animals with color spots were distributed randomly among litters, and per animal only one color spot could be observed. Mutagen treatment did not only lead to the appearance of color spots, but also to a reduction of litter-size and to an increased loss of litters before and after birth.

Discussion

In Vitro *Test with* S. cerevisiae

The result, that chlorophenols as well as the predioxins show activity in one or two of the genetic systems of MP-1, but no substance in all three, becomes perhaps more intelligible if the complexity of the dose/effect-relationship of Cl_5-predioxin in comparison to other mutagens (Fahrig, 1976; unpublished results) is taken into consideration. The dose/effect-curves of all poly- and all monofunctional alkylating agents tested up till now, of 4-Nitroquinoline 1-oxide as well as of $NaNO_2$ are linear in a double logarithmic scale with a slope of one or appreciably greater than one, but in nontoxic concentration ranges parallel for induced mutations, intragenic and intergenic recombinations. In contrast to this, with Cl_5-predioxin only at low concentrations a linear increase in the frequency of mutants and intergenic recombinants can be observed, followed at higher concentrations by a decrease leading even to a reduction of the spontaneous frequency. In the intragenic recombination system this decrease can be observed even at low concentrations.

One possible explanation for these results is, that chlorophenols and predioxins have not only the ability to damage DNA, a situation for repair synthesis resulting in mutations as well as recombinations (Fahrig, 1976), but also have the ability to inhibit repair synthesis itself at higher concentrations. On the one hand, this inhibition would prevent fixation of induced DNA damage as recombination or mutation, but would lead in many cases to lethality. And indeed, it is apparent that the lethal effect of Cl_5-predioxins begins in the same concentration range where a decrease of mutation and intergenic recombination can be observed (Figs. 1,2). On the other hand, inhibition of repair synthesis would lead to a reduction of all spontaneously occurring DNA-alterations of which fixation is dependant on repair. The repair systems responsible for fixation of intragenic recombination seem to be more sensitive to Cl_5-predioxin than those responsible for fixation of intergenic recombination or mutation.

The discussion presented here offers only one of perhaps several possible explanations for the complexity of the dose/effect-relationship of chlorophenols and predioxins. One of these possibilities, namely selective inactivation of convertants by Cl_5-predioxins, can be excluded (Fahrig, unpublished results), but further data are needed to come to well-grounded statements concerning the molecular mechanism of chlorophenols and predioxins.

Spot Test with Mice

The expression of the recessive color genes under study on an *a/a* background (uniform dark color apart from some yellow-gray hair near the ears, mammae and genitalia) would be expected at the *b*-locus to be chocolate brown, at the *c*-locus to be light brown to cream, at the *p*-locus to be blue lilac to light gray, at the *d*-locus to be blue gray to gray, and at the linked loci *p* and *c^{ch}* to be white (Searle, 1968), i.e., the spot colors should be either white, brownish or grayish.

From the comparison of coat color expectations and the histological examination of induced color spots there is good reason for believing that light gray spots are due to expression of *p* or perhaps also *c,* gray spots of *p* or *d,* light brown spots of *c* and brown spots of *b,* but the possibility of dominant mutations affecting the melanocyte cannot be forgotten. The mechanism by which a heterozygous recessive gene comes to expression should be either gene mutation or a recombinational process. Of the numerical and structural chromosome aberrations produced only the few which are able to pass the filter of several mitoses come into consideration. It is not clear which genetic alteration is responsible for the expression of a heterozygous recessive gene, but we can suppose that at least the same spectrum of heritable DNA-alterations is covered as with the multipurpose strain of *S. cerevisiae.*

For white-gray spots it is not clear whether they express genetic alterations or pigment cell killing. If the traces of pigment found in hairs of this type are *pp*-granules, then they are presumably the ventral expression of *p* $c^{ch}/p+$, and all possible DNA-alterations could be responsible for their appearance.

White spots could be the result of a genetic alteration leading to *p* c^{ch}/p $c^{ch}/$ or *p*c^{ch}/deletion. The possibility of gene mutations can be excluded, because a simultaneous mutation at the linked loci both *p* and *c* would be a too rare event to be likely. There is also no proof that large deletions covering both the *c* and *p* loci (14 units apart) are sufficiently viable when heterozygous to form large clones of melanocytes. *A priori* one would expect such deletions to be induced much less frequently than small ones covering either locus alone. Other possibilities are mitotic nondisjunction leading to monosomy or intergenic recombination (mitotic crossing over) leading to the genotype *p* c^{ch}/p c^{ch}. Viable cells would be produced at least by intergenic recombination.

As far as white midventral spots are concerned, it is likely that they are   not the result of genetic alterations, because they can be induced at least by X-rays in C57BL alone (Russell and Major, 1957; Fahrig, 1975), where *p* c^{ch} cannot show up.

As white and white-gray spots can result as well from pigment cell killing as from genetic alterations, spots of this color can be considered to be of questionable genetic relevance.

Conclusions

The following can be said concerning the activity of chlorophenols and their impurities in comparison to that of other mutagens in the spot test: At a control frequency of 0.1%, 50 mg of the chlorophenols/kg induced about 0.6%, 50 mg Cl$_3$-predioxin/kg about 2.4%, and 50 mg Cl$_5$-predioxin/kg about 1.3% color spots of genetic relevance. Comparing the frequency of spots induced with the dose applied, the chlorophenols belong to the weakest mutagens tested up till now in the spot test (Fahrig, 1977), while the activity of Cl$_3$-predioxin is comparable to that of N-methyl-N'-nitro-N-nitrosoguanidine (MNNG) and acridine orange (AO), the activity of Cl$_5$-predioxin to that of Dimethylnitrosamine (DMN) and trichloroethylene (Fahrig, 1977). This means, chlorophenols have definite but weak mutagenic activity, while some of their impurities have much stronger mutagenic activity.

ACKNOWLEDGMENT

The expert technical assistance of Mrs. B. Fahrig, Mrs. A. Forstmeier, Mrs. H. Schlegel and Mrs. B. Velden is gratefully acknowledged. We are indebted to Dr. Åke Norström for the dibenzofuran and the diphenyl ethers.

References

Brusick, D.J., 1972. Induction of cycloheximide-resistant mutants in *Saccharomyces cerevisiae* with N-methyl-N'-nitro-N-nitrosoguanidine and ICR-170. *J. Bacteriol.,* **109:** 1134-1138.

Buser, H.-R., 1975. Analysis of polychlorinated dibenzo-*p*-dioxins and dibenzofurans in chlorinated phenols by mass fragmentography. *J. Chromatogr.,* **107:** 295-310.

Buser H.-R., and H.P. Bosshardt, 1976. Determination of polychlorinated dibenzo-*p*-dioxins and dibenzofurans in commercial pentachlorophenols by combined gas chromatography-mass spectrometry. *J. Ass. Offic. Anal. Chem.,* **59:** 562-569.

Fahrig, R., 1975. Development of host-mediated mutagenicity tests - yeast systems. II. Recovery of yeast cells out of testes, liver, lung and peritoneum of rats. *Mutation Res.,* **31:** 381-394.

Fahrig, R., 1975. A mammalian spot test: Induction of genetic alterations in pigment cells of mouse embryos with X-rays and chemical mutagens. *Molec. gen. Genet.,* **138:** 309-314.

Fahrig, R., 1976. The effect of dose and time on the induction of genetic alterations in *Saccharomyces cerevisiae* by aminoacridines in the presence and absence of visible light irradiation in comparison with the dose-effect-curves of mutagens with other type of action. *Molec. gen. Genet.,* **144:** 131-140.

Fahrig, R., 1977. The mammalian spot test (Fellfleckentest) with mice. *Arch. Toxicol.,* **38:** 87-98.

Kimbrough, R.D., 1974. The toxicity of polychlorinated polycyclic compounds and related chemicals. *CRC Critical Reviews in Toxicology,* 445.

Levin, J.-O., C. Rappe, and C.-A. Nilsson, 1976. Use of chlorophenols as fungicides in sawmills. *Scandinavian J. Work Environ. Hlth.,* **2:** 71-81.

Lyman, F.L., and T. Furia, 1969. Toxicology of 2,4,4'-trichloro-2'-hydroxy-diphenyl ether. *Ind. med.,* **38:** 45-52.

McConnel, E.E., 1975. Memorandum Sept. 22, 1975 U.S. Department of Health, Education and Welfare.

Menon, J.A., 1958. Tropical hazards associated with the use of pentachlorophenol. *British Med. J.,* 1156-1158.

Nilsson, C.-A., and L. Renberg, 1974. Further studies on impurities in chlorophenols. *J. Chromatogr.,* **89:** 325-333.

Russell, L.B., and M.H. Major, 1957. Radiation-induced presumed somatic mutations in the house mouse. *Genetics,* **42:** 161-175.

Searle, A.G., 1968. *Comparative genetics of coat color in mammals.* Logos/Academic, London.

Truhart, R., P. L'Epee, and E. Boussemart, 1952. Recherches sur la toxicologie du pentachlorophenol: II. Intoxications professionelles dans l'industrie du bois. Observations de deux cas mortels. *Arch. mal. prof. med. trav. secur. soc.,* **13:** 567-569.

Zimmermann, F.K., 1971. Induction of mitotic gene conversion by mutagens. *Mutation Res.,* **11:** 327-337.

Negative Chemical Ionization Mass Spectrometry: Applications in Environmental Analytical Chemistry

RALPH C. DOUGHERTY and E.A. HETT

Abstract—Negative Chemical Ionization (NCI) mass spectrometry is unusually sensitive to non-specific toxic substances and is generally much less sensitive to biomolecules. These features make NCI mass spectra particularly attractive for screening environmental substrates for environmental contamination. This paper discusses the NCI technique as well as the levels of contamination that should be of concern. NCI spectra of several classes of toxic substances are presented. Application of the procedure to screening for environmental contamination is reviewed.

Introduction

Human cancer is a problem that engages a number of environmental scientists. The "Atlas of Cancer Mortality" (Mason *et al.*, 1975) provides the strongest evidence that cancer in humans is related to environmental factors. Figures 1 and 2 present the age adjusted frequency data for cancer mortality (all sites combined) in the years between 1950-1969 for white males and females. There are a few direct observations that one can make by examining these pages of the atlas. Counties of high mortality rates for males in the West often are counties that contain a heavy metal smelter. Metal refining is also associated with the number of Northeastern counties that appear in the highest decile. Quite a number of other counties in the highest decile in the Northeast are counties that contain substantial chemical industries.

RALPH C. DOUGHERTY and E.A. HETT • Department of Chemistry, Florida State University, Tallahassee, Florida 32306, U.S.A.

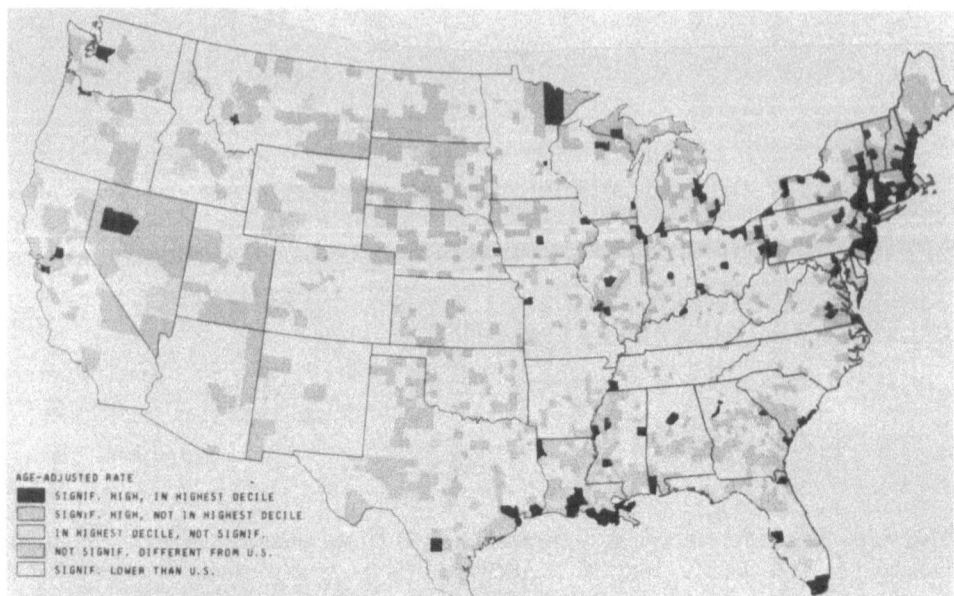

Figure 1. Cancer mortality in white males during 1950-1969, by county; all sites combined (Mason *et al.*, 1975).

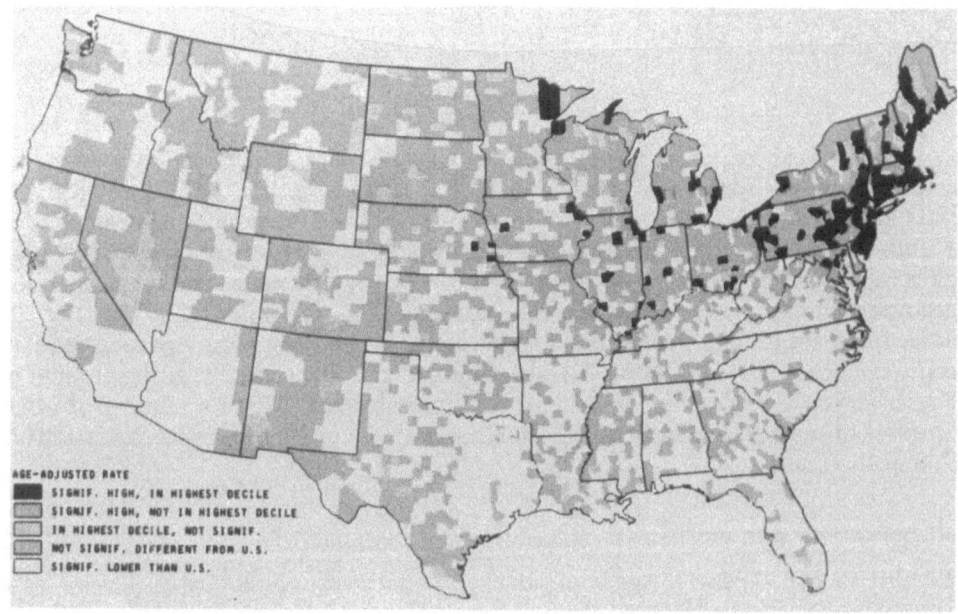

Figure 2. Cancer mortality in white females during 1950-1969, by county; all sites combined (Mason *et al.*, 1975).

It is our opinion that the most effective cancer control strategy, in the short run at least, will be the control of human exposure to cancer causing substances. Development of the control strategy will require the development of analytical techniques that will allow us to detect any unknown member of broad classes of toxic substances in environmental substrates when the presence of that specific compound is not suspected.

Criteria for Development of Analytical Procedures

There are two very general questions that must be addressed prior to development of analytical techniques. The first is what are appropriate levels of concern for toxic substances. I believe that the answer to this question may be found at least in part in the data contained in Tables 1 and 2. Table 1 lists the approximate number of cellular constituents in an average human cell. At any given time there are approximately 150,000 RNA's and proteins in a given cell. The turnover time for the RNA's is between 5-10 hours. The number of control molecules per cell, the molecules that set the balance between growth and homeostasis, has been estimated to be less than 20 per cell. One view of carcinogenesis holds that the induction of cancer requires a critical mutation or chemical reaction in more than one of the control molecules. In a very simple minded view the probability of this occurring through action of two isolated molecules in a cell would be something of the order of one part in 10^{10}. This must be considered to be an upper estimate.

TABLE 1. Cellular Constituents

DNA's (2)	code for 10^6 RNA's Proteins	
RNA's $\sim$ 15,000/cell	$\sim$ 1-10 each Larger for specialized cells	t ± $\sim$ 5-10 hr some longer
Proteins $\sim$ 15,000/cell	$\sim$ 1-10 each	
Control Molecules- RNA's/Proteins	< 20/cell	

If we take the average mass of a carcinogenic substance to be something around 200 daltons it is possible to calculate the number of molecules that will reside in a given cell at various levels of contamination. This data is given in Table 2. From the data in Table 2 it is clear that atomolar concentrations, that is, subfemtogram amounts of toxic substances, will rarely be significant.

TABLE 2. Number of Molecules Per Cell

Level	Prefix	Level	Number of Molecules Per Cell
10^{-6}	micro-	ppm	3×10^6 molecules/cell (220 daltons)
10^{-9}	nano-	ppb	3×10^3
10^{-12}	pico-	ppt	3
10^{-15}	femto-	ppquad	3×10^{-3}
10^{-18}	atto-	ppquint	3×10^{-6}

Parts per trillion concentrations of toxic substances may be important if the molecules are exceptionally toxic. One class of molecules that falls into this category are the polychlorodioxins. Tetrachlorodioxin is lethal at parts per billion concentrations and this suggests that parts per trillion concentrations should be of concern for long term contamination. For most molecules, however, parts per billion concentrations are the lowest limits that one should concern oneself with in terms of long term toxicology. This means that the screening techniques that one should develop should be sensitive to nanogram quantities of toxic materials in the presence of orders of magnitude larger quantities of biomolecules.

The second general question that must be addressed is what are the restrictions to be placed on the methodology by the requirement that the screening procedures detect any unknown member of a broad class of toxic substances. That restriction is actually quite severe. It means that the methodology cannot involve adsorption chromatography nor can it involve gas chromatography. In the case of adsorption chromatography there are many materials that will simply not pass through an adsorption chromatography step. In the case of gas chromatography, column selection and the requirement for derivatization requires that the analyst have some knowledge of the substance he is looking for prior to analysis. The cleanup procedures that can be used are thus restricted to solvent-solvent extraction and gel permeation chromatography which does not involve adsorption of molecular species.

Negative Chemical Ionization Mass Spectrometry

Negative chemical ionization (NCI) mass spectrometry seems uniquely suited for screening partially cleaned up samples of environmental substrates for contamination with toxic substances. NCI mass spectra can very generally be obtained from toxic materials with nanogram sensitivities. The sensitivity of the technique to the biomolecules that usually dominate environmental samples is often three or four orders of magnitude lower, that is, 10 micrograms of material would be required to obtain a reproducible spectrum. The reason for this differential sensitivity is easy to understand on the basis of two observations. (1) The biomolecules are highly reduced. With the exception of free carboxylic acids and some of the prosthetic groups from the electron transport chain the biomolecules are virtually transparent to negative chemical ionization. These molecules have negative electron affinities and only very weakly attach gas phase nucleophiles. (2) In contrast, carcinogenic substances can be very generally classed as either oxidizing agents, alkylating agents or molecules that can be converted readily into alkylating agents. These molecules generally have positive electron affinities and they generally attach gas phase nucleophiles to form stable complexes.

The analytical potential of NCI mass spectrometry is broader but quite similar to that of gas chromatography with electron capture detection. Compounds that give intense electron capture responses will virtually always be sensitive to NCI mass spectra. Mass resolution in NCI spectra serves the same function as time resolution in gas chromatography. By using a high resolution mass spectrometer it is possible to obtain mass resolution of the order of one part in 10^4 which is considerably higher resolution than is available with a conventional gas chromatograph. The reduction in chemical noise by mass analysis also contributes to a lower detection limit for the NCI system as compared to gas chromatography.

NCI mass spectra are obtained by operating a chemical ionization mass spectrometer (ion source pressure ca. 1 torr) in the negative ion mode. Under these conditions there are five major negative ion forming reactions that occur, assuming that the ion source contains oxygen and organochlorine compounds. These reactions are the following:

$$
\begin{aligned}
M + e_s^- &\longrightarrow M^{*\pm} \xrightarrow{\ N\ } M^{\pm} & (1)\\
M + Cl^- &\longrightarrow (M + Cl)^- & (2)\\
M + Cl^- &\longrightarrow (M - H)^- + HCl & (3)\\
M + O_2^{\pm} &\longrightarrow (M - Cl + O)^- + ClO^{\cdot} & (4)\\
M + e_s^- &\longrightarrow A^- + B^{\cdot} & (5)
\end{aligned}
$$

Resonance capture of an electron to give an excited molecule anion, reaction 1, will occur for all species that have positive electron affinities. This group includes many toxic substances. Chloride association, reaction 2, will occur for all hydrogen bonding or carbon bonding substrates. The dominant ion in the NCI mass spectra of most organic polychlorides is $(M \pm Cl)^-$. Acid-base exchange reactions, 3, or disassociative electron capture, 5, can give $(M - H)^-$ ions. This is the mode of ionization for pentachlorophenol (PCP). Both aromatic and aliphatic polychlorides will exchange chloride for oxygen in a reaction that requires either superoxide, 4, or the molecule ion of the substrate.

The choice of reagent gases is critical to the development of the library of standard NCI mass spectra. The gas mixture that we are now routinely using consists of 5-4-1 mixture of isobutane, methylene chloride and oxygen. There are three reasons for choosing this gas mixture. (1) Isobutane is present as a diluent because its presence will increase the filament life. (2) Methylene chloride is a source of chloride ions. Chloride is a sufficiently good nucleophile in the gas phase that it will attach to a large number of alkylating and oxidizing agents that would otherwise not produce NCI mass spectra. (3) Oxygen is exceptionally reactive in anionic mass spectra and it is virtually impossible to rigorously exclude it from an NCI experiment. In order to make the spectra reproducible for samples run at different times with different instruments we add controlled quantities of oxygen to gas mixture and thus the oxygen reaction ions will be present in a controlled way.

Negative chemical ionization mass spectrometry was first used to obtain the spectra of polycyclic chlorinated pesticides (Dougherty *et al.*, 1972). Polycyclic chlorinated insecticides of the dieldrin class uniformly gave intense ions which corresponded to chloride adducts of the molecule. The sample requirements for these spectra were small, of the order of nanograms, and the selectivity was high. We decided to use the technique to look for organic polychlorides in extracts of environmental substrates on the basis of these initial observations. When we applied NCI mass spectrometry to a series of extracts from sources ranging from birds to human tissues, we detected and identified polychlorobiphenyls (Dougherty *et al.*, 1973), mirex, dieldrin, dichlorodiphenyl acetic acid (Dougherty and Piotrowska, 1976a), other DDT metabolites and aldrin. In these preliminary studies the samples that we examined were generally known to contain the substances that were found.

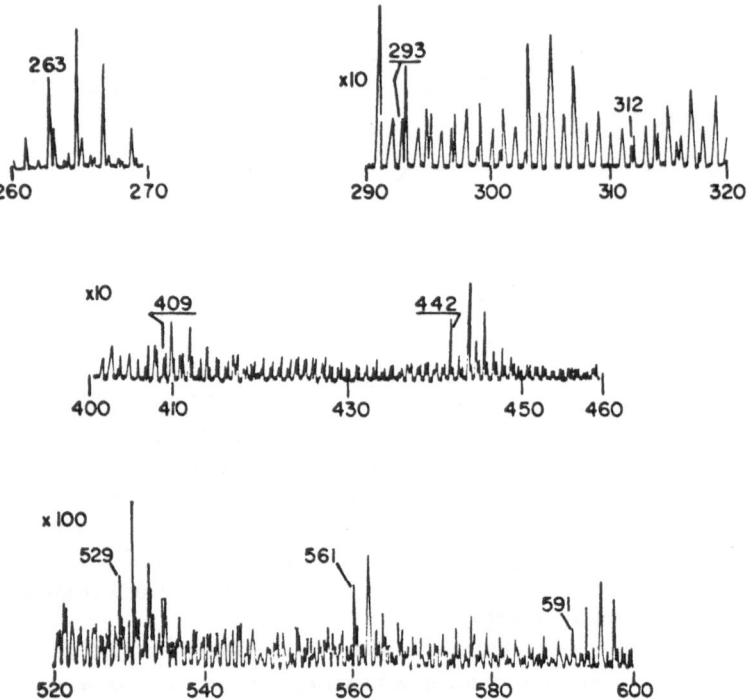

Figure 3. Methylene chloride NCI, 0.5 torr, methylene chloride extract of human urine adjusted to pH 6, high resolution was required to separate the polyhalogen ions from the positive mass defect ions adjacent to them (Dougherty *et al.*, 1976).

In order to extend the utility of NCI screening of environmental substrates for toxic substances we applied a rudimentary screening technique to human urines (Dougherty et al., 1976). Extraction of urines with methylene chloride followed by drying and evaporation gave residues that produced exceptionally rich NCI mass spectra using methylene chloride as the reagent gas. One example of the kind of spectra that we obtained is shown in Figure 3. Although the spectrum in Figure 3 is obviously rich in organic polyhalides as shown by ion clusters starting at m/z 263, 312 and 409 as well as many others, the sample matrix was so rich in polar biomolecules that examination of only a handful of spectra deteriorated the instruments' performance and the source had to be cleaned and rebuilt. In order to exclude polar biomolecules from the extract that we finally examined we settled upon extraction with pesticide quality hexane followed by drying and evaporation. Hexane extraction gives good recoveries for most nonpolar or slightly polar toxic substances and it leaves the polar biomolecules in the aqueous phase. Since pentachlorophenol appeared in every urine sample that we examined as pentachloro -phenoxide, m/z 263, we decided to determine order of magnitude concentrations of PCP in human urines. The calibration standard that we selected was p-chlorobenzophenone, a compound which gives intense NCI spectra that are dominated by odd electron even mass ions.

In a survey of the organic polychlorides in the urines of college-age men we found PCP in concentrations ranging from 1-80 ppb and a series of ions that corresponded to chloride adducts of the 2,4,5-trichlorophenoxyacetic acid (2,4,5-T) family of herbicides (Dougherty and Piotrowska, 1976b). We also routinely found ions corresponding to the chloride adduct of trichlorophenol, a compound that is known to be a metabolite of both 2,4,5-T and PCP. A limited longitudinal study of two individuals indicated that concentration of PCP in urine varied with time and that the appearance of other organopolyhalides like 2,4,5-T varied dramatically with time. This information suggested that the diet was one possible source of these toxic substances. In order to investigate this possibility we conducted a spot survey on selected items from the food chain. In spite of the fact that our initial cleanup procedure for lipid rich substrates (Dougherty and Piotrowska, 1976b) was unsatisfactory we were able to obtain evidence of the presence of PCP and a number of other polychlorinated organics in concentrations ranging from 1 to 100 ppb. PCP appeared in most of the grain and sugar products that we examined. We also found traces of this molecule in powdered milk.

We are presently involved in expanding the NCI screening program in two directions. We are attempting to include a number of classes of toxic substances other than organic polychlorides in the screening effort. We are also developing techniques for screening microsamples and lipid rich substrates for contamination.

The inclusion of new classes of toxic substances requires a demonstration that NCI mass spectrometry will produce reproducible mass spectra for these molecules at very low concentration. Figures 4 through 9 present the NCI mass spectra of a series of structurally distinct toxic substances obtained with the isobutane, methylene chloride and oxygen reagent gas mixture. The spectra were obtained under conditions which gave the reagent gas spectrum at 10 times the abundance of the spectrum of the substrate; this is necessary to insure that the ion molecule

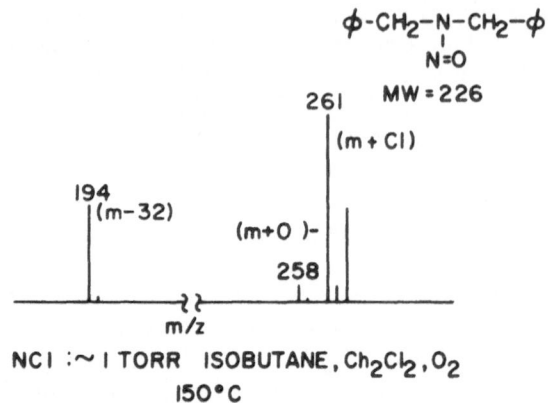

Figure 4. NCI mass spectrum of dibenzylnitrosamine.

reactions that produce the ions in the spectrum will be reproducible from one sample to the next. The spectra in the figures do not include the spectra of the reagent gases. Figure 4 presents the NCI mass spectrum of dibenzylnitrosamine. Nitrosamines as a class give very weak responses to electron capture detection. The results of this spectrum illustrate the fact that NCI mass spectra can be obtained at high sensitivity for compounds that are insensitive to electron capture detection. The reason for this stems from the fact that molecules that will readily associate gas phase nucleophiles such as chloride will give intense NCI mass spectra in spite of the fact that they are not electron capturing substances.

Malathion (Fig. 5) is one of the widely used substitutes for DDT. Under the conditions of NCI mass spectrometry it undergoes disassociative electron capture to give ions at m/z 315 and m/z 157. These ions correspond directly to cleavage at one

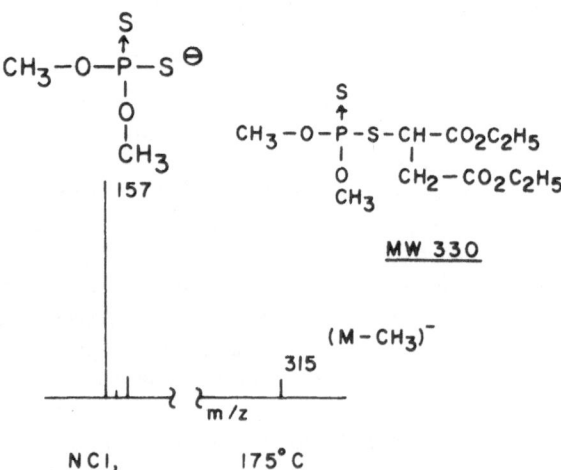

Figure 5. NCI mass spectrum of Malathion.

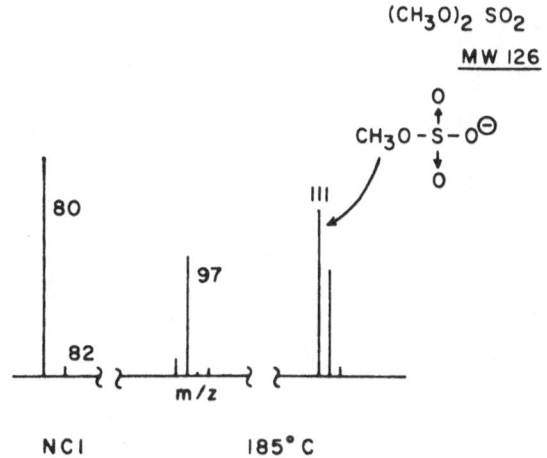

Figure 6. NCI mass spectrum of dimethylsulfate.

of the side chains on the thiophosphate ester. Dimethylsulfate, (Fig. 6), another potent mutagen, responds very similarly to the phosphate esters in that it gives an intense ion for disassociative capture with cleavage of the methyl group at m/z 111.

Polynuclear aromatic hydrocarbons are a widely recognized class of carcinogenic substances. These compounds routinely give intense NCI mass spectra. Most polynuclear aromatic hydrocarbons require activation by liver areneoxidase before they are capable of transforming cells. Trans-stilbene oxide (Fig. 7) is an example of an aromatic epoxide, and indeed it produces an intense NCI mass spectrum that is dominated by the M-H ion at m/z 195.

Carbamates are a widely used class of antifungal substances that are also

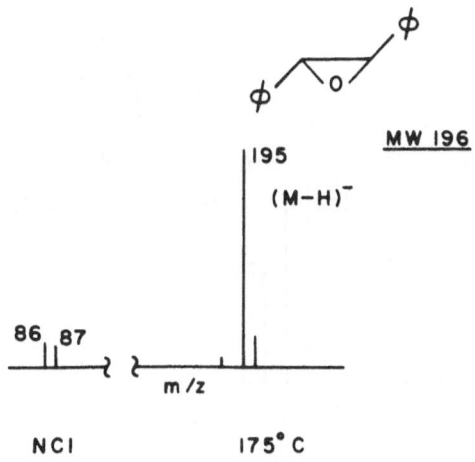

Figure 7. NCI mass spectrum of trans-stilbene oxide.

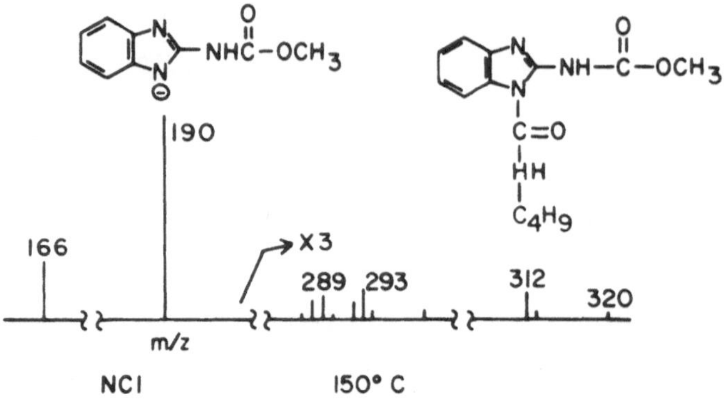

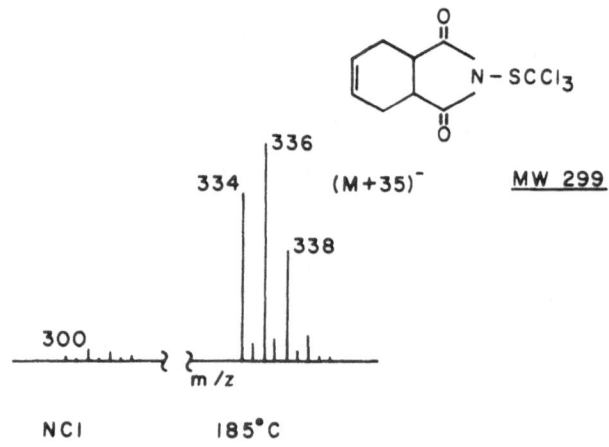

Figure 8. NCI mass spectrum of Benomyl.

potentially mutagenic. The NCI mass spectrum of Benomyl is presented in Figure 8. The dominant ion in the spectrum results from disassociative capture of an electron to give m/z 190. Captan is another widely used fungicide and like other organic polychlorides it attaches chloride to give an ion m/z 334 (Fig. 9).

The spectra in Figures 4 through 9 are representative of the spectra obtained for more than 100 different substances that form the nucleus of our library of NCI mass spectra. The fact that compounds as diverse as polychlorinated organics, nitrosamines or epoxides will produce intense NCI mass spectra suggests that we can anticipate broad general utility of NCI mass spectrometry in screening environmental substrates for contamination with toxic substances.

Lipid rich substrates have always posed serious problems for toxic substance analysis at ppm and lower levels. The reason for this is that separation of the toxic substances of interest from the matrix is considerably more difficult than in a case of aqueous substrates. The utility of gel permeation chromatography for separating

Figure 9. NCI mass spectrum of Captan.

toxic substances from lipid rich substrates has been demonstrated by Stalling *et al.* (1972). The basis for the application of this procedure is the fact that neutral and polar lipids generally have significantly higher molecular weights than those of nonspecific toxic substances. This means that the neutral lipids will elute from a gel permeation column significantly ahead of the toxic compounds of interest.

We have developed procedures for obtaining small samples of breast fluid from nonlactating females. The reason for our interest in breast fluid stems from the fact that breast cancer incidence among women who have lactated is significantly lower than women who have not lactated. Breast fluid is lipid rich and has approximately the composition of human milk. Approximately 100-200 microliters of breast fluid can be obtained by use of a breast pump from approximately 70% of nonlactating caucasian females. Our preliminary screening of breast fluid has revealed the presence of a number of organic polyhalides, these include m/z 286, Cl_2; m/z 315, Cl_4; m/z 321, Cl_5; and m/z 425, Cl_5. It remains to identify the compounds that are the sources of these ions. In a survey of nearly 1500 subjects the Environmental Protection Agency has detected low levels of 3 polychlorinated insecticides in human milk. Their survey detected dieldrin in 80% of the samples, heptachloroepoxide in 63% and oxychlorodane in 74%. The EPA survey depended upon gas chromatography and only screened for the presence of 6 different compounds.

Conclusions

A substantial amount of evidence has been collected which suggests that negative chemical ionization mass spectrometry will be an exceptionally valuable tool in environmental and analytical chemistry in the next decade. Negative chemical ionization is highly selective for compounds that have positive electron affinities or compounds that will readily associate with gas phase nucleophiles. Virtually all known classes of carcinogenic substances fit this description. NCI mass spectra are exceptionally unsensitive to biomolecules, the kinds of molecules that one encounters in environmental substrates. When combined with cleanup procedures that do not involve adsorption chromatography or gas chromatography, NCI mass spectra should provide the basis for the development of a screening procedure that will reliably detect any unknown member of broad classes of toxic substances at ppb levels when the presence of that toxic substance is not suspected.

References

Dougherty, R.C., A. Bergner, P. Levonowich, and J.D. Roberts, 1976. Positive and negative chemical ionization mass spectra for pesticide screening. In, *Advances in Mass Spectroscopy in Biochemistry and Medicine* (A. Friterio and N. Castagnoli, Eds.) Vol. **1**, pp. 181-192, Spectrum Publ. Inc., New York.

Dougherty, R.C., J. Dalton and F.J. Biros, 1972. Negative chemical ionization mass spectra of polycyclic chlorinated insecticides. *Org. Mass Spectrom.*, **6**: 1171-1181.

Dougherty, R.C., and K. Piotrowska, 1976a. Multiresidue screening by negative chemical ionization mass spectrometry of organic polychlorides. *J. Ass. Offic. Anal. Chem.,* **59:** 1023-1027.

Dougherty, R.C., and K. Piotrowska, 1976b. Screening by negative chemical ionization mass spectrometry for environmental contamination with toxic residues: application to human urines. *Proc. Natl. Acad. Sci., U.S.A.,* **73:** 1777-1781.

Dougherty, R.C., J.D. Roberts, H.P. Tannenbaum, and F.J. Biros, 1973. Positive and negative chemical ionization mass spectra of polychlorinated pesticides. In, *Mass Spectroscopy and NMR Spectroscopy in Pesticide Chemistry* (R. Haque and F.J. Biros, eds.), pp. 33-47, Plenum, New York.

Mason, T.J., F.W. McKay, R. Hoover, W.J. Blot, and J.F. Fraumeni, Jr., 1975. *Atlas of Cancer Mortality for U.S. Counties: 1950-1969.* United States Department of Health, Education and Welfare, Publ. No. (NIH) 75-780.

Stalling, D.L., R.C. Tindle and J.L. Johnson, 1972. Cleanup of pesticides and polychlorinated biphenyl residues in fish extracts by gel permeation chromatography. *J. Ass. Offic. Anal. Chem.,* **55:** 32-38.

Human Exposure to Pentachlorophenol

RALPH C. DOUGHERTY

Abstract—Several reports have suggested that people living in industrialized societies will generally be contaminated with pentachlorophenol (PCP) at concentrations of 1 ppb and higher. The most likely source for this contamination appears to be food chain exposure to PCP treated wood products. Another possible source of exposure is metabolism of hexachlorobenzene to pentachlorophenol. Pentachlorophenol has been shown to be a mutagen and commercial preparations of the compound are often contaminated with dioxins. Both of these observations may be significant to the toxicological evaluation of the present hazard from pentachlorophenol exposure.

Introduction

Pentachlorophenol (PCP) is a widely used, versatile and very persistent pesticide. The literature prior to 1967 concerning the properties and occurrence of PCP in environmental substrates has been carefully reviewed by Bevenue and Beckman (1967). This review will focus on the literature subsequent to 1967 and will discuss the following areas: methodology for PCP analysis; chronic and acute human exposure to PCP; sources of exposure to PCP; contamination of PCP with dioxins; and toxic effects of PCP.

RALPH C. DOUGHERTY • Department of Chemistry, Florida State University, Tallahassee, Florida 32306, U.S.A.

Methodology for Pentachlorophenol Analysis

The most widely used technique for measuring PCP in environmental substrates involves an initial cleanup, followed by derivatization and analysis using electron capture gas chromatography. Other methods of analysis that have been used in recent past include mass spectrometry and ultraviolet ratio spectrophotometry. Gas chromatographic methods that are now in use are closely related to the original procedures used by Bevenue *et al.* (1966, 1967b) in their investigation of PCP in human urine.

Rivers (1972) introduced a new procedure for determination of PCP in human blood and urine that maintained the original diazomethane derivatization of PCP and gas chromatography with electron capture detection of the resultant pentachloroanisole that had been developed by Bevenue *et al.* (1966, 1968). The major changes that were introduced by Rivers (1972) involved a simplification of the cleanup procedure for both blood and urine samples. Recoveries of PCP using the River's method fell between 89 and 99%. PCP concentrations in six human plasma samples determined by this method ranged from 0.07 to 45.4 ppm.

Shafik (1973) developed a procedure for analysis of PCP and hexachlorophene in human adipose tissue. The analytical protocol in the Shafik method differed from those above in that he used diazoethane rather than diazomethane to derivatize the PCP. Average recoveries of PCP from human fat were 75%. In a series of nineteen samples of adipose tissue taken from the general population Shafik found levels that varied from 5 to 52 ppb of PCP. The average value for the nineteen measurements was 24 ppb. Although Shafik's method depended upon the extraction of PCP into 10% sodium hydroxide it is not clear that the conditions used would result in the hydrolysis of lipid soluble PCP conjugates; the results must, therefore, be considered as lower limits of the PCP concentration in the samples examined.

Hoben *et al.* (1976) used a very similar procedure to determine PCP levels in rat plasma, urine, tissue and in aerosol samples. In this study PCP from urine was derivatized by reaction with diazoethane and PCP from all other sources was derivatized by a reaction with diazomethane. Quantitation once again involved electron capture gas chromatography. PCP recovery from urine ranged between 92 and 99% while recoveries from tissue samples ranged between 87 and 96%. PCP in aerosol mixtures was determined after collection of the sample with a standard aerosol impinger.

Renberg (1974) has described a method for analysis of PCP, other chlorinated phenols and phenoxy acids in organic tissues, soil and water based on ion exchange separation using Sephadex QAE, A25 as the ion exchange bed. An organic extract of the substrate was subjected to ion exchange chromatography and the chlorinated phenols and phenoxy acids were later eluted from the column and derivatized with diazomethane prior to gas chromatography with electron capture detection. Ion exchange is a convenient procedure for separating acidic substances from normal biomolecules. Recoveries for PCP ranged from 92 to greater than 97% in samples of fish tissue, water and soil. PCP levels in fish found dead in a contaminated river ranged from 0.35 to 26 ppm. The PCP level in extractable fat from the same animals ranged from 1.9 to 1200 ppm. A water sample from the river taken two days after the discharge contained only 0.35 ppb of PCP.

Acetylation of PCP with acetic anhydride to produce pentachlorophenylacetate has also been used as a derivatization procedure prior to gas chromatographic analysis. Krijgsman and Van De Kamp (1977) have reported a capillary gas chromatographic method which can be used to identify the position isomers of the halophenols as well as PCP. Gas chromatographic analysis of PCP using the acetate derivative was first used by Rudling (1970).

The laurate ester of PCP is a component of some commercial wood preservative solutions. This substance can be conveniently analyzed by direct gas chromatography with electron capture detection (Dunn and Kelso, 1968).

It is possible to obtain analyses of PCP by gas chromatography without derivatization prior to analysis. A method for doing this was developed by Barthel *et al.* (1969) using ¼ inch packed columns with 3% of diethyleneglycol succinate and 2% phosphoric acid on 60 to 80 mesh chromasorb G. Detection limits for this column using a ^{63}Ni detector were between 0.1 and 0.02 ng of PCP. Gas chromatography without derivatization was also used by Canada and Regnier (1977) for determination of PCP using isotope ratio mass chromatography. The column used in this case was an 8 foot, ¼ inch packed column with 3% dexsil 300 on Gaschrom Q. By using isotope ratio mass chromatography a technique in which a characteristic chlorine isotope pattern is found in the mass spectra obtained by a GC mass spectrometer, these authors were able to detect approximately 1.2 μg of PCP in a complex mixture.

Negative chemical ionization mass spectrometry produces a very small number of ions from any given substrate and is generally very selective for toxic substances as compared to biomolecules. This technique, which involves operating a mass spectrometer with a source pressure of approximately 1 torr in the negative ion mode, has been used for both qualitative and quantitative identification of PCP in biological mixtures (Dougherty and Piotrowska, 1976). The analytical procedure for human urines involved hydrolysis with sulfuric acid to cleave urinary conjugates, extraction with pesticide grade hexane, evaporation of the extract after it has been washed with water and dried with magnesium sulfate, and analysis of the residue by negative chemical ionization mass spectrometry. Standardization was accomplished through use of internal *p*-chlorobenzophenone. Sensitivities exceeded 1 ng. Recoveries at the level of one ppm exceeded 98%; recoveries at the level of 10 ppb exceeded 80%.

Direct use of GC-mass spectrometry and gas chromatography with electron capture detection for determination of PCP in river water has been described by Buhler *et al.* (1973). Sensitivity of the overall procedure exceeded 50 ppb using 900 ml samples. PCP was methylated with diazomethane prior to analysis.

One of the simplest and least expensive procedures for determining PCP has been developed by Fountaine *et al.* (1975). The procedure is based on ultraviolet ratio spectrophotometry which is obtained as a function of pH for a solution of PCP and other substances in water. The absorption spectrum of PCP and other phenols depends strongly upon pH. By observing absorbances at 342.7 nm (a resonance line from a copper hollow cathode light source) at pH 2 and pH 10 the authors were able to demonstrate a linear calibration for the system in the region 10 ppb to 500 ppb for PCP in water. They examined more than twenty compounds, including many of the polyhalogenated phenols, for interferences and they found none.

Chronic and Acute Human Exposure to Pentachlorophenol

Bevenue *et al.* (1967b) conducted the first and by far the most extensive study of PCP levels in human biofluids for nonoccupationally exposed individuals. The Bevenue group found an average of 40 ppb of PCP (range: sub ppb to 1.8 ppm) in the urines of 117 nonoccupationally exposed individuals. In another group of 173 individuals Bevenue *et al.* found an average of 44 ppb of PCP in urine with a high of 570 and a low of 3 ppb. Akisada (1965) reported a range of 10 to 50 ppb for PCP in the urines of 20 individuals in Japan. Cranmer and Freal (1970) reported a range of 2 to 11 ppb for PCP in the urine of 6 nonoccupationally exposed individuals in Great Britain. Our studies of PCP in the urines of 60 university students in Florida found a range of 9 to 80 ppb of PCP in urine with an average of approximately 20 ppb (Dougherty and Piotrowska, 1976). The EPA human monitoring program has detected a range of 1 to 193 ppb of PCP in the urines of 34 nonoccupationally exposed individuals from different parts of the United States (Kutz *et al.*, 1976a).

In the study of plasma levels of PCP in controls and patients on chronic hemodialysis in New Orleans, Louisiana, Pearson *et al.* (1976) found an average PCP plasma level of 16 ppb in 23 dialysis patients and an average of 15 ppb in the plasma of 14 controls. These values are quite close to the urine values and to the value for the PCP concentration in human adipose tissue obtained by Shafik (1973). Shafik found an average concentration of 25 ppb of PCP in human fat with a range of 5 to 52 ppb. These fat samples were said to come from the general public, and one must assume that the donors were residents of south Florida.

The combined results of all of these studies suggest that contamination of human populations with PCP at a level of 10 to 20 ppb is quite general in industrialized societies.

Acute intoxication with PCP or its salts is an uncommon occurrence; however, a number of fatalities have been reported. Bevenue and Beckman reviewed the acute toxicology of PCP prior to 1967. Postmortem samples of serum, tissue and urine from individuals who died from PCP intoxication have contained between 20 and 140 ppm of PCP in tissue and 28 to 96 ppm of PCP in urine. Bevenue *et al.* (1967b) reported 6 cases of occupationally exposed pest control operators in Hawaii whose urinary PCP exceeded 10 ppm. The highest value recorded was 36 ppm. It is possible that tolerance to PCP might develop with continued exposure, which would account for a lack of obvious symptoms in the heavily exposed pest control operators.

There was an unfortunate incident in which 20 infants in a small hospital in St. Louis developed an unusual illness which was severe in 9 cases and lethal in 2 cases. The illness was traced to the misuse of a laundry product which contained sodium pentachlorophenate (Barthel *et al.*, 1969; Armstrong *et al.*, 1969). Postmortem tissue samples of one of the children who died contained 21 to 33 ppm pentachlorophenol. The serum levels of PCP in another infant ranged from 118 ppm prior to a blood transfusion to 31 ppm on the following day. Serum levels of PCP in 6 exposed, asymptomatic infants ranged 7 to 26 ppm. Concentrations of PCP in diapers used in the nursery ranged from 109 to 172 ppm. Serum values of PCP in two healthy control infants were 69 and 459 ppb respectively.

Jirasek *et al.* (1976) reported a study of 80 industrial workers who were exposed to sodium pentachlorophenate and 2,4,5-trichlorophenoxyacetic acid. These workers showed a variety of symptoms including chloracne, porphyria cutanea tarda, disorders in porphyrin metabolism, fat metabolism and carbohydrate metabolism, as well as neurological lesions. The workers were exposed to a number of compounds in the industrial environment. Although some of the effects may have been due to PCP the major causative agent was believed to be 2,3,7,8-tetrachlorodibenzo-*p*-dioxin.

There have been two reports of PCP intoxication intermediate between general chronic exposure and lethal or nearly lethal exposure. Ueda *et al.* (1962) reported that four families in Japan became weak and suffered from throat and skin irritation after drinking and bathing in water that contained 12.5 ppm PCP. Bevenue *et al.* (1967a) reported the results of a case in which a man bathed his hands in a PCP solution for 10 minutes while cleaning a paint brush. Pain in the man's hands caused him to stop. Two days later his urine showed 236 ppb of PCP. One month elapsed before the urinary PCP for this individual had returned to "background" levels, i.e. 17 ppb.

Sources of Chronic Human Exposure to Pentachlorophenol

Detrick (1977) has analyzed possible sources in human exposure to PCP and suggested four possible sources. These were: (1) natural formation of PCP in the environment, (2) PCP appearing as the result of metabolism of other chlorinated compounds, (3) formation of PCP in water chlorination systems and (4) intake of PCP as a result of human activities.

The suggestion that PCP is a natural compound in the environment seems reasonably unlikely. There is no evidence that PCP is a natural metabolite of any organism. The identification of the fungal antibiotic Drosophilin *A* as *p*-methoxytetrachlorophenol (Anchel, 1952) was cited by Detrick as evidence that PCP may be a natural part of the biosphere. A large number of polychlorinated organics including carbon tetrachloride are known to be natural metabolites of microorganisms particularly those living in the ocean. It seems quite likely that if PCP were a natural metabolite, it would have been identified as such because of its widespread presence.

Humans in the United States are exposed to a substantial number of organochlorine pesticides and human adipose tissue contains residues of these chemicals. The EPA National Human Monitoring Program presently monitors human adipose tissue for a number of polychlorinated organics. These include DDT and its metabolites, BHC and its isomers, aldrin, dieldrin, heptachlor, heptachlorepoxide, endrin, mirex, oxychlorodane, trans-nonachlor, polychlorinated biphenyls and hexachlorobenzene. The fact that hexachlorobenzene residues are not discussed in the 1976 EPA report (Kutz *et al.*, 1976b) suggests that these residues in human tissues, if present, exist at only low levels.

Pentachlorophenol was the major metabolite of hexachlorobenzene in rats (Engst *et al.*, 1976). In rat urine there was nearly twice as much PCP as hexachlorobenzene. Fecal excretion was dominated by hexachlorobenzene with some pentachlorobenzene. The fact that hexachlorobenzene has not been detected in human urine suggests that metabolism of hexachlorobenzene should be only a minor contributor to the human PCP load.

Detrick (1977) has suggested that chlorination of phenol in water supplies in sewage effluents may be responsible for the wide occurrence of PCP. He reports that chlorination of 1 ppm of phenol in water with 10 ppm of chlorine leads to the production of 0.2 ppb of PCP. Although these results are interesting, it seems quite unlikely that this route would substantially contribute to the human PCP burden. A concentration of 1 ppm of phenol in water is intolerable from the point of view of taste, in fact humans can detect phenol in water in the ppb range. The perchlorination of phenol with hypochlorite must require photochemical activation, a situation which is not generally obtainable in water treatment or sewage chlorinators.

A longitudinal study conducted by Bevenue *et al.* (1976b) and the limited longitudinal study conducted in this laboratory (Dougherty and Piotrowska, 1976) have indicated substantial variations in PCP concentrations in urine as a function of time. Studies of PCP metabolism have indicated that roughly half of the ingested PCP is excreted unchanged (Ahlborg *et al.*, 1974). Furthermore the half life for excretion of PCP in mammals appears to be fairly short. The variation in PCP levels in human urines would thus suggest a variation in exposure with time. The most likely source of this variation is the food chain.

PCP has been found as a residue in the total diet (Manske and Corneliassen, 1974) by the Food and Drug Administration. The number of composite samples that contained PCP was only two out of a total of 360 composite samples, the maximum concentration was 11 ppb. PCP has been found at concentrations ranging from 1 to 4 ppb in food fish (Zitko *et al.*, 1974). A microbial PCP metabolite, pentachloroanisole has been found in concentrations ranging between 1 and 18 ppb in broiler chicken tissues (Harper and Banave, 1975). The source of pentachloroanisole in broiler tissues appears to be wood shavings from PCP-treated lumber.

In a spot survey of selected items from the food chain we found PCP residues in powdered dry milk, soft drinks, bread, candy bars, cereal, noodles, rice, sugar, and wheat (Dougherty and Piotrowska, 1976). The concentrations ranged from 1 ppb to 0.1 ppm. The presence of PCP residues in all of the grain and sugar products that we examined would be consistent with the storage of these products in PCP-treated wooden storage containers.

The estimated annual production of PCP is 200 million pounds world-wide (Nilsson *et al.*, 1974). Roughly 80% of the annual production is used for preservation of wood. Surface treatment of wood with PCP is used for preservation during shipment. Pressure treatment with PCP and its derivatives is used to induce long term stability of wood that is used in exposed or wet environments. Other uses of pentachlorophenol take advantage of its antifungal, antibacterial and contact defoliant properties.

Pentachlorophenol Contamination with Dioxins

Most pentachlorophenol preparations are contaminated to a greater or lesser extent with hexachloro-, heptachloro-, and octachlorodibenzo-p-dioxins. Concern about dioxin contamination is based primarily on the extremely high toxicity of 2,3,7,8-tetrachlorodibenzo-p-dioxin (TCDD). 1,2,3,4,7,8-hexachlorodibenzo-p-dioxin (HCDD) is ten times less effective in inducing arylhydrocarbon hydroxylase activity than TCDD (Bradlow *et al.*, 1975). Although heptachloro- and octachlorodioxin (OCDD) are both less toxic than HCDD, their presence in materials that are produced in large quantities and distributed throughout the environment are of some concern. The concern stems not only from their toxicity but also from their persistence and potential for biological magnification.

One of the first records of dioxin concentrations in commercial PCP samples indicated that all samples examined (a total of 8) contained HCDD at levels ranging from 0.17 to 39 ppm (Firestone *et al.*, 1972). The same authors also reported the presence of hepta and octachlorodioxins in the PCP samples they examined. The examination of a technical grade PCP sample in 1973 indicated that it contained hexa, hepta, and octachlorodioxins at concentrations of 42, 24 and 11 ppm, respectively (Villanueva *et al.*, 1973).

The amount of polychlorodioxins that one finds in a given sample depends strongly on the procedure that is used to cleanup the sample prior to analysis (Villanueva *et al.*, 1975). Using four different procedures from the literature Villanueva *et al.* obtained recoveries that varied from 57 to 99% for samples that contained 100 ppm of OCDD. Electron capture analysis for hexachlorodioxin in pentachlorophenol using three different methods gave values ranging from 109 to 551 ppm, corresponding values for OCDD ranged from 204 to 8,042 ppm.

Buser and Bosshardt (1976) developed a new method for analysis of tetra-, hexa- and octachlorodioxins in pentachlorophenol. This gas chromatographic, mass spectrometric method gave recoveries for tetra-, hexa-, and octachlorodioxin that were 80 to 95%, 80 to 95% and 95% respectively. The concentrations of tetra-, hexa-, and octachlorodioxin found in samples of pentachlorophenol obtained from Swiss manufacturers in 1973 ranged from less than 10 to 250 ppb for TCDD, less than 30 ppb to 10 ppm for HCDD, and 1.5 to 370 ppm for OCDD.

Manufacturers of pentachlorophenol have been aware of problems associated with dioxin contamination of their products and have gone to considerable lengths to decrease the concentrations of dioxins in the commercial material. Two reports from Dow Chemical Company indicate that their recent products contain less than half a part per million of HCDD and the concentration of OCDD ranged between 2 and 16 ppm (Blaser *et al.*, 1976; Pfiffler, 1976).

Toxicity of Pentachlorophenol

The fact that lethal concentrations of pentachlorophenol in human biofluids have ranged from 20 to 190 ppm is consistent with observations of pentachlorophenol toxicity in other mammalian species (Bevenue and Beckman, 1967). The question of an appropriate safety margin for long term exposure to toxic substances is a subject of considerable debate. If an appropriate safety margin were a factor of 10^{-3}, current human exposure to pentachlorophenol appears to be right at that safety margin. A number of scientists maintain that in the case of mutagenic substances there is no appropriate safety margin and permissible concentrations of mutagens in the environment should be below the parts per trillion level, that is, in the concentration region corresponding to only a few molecules per cell.

There are conflicting reports in the literature concerning the mutagenicity of pentachlorophenol. Anderson *et al.* (1972) found that PCP was incapable of inducing point mutations in four different microbial systems. The microbes used in these studies included histidine requiring mutants of *Salmonella typhimurium*, T_4 bacteriophage, rII mutants and *E. coli* mutants. The tests did not employ activation using liver microsomes.

Masurekar *et al.* (1972) reported that treatment of *Penicillium chrysogenum* colonies, that had previously been treated with ethylmethane sulfonate, with 25 ppm of PCP in water for 7.5 minutes increased the auxotroph frequency in the mutant population by up to a factor of ten.

Fahrig (1975) reported that PCP at a concentration of 1.9 mM and a contact time of 6 hours increased the mitotic gene conversion rate at 2 different loci of *S. cerevisiae* by a factor of 15 and 12 respectively. Fahrig also noted Schoeller's finding that PCP gave weakly positive results for chromosome aberrations in human lymphocytes cultured *in vitro*.

Innes *et al.* (1969) concluded that PCP, when administered at the maximum tolerable dose (46.4 mg/kg/day), gave no significant indication of tumorigenicity after oral administration in selected strains of mice.

The major toxic effect of PCP on biological systems is not directly related to its mutagenicity. PCP is known to be an uncoupler of oxidative phosphorylation and it has been shown to alter the electrical conductivity of membranes (Smejtek *et al.*, 1976; Kolesova *et al.*, 1973). The fact that PCP is a broad spectrum biocide may be largely due to its influence on membrane properties and oxidative phosphorylation.

The toxic effects of PCP can not be divorced from the effects of the polychlorodioxins that contaminate commercial formulations. Goldstein *et al.* (1977) have shown that commercial PCP had a significantly larger effect on induction of liver enzymes, specifically aryl hydrocarbon hydroxylase, in rats than the reagent grade compound. The difference was ascribed to the presence of polychlorodioxins in the commercial PCP. The dioxin concentrations in the commercial PCP used were 8 ppm HCDD, 520 ppm heptachlorodioxin and 1380 ppm OCDD. The fact that the dioxins are considerably less water soluble than PCP and thus subject to a much larger biomagnification (Neely *et al.*, 1974) makes this finding ominous for exposure to PCP from commercial sources.

Conclusions

Studies cited here have indicated a very general human exposure to pentachlorophenol at concentrations ranging between 1 and 100 ppb. These concentrations are 10^{-3} to 10^{-4} of the lethal concentration for this compound in the enviroment. The most likely sources for human exposure to pentachlorophenol are the food chain and direct contact with PCP-treated wood products. Contamination of the food chain is probably related to pentachlorophenol treatment of storage structures for food products. Another possible source of PCP contamination is the metabolism of hexachlorobenzene obtained from environmental sources.

In view of studies that suggest that pentachlorophenol is mutagenic or at least a comutagen, it seems likely that current human exposure to pentachlorophenol poses a significant health hazard.

References

Ahlborg, U.G., J.E. Lindgren, and M. Mercer, 1974. Metabolism of pentachlorophenol. *Arch. Toxicol.,* **32:** 271-281.

Akisada, T., 1965. Simultaneous determination of pentachlorophenol and tetrachlorophenol in air and urine. *Bunseki Kagaku,* **14:** 101-105.

Anchel, M., 1952. Identification of Drosophilin A as p-methoxytetrachlorophenol. *J. Am. Chem. Soc.,* **74:** 2943.

Anderson, K.J., E.G. Leighty, and M.T. Takahashi, 1972. Evaluation of herbicides for possible mutagenic properties, *J. Agr. Food Chem.,* **20:** 649-656.

Armstrong, R.W., E.R. Eichner, D.E. Klein, W.F. Barthel, J.V. Bennett, V. Jonsson, H. Bruc, and L.E. Loveless, 1969. Pentachlorophenol poisoning in a nursery for newborn infants. II. Epidemiologic and toxicologic studies. *J. Pediatrics,* **75:** 317-325.

Barthel, W.F., A. Curley, C.L. Thrasher, V.A. Sedlak, and R. Armstrong, 1969. Determination of pentachlorophenol in blood, urine, tissue and clothing. *J. Ass. Offic. Anal. Chem.,* **52:** 294-298.

Bevenue, A., and H. Beckman, 1967. Pentachlorophenol: A discussion of its properties and its occurrence as a residue in human and animal tissue. *Residue Rev.,* **19:** 83-129.

Bevenue, A., M.L. Emerson, L.J. Casarett, and W.L. Yauger, 1968. A sensitive gas chromatographic method for the determination of pentachlorophenol in human blood. *J. Chromatogr.,* **38:** 467-472.

Bevenue, A., T.J. Haley, and H.W. Klemmer, 1967a. A note on the effects of a temporary exposure of an individual to pentachlorophenol. *Bull. Environ. Contam. Toxicol.,* **2:** 293-296.

Bevenue, A., J. Wilson, L.J. Casarett, and H.W. Klemmer, 1967b. A survey of pentachlorophenol content in human urine. *Bull. Environ. Contam. Toxicol.,* **2:** 319-333.

Bevenue, A., J.R. Wilson, E.F. Porter, M.K. Song, H. Beckman, and G. Mallett, 1966. A method for determination of pentachlorophenol in human urine in picogram quantities. *Bull. Environ. Contam. Toxicol.,* **1:** 257-266 (1966).

Blaser, W.W., R.A. Bredeweg, L.A. Shadoff, and R.H. Stehl, 1976. Determination of chlorinated dibenzo-p-dioxins in pentachlorophenol by gas chromatography — mass spectrometry. *Anal. Chem.,* **48:** 984-986.

Bradlow, J.A., L.H. Garthoff, D.M. Graff, and N.E. Harley, 1975. Detection of chlorinated dioxins: induction of aryl hydrocarbon hydroxylase activity in rat hepatoma cell culture. *Toxicol. Appl. Pharmacol.,* **33:** 166.

Buhler, D.R., M.E. Rasmusson, and H.S. Nakave, 1973. Occurrence of hexachlorophene and pentachlorophenol in sewage and water. *Environ. Sci. Tech.,* **7:** 929-934.

Buser, H.R., and H.P. Bosshardt, 1976. Determination of polychlorinated dibenzo-*p*-dioxins and dibenzofurans in commercial pentachlorophenols by combined gas chromatography — mass spectrometry. *J. Ass. Offic. Anal. Chem.,* **59:** 562-569.

Canada, D.C., and F.E. Regnier, 1976. Isotope ratios as a characteristic selection technique for mass chromatography. *J. Chromatogr. Sci.,* **14:** 149-154.

Cranmer, M.F., and J. Freal, 1970. Gas chromatographic analysis of pentachlorophenol in human urine. *Life Sci.,* **9:** 121-128.

Detrick, R.S., 1977. Pentachlorophenol, possible sources of human exposure. *Forest Prod. J.,* **27:** 13-16.

Dougherty, R.C., and K. Piotrowska, 1976. Screening by negative chemical ionization mass spectrometry for environmental contamination with toxic residues: application to human urines. *Proc. Natl. Acad. Sci. U.S.A.,* **73:** 1777-1781.

Dunn, P., and A.G. Kelso, 1968. The determination of pentachlorophenol laurate using electron capture gas chromatography. *J. Gas Chromatogr.,* **6:** 113-114.

Engst, R., R.M. Macholz, and M. Kujawa, 1976. The metabolism of hexachlorobenzene (HCB) in rats. *Bull. Environ. Contam. Toxicol.,* **16:** 248-252.

Fahrig, R., 1975. Comparative mutagenicity studies with pesticides. IARC Scientific Publications No. **10:** 161-181.

Firestone, D., J. Ress, N.L. Brown, R.P. Barron and J.N. Domico, 1972. Determination of polychlorodibenz-*p*-dioxins and related compounds in commercial chlorophenols. *J. Ass. Offic. Anal. Chem.,* **55:** 85-93.

Fountaine, J.E., P.B. Joshipura, P.N. Keliher and J.D. Johnson, 1975. Determination of pentachlorophenol by ultraviolet ratio spectrophotometry. *Anal. Chem.,* **47:** 157-159.

Goldstein, J.A., M. Friesen, R.E. Linder, P. Hickman, J.R. Hass, and H. Bergman, 1977. Effects of pentachlorophenol on hepatic drug-metabolising enzymes and porphyria related to contamination with chlorinated dibenzo-*p*-dioxins. *Biochem. Pharm.,* **26:** 1549-1557.

Harper, D.B., and D. Banave, 1975. Chloroanisole residues in broiler tissues. *Pesticide Sci.,* **6:** 159-163.

Hoben, H.J., S.A. Ching, L.J. Casarett, and R.A. Young, 1976. A study of the inhalation of pentachlorophenol by rats, part I. A method for the determination of pentachlorophenol in rat plasma, urine and tissue and in aerosol samples. *Bull. Environ. Contam. Toxicol.,* **15:** 78-85.

Innes, J.R.M., B. Mulland, M.G. Valerio, L. Petrucelli, L. Fishbein, E.R. Hart, A.J. Pallotta, R.R. Bates, H.L. Falk, J.J. Gart, M. Klein, I. Mitchell, and J. Peters, 1969. Bioassay of pesticides and industrial chemicals for tumorigenicity in mice: a preliminary note. *J. Nat. Cancer Inst.,* **42:** 1101-1114.

Jirasek, L., J. Kalensky, K. Kubec, J. Pazderova, and E. Lukas, 1976. Chlorakne, parphyria cutanea tarda and andere intoxikationen durch herbizich. *Hautartt,* **27:** 328-333.

Kolesova, G.M., L.I. Boguslavskii, and L.S. Yaguzhinskii, 1973. Interaction of pentachlorophenol — an uncoupler of oxidative phosphorylation — with lecithin. *Dokl. Akad. Nauk. SSSR,* **210:** 1453-1456.

Krijgsman, W., and C.G. van de Kamp, 1977. Determination of chlorophenols by capillary gas chromatography. *J. Chromatogr.,* **131:** 412-416.

Kutz, F.W., R.S. Murphy, and S.C. Strassman, 1976a. Estimation of exposure to specific pesticides in the U.S. population: a national survey approach. Paper presented at the 104th Annual Meeting of the American Public Health Association, Miami Beach, Florida.

Kutz, F.W., A.R. Yobs, and S.C. Strassman, 1976b. Organochlorine pesticide residues in human adipose tissue. *Bull. Soc. Pharm. Environ. Pathologists,* 4: 17-19.

Manske, D.D., and P.E. Corneliassen, 1974. Residues in food and feed, pesticide residues in total diet samples (VIII). *Pesticide Monit. J.,* 8: 110-124.

Masurekar, P.S., M.P. Kahgan, and A.L. Demain, 1972. Mutagenesis and enrichment of auxotrophs in *Penicillium chrysogenum. Appl. Microbiol.,* 24: 995-996.

Neely, W.B., D.R. Branson, and G.E. Blau, 1974. Partition Coefficient to measure bioconcentration potential of organic chemicals in fish. *Environ. Sci. Tech.,* 8: 1113-1115.

Nilsson, C.A., K. Andersson, C. Rappe and S.O. Westermark, 1974. Chromatographic evidence for the formation of chlorodioxins from chloro-2-phenoxyphenols. *J. Chromatogr.,* 96: 137-147.

Pearson, J.E., C.D. Schultz, J.E. Rivers and F.M. Gonzalez, 1976. Pesticide levels of patients on chronic hemodialysis. *Bull. Environ. Contam. Toxicol.,* 16: 556-558.

Pfiffler, C.D., 1976. Determination of chlorinated dibenzo-*p*-dioxins in pentachlorophenol by liquid chromatography. *J. Chromatogr. Sci.,* 14: 386-391.

Renberg, L., 1974. Ion exchange technique for the determination of chlorinated phenols and phenoxy acids in organic tissue, soil and water. *Anal. Chem.,* 46: 459-461.

Rivers, J.B., 1972. Gas chromatographic determination of pentachlorophenol in human blood and urine. *Bull. Environ. Contam. Toxicol.,* 8: 294-296.

Rudling, L., 1970. Determination of pentachlorophenol in organic tissues and water. *Water Res.,* 4: 533-537.

Shafik, T.M., 1973. The determination of pentachlorophenol and hexachlorophene in human adipose tissue. *Bull. Environ. Contam. Toxicol.,* 10: 57-63.

Smejtek, S., K. Hsu, and W.H. Perman, 1976. Electrical conductivity in lipid bilayer membranes induced by pentachlorophenol. *Biophys. J.,* 16: 319-335.

Ueda, K., M. Nagai, and M. Osafune, 1962. Contamination of drinking water with PCP (pentachlorophenol). *Osaka Shiritsu Eisei Kenkyu-sho Kenkyu Hokoku,* 7: 19-22.

Villanueva, E.C., V.W. Burse, and R.W. Jennings, 1973. Chlorodibenzo-*p*-dioxin contamination of two commercially available pentachlorophenols. *J. Agr. Food Chem.,* 21: 739-740.

Villanueva, E.C., R.W. Jennings, V.W. Burse, and R.C. Kimbrough, 1975. A comparison of analytical methods for chlorodibenzo-*p*-dioxins in pentachlorophenol. *J. Agr. Food Chem.,* 23: 1089-1091.

Zitko, V., O. Hutzinger, and P.M.K. Choi, 1974. Determination of pentachlorophenol and chlorobiphenols in biological samples. *Bull. Environ. Contam. Toxicol.,* 12: 649-652.

Survey of Pesticide Residues and Their Metabolites in Urine from the General Population

FREDERICK W. KUTZ, ROBERT S. MURPHY, and
SANDRA C. STRASSMAN

Abstract—The National Human Monitoring Program for Pesticides is collaborating with the U.S. Public Health Service in a three year study to assess the exposure of the general population of the United States of America through analysis of human urine for residues of selected pesticides and their specific metabolites. Samples are collected on a national probability design and analyzed for selected organochlorine, carbamate, chlorophenoxy and organophosphorous pesticides. Medical, nutritional and pesticide usage data are available from each donor. Preliminary results of the urine analyses indicate that the general population is being exposed to some of these types of pesticides.

Introduction

Residues of pesticides and their metabolites in various human tissues and fluids collected from the general population are indicative of the total body burden of these pesticides and of past and present exposure to them. It should be emphasized that most members of the general population are not occupationally exposed to pesticides; therefore, their contact comes from other more covert sources. Pesticides may gain entrance to the human body through the intestine subsequent

FREDERICK W. KUTZ and SANDRA C. STRASSMAN • Ecological Monitoring Branch (WH-569), United States Environmental Protection Agency, Washington, D.C. 20460, U.S.A.

ROBERT S. MURPHY • National Center for Health Statistics, United States Public Health Service, Rockville, Maryland, U.S.A.

to ingestion; through the lungs as a result of airborne pesticide-laden dusts, vapors and aerosols; by penetration through the intact skin; and (rarely) by absorption directly into the bloodstream through broken skin.

Once within the human body, the residue is subjected to numerous metabolic processes. In the case of certain lipophilic organochlorine pesticides, residues of the parent compound or metabolites are assimilated and stored in the lipid portion of adipose tissues. Residues of these chemicals also may be detected in the lipid portion of such fluids as milk and blood serum. On the other extreme are the pesticides which are rapidly metabolized and excreted. Certain of the organophosphorous and carbamate pesticides undergo these dynamic changes. Other chemicals such as certain organochlorine and chlorophenoxy herbicides are capable of passing directly through the human body virtually intact, and then excreted.

From a regulatory perspective, findings of pesticide residues in humans representative of the general population provide a major element in pesticide policy decision-making. These residues are demonstrative of the extent of the environmental distribution of the particular pesticide, and when coupled with laboratory animal data showing adverse biological effects, signal a potential public health hazard. In recent decisions regarding the uses of aldrin, dieldrin, heptachlor and chlordane, the Administrator of the U.S. Environmental Protection Agency cited and considered the findings of these chemicals in human tissue.

One program which examined humans for pesticide residues and associated chemicals is the National Human Monitoring Program for Pesticides. This program is operated by the U.S. Environmental Protection Agency and functions to determine, on a national scale, the incidence and level of exposure to pesticides experienced by the general population and to identify trends in these factors when they occur. At present, the human monitoring survey examines adipose tissue, urine and blood serum. The adipose tissue samples are collected by the cooperating pathologists while the blood serum and urine are collected under a cooperative arrangement with the National Center for Health Statistics (NCHS) of the U.S. Public Health Service. This paper briefly describes the ongoing collaborative study and indicates the types of results from urine analyses which will be available at the end of the study in 1979.

Health and Nutritional Examination Survey

The Health and Nutritional Examination Survey II (HANES II) of the National Center for Health Statistics, entails collecting data from a scientifically selected segment of the general population. Survey personnel will collect samples of blood serum and urine from a probability sample of persons 12 to 74 years old for pesticide residue and metabolite determination. The results of this cooperative arrangement will establish baseline data on the exposure of the general population to organophosphate, carbamate, chlorophenoxy and certain organochlorine pesticides; correlate residue and metabolite data with various medical and nutritional parameters; and collect some information on the pesticide use patterns of the general population.

The purpose of HANES II is to measure the prevalence of certain health and nutritional indicators, and to monitor change in them over time. Major conditions of interest are anemia, diabetes, kidney disease, heart disease, liver disease, hypertension, hearing problems, allergies, osteoarthritis and disc degeneration in the cervical and lumbar spines, respiratory function, and a limited assessment of some speech problems in young school children. Evaluation of nutritional status will include dietary intake and food frequency data interrelated with physical examination, medical history, and biochemical assessment data.

The following medical assessments are being made by the Health and Nutritional Examination Survey II:

a) Nutritional and general health assessment: a national probability sample of the civilian, non-institutionalized population are being selected in such a manner that estimates of health and nutritional status can be made for persons six months through 74 years of age, and specifically for population groups at high risk of poor nutrition, e.g., preschool children, the aged, the poor and women of child-bearing age. It will be possible to make some estimates for minority persons within these groups. Hawaii, Alaska, and American Indian Reservation lands are included in the target universe.

b) Diabetes is being assessed on adults 20-74 years of age.

c) Liver disease is being assessed on adults 35-74 years of age.

d) Arthritis of the back and spine and heart disease is being assessed on adults 25-74 years of age.

e) Kidney disease is being assessed in the group 12-74 years of age.

f) Respiratory function and hypertension is being assessed on sample persons 6-24 years of age.

g) Allergies are being assessed on sample persons 6-74 years of age.

h) Speech defects are being assessed on 4, 5 and 6 year olds. Audiometry testing is being done on children 4-19 years of age.

Sample Size and Locations

Approximately 28,000 persons are being selected in 64 communities throughout the four broad census regions of the conterminous United States. It is estimated that 22,000 people representing a national probability sample will be examined between February 1976 and June 1979.

Each community in the conterminous United States is grouped into one of 64 strata on the basis of similar characteristics. One community is selected from each stratum. Some communities such as New York and Los Angeles are populous enough to define a unique stratum and, hence, are selected into the sample with a probability of one. Within a community, a systematic sample of clusters of 16 households is selected; HANES sample units are a random selection of 8 of the 16 households of each cluster. A Census Bureau interviewer conducts a household interview and selects the HANES sample persons from the household roster according to the following rules:

¾ of those persons less than six years and 60 years and over.

¼ of those persons 6-59 years.

Upon completing the household interview and the required medical histories, the Census interviewer arranges an appointment for examination in the HANES mobile examination centers. Included in the medical history are several questions relating to pesticide usage practices.

The types of tabulations and publications planned are similar to those used in publications from earlier Health Examination Survey programs. The publications will include:

a) Publication of national estimates of the *prevalence* of target conditions or states such as persons with specific pesticides levels in excess of a given amount, diabetes, kidney disease, heart disease, anemias, liver disease, hypertension, back and neck pathology, etc., cross-classified by demographic and socio-economic variables;

b) Publication of the results of *monitoring* some indicators of health and nutritional status such as changes in the level of iron deficiency in the U.S. population for HANES I to HANES II;

c) Publication of *normative data* on such characteristics as pesticide levels, body measures, spironmetry, dietary intake, and laboratory assessments such as vitamin, trace element and chloesterol levels in the blood or urine; and

d) Publications which explore the *interrelationships* among different health characteristics, such as the interrelationship of diabetes, kidney disease and hypertension, and the relationship of body measures or anemias to other indicators of nutritional status, e.g., dietary intake.

All pesticide residue analyses are being conducted by contract laboratories using only methodologies specified by the program. These laboratories are equipped with gas-liquid chromatographs with electron capture and flame photometric detectors. All laboratories are required to maintain acceptable performance levels in the interlaboratory quality assurance program, established and moderated by the EPA Environmental Toxicology Division, Research Triangle Park, North Carolina. This laboratory also serves as a source of technical consultants for the analytical portion of the program.

Pesticide Residues in Human Urine Samples

The multi-residue approach used to analyze the urine samples detects the chemicals as shown in Table 1. The pesticide origins of these chemicals are also presented. Thin layer chromatography, electrolytic conductivity detectors and in some cases, combined gas chromatography-mass spectrometry are employed as confirmatory analytical techniques.

The results of the urine analyses for pesticide residues and metabolites are shown in Tables 2 to 5. It should be emphasized here that these are preliminary results and should not be construed as representative of the general population. As can be seen, residues of pesticides and their metabolites are being detected in human urine. Among the pesticides detected, pentachlorophenol was the most ubiquitous compound encountered as it was found in 85% of the urine samples analyzed (Table 2). When the survey is completed, estimates of levels and

TABLE 1. List of Chemicals Detectable in Human Urine and their Pesticide Origin

Chemical	Pesticide Origin
(Carbamate Method)	
α-Naphthol	Carbaryl and Naphthalene
Isopropoxyphenol	Propaxur
Carbofuran phenol and 3-Ketocarbofuran	Carbofuran
(Malathion Method)	
α-Monocarboxylic acid (MCA) and Dicarboxylic acid (DCA)	Malathion
(Multiphenol Method)	
Pentachlorophenol	Pentachlorophenol, Lindane and Hexachlorobenzene
p-Nitrophenol	Methyl and Ethyl Parathion
2,4-D	2,4-D
2,4,5-T	2,4,5-T
Silvex	Silvex
3,5,6-Trichloro-2-pyridinol (3,5,6-TC-2-P)	Chloropyrifos
2,4,5-Trichlorophenol (2,4,5-TCP)	2,4,5-Trichlorophenol used as a disinfectant; or a metabolite of certain organochlorine insecticides
Dicamba	Dicamba
(Dialkyl Phosphate Method)	
Dimethyl phosphate (DMP)	Any organophosphorus insecticide
Diethyl phosphate (DEP)	containing these phosphate or
Dimethyl phosphorothionate (DMTP)	thiosulphate molecules
Diethyl phosphorothionate (DETP)	
Dimethyl phosphorodithionate (DMDTP)	
Diethyl phosphorodithionate (DEDTP)	

frequencies can be made on many epidemiological, medical and geographic parameters.

A mechanism has been established between EPA and NCHS so that laboratory findings indicative of acute effects are reported to the volunteers' primary health care provider.

TABLE 2. Occurrence of Pesticide-related Phenolic Residues in Human Urine[a]

Residue[b]	Percent positive	Arithmetic mean (ppb)	Maximum value (ppb)
Pentachlorophenol	84.8	6.3	193.0
3,5,6,-TC-2-P	16.1	<5.0	31.7
2,4,5-TCP	1.7	< 5.0	32.4
p-Nitrophenol	1.7	< 10.0	113.0
Silvex	0.2	< 5.0	3.2
2,4-D	0	–	Trace
2,4,5-T	0	–	Trace
Dicamba	0	–	–

[a]Based on the analysis of 416-418 samples collected from the general population via the Health and Nutritional Examination Survey II, National Center for Health Statistics.
[b]Limits of detection range from 5 to 30 ppb (5 to 30 μg/l).

TABLE 3. Residues of Carbamate Pesticide Metabolites in Human Urine[a]

Chemical[b]	Percent positive	Arithmetic mean (ppb)	Maximum value (ppb)
α-Naphthol	13.8	<10	230
Isopropoxyphenol	1.3	< 40	150
Carbofuranphenol	0.7	< 40	70
3-Ketocarbofuran	1.0	< 30	140

[a]Based on the analysis of 302-370 samples collected from the general population via the Health and Nutritional Examination Survey II, National Center for Health Statistics.
[b]Limits of detection range from 10 to 40 ppb (10 to 40 μg/ l).

TABLE 4. Residues of Malathion Metabolites in Human Urine[a]

Metabolite[b]	Percent positive	Arithmetic mean (ppb)	Maximum value (ppb)
α-Monocarboxylic acid	9.4	< 30	970
Dicarboxylic acid	2.8	< 30	280

[a]Based on the analysis of 360-370 samples collected from the general population via the Health and Nutritional Examination Survey II, National Center for Health Statistics.
[b]Limits of detection = 30 ppb (30 μg/l).

TABLE 5. Occurrence of Dialkyl Phosphate Residues in Human Urine[a]

Dialkyl Phosphate Residue[b]	Percent positive	Arithmetic mean (ppb)	Maximum value (ppb)
Dimethyl Phosphate (DMP)	11.5	< 20	80
Diethyl Phosphate (DEP)	7.9	< 20	420
Dimethyl Phosphothionate (DMTP)	6.5	< 20	360
Diethyl Phosphothionate (DETP)	10.8	< 20	900
Dimethyl Phosphodithionate (DMDTP)	0	–	Trace
Diethyl Phosphodithionate (DEDTP)	0.2	< 20	20

[a]Based on the analysis of 418 samples collected from the general population via the Health and Nutritional Examination Survey II, National Center for Health Statistics.
[b]Limit of Detection for all residues = 20 ppb (20 μg/l).

Health Implications of 2,3,7,8-Tetrachlorodibenzo-p-dioxin Exposure in Primates

J.R. ALLEN and J.P. van MILLER

Abstract—Tetrachlorodibenzo-p-dioxin (TCDD) was administered in the diet at a level of 500 parts per trillion to eight adult female rhesus monkeys for nine months. Effects included acne, loss of hair and eyelashes and periorbital edema during the initial three months of exposure. There was also a decrease in 17β-estradiol and progesterone levels as well as the development of irregularities in their menstrual cycles. Inability to conceive and early abortions occurred in the animals following six months exposure to TCDD. Death in 5 of the 8 experimental animals was preceded by a severe pancytopenia. Microscopic changes included suppression of hematopoiesis in the bone marrow, hypocellularity of the lymphoid tissue, widespread hemorrhage, cardiac hypertrophy, edema and dilatation, hyperplasia and metaplasia of ductal tissue throughout the body, hypertrophic gastritis with ulceration, and dilatation of the biliary system. Death in most instances was attributed to the extensive hemorrhage that was associated with the thrombocytopenia.

Introduction

2,3,7,8-Tetrachlorodibenzo-p-dioxin (TCDD) is a contaminant in the widely used industrial chemical 2,4,5-trichlorophenol (Sparschu *et al.,* 1971). The production of trichlorophenol is the initial step in the production of 2,4,5-T, a widely used herbicide and hexachlorophene and trichlorophene, antibacterial agents that have

J.R. ALLEN and J.P. van MILLER • University of Wisconsin Medical Center, Department of Pathology and Regional Primate Research Center, Madison, Wisconsin 53706, U.S.A.

been incorporated in a number of soaps and cosmetics (Rawls and O'Sullivan, 1976). In the synthesis of trichlorophenol, minute amounts of TCDD are normally produced. However, excessively high temperature markedly increases the amount of TCDD that is formed (May, 1973).

While the environmental persistence of TCDD and many of its toxic manifestations are similar to other chlorinated hydrocarbons, this extremely toxic compound is capable of producing effects at levels thousands of times less than the other compounds (Neubert et al., 1973). Our knowledge of the effects of TCDD exposure has come from three different sources. Industrial accidents have resulted in severe human exposure. The use of synthetic compounds containing TCDD, particularly the herbicide 2,4,5-trichlorophenoxyacetic acid (2,4,5-T) has resulted in human and animal exposure to the toxin (Rose and Rose, 1972). Furthermore, numerous laboratory investigations have been done over the past 10 years in an attempt to determine the toxic effects of TCDD.

Industrial Accidents

Several incidents of accidental exposure have afforded us an insight into the ramifications of human exposure to TCDD (May, 1973). Although these incidents involved a small number of people exposed to a large quantity of the compound, many of the effects may indicate what might be experienced by the general population exposed to minute quantities of TCDD over a long period.

The high temperatures which increase the amount of TCDD produced in the synthesis of trichlorophenol may also cause a reaction product that generates its own heat. This intense heat may subsequently result in an explosion disseminating large quantities of phenol and TCDD over the surrounding area (May, 1973). Explosions of this origin have occurred in England, France, Germany, Italy, the Netherlands, Czechoslovakia, Austria, and the United States. With the exception of the Italian incident, the exposure to TCDD and other toxic chemicals has been limited to the people working in the immediate area of the explosion. While these explosions required extreme clean-up measures, such as complete destruction of the contaminated buildings, the number of humans exposed was small. However, in July of 1976 an industrial explosion occurred in northern Italy that released a mixture of materials containing TCDD over a large area of land being used for industrial, urban and agricultural purposes. In this incident, several thousand people as well as large numbers of domestic animals were directly or indirectly exposed to various substances, the most toxic of which was TCDD. Those persons closest to the source of contamination became ill almost immediately with headaches, nausea and vomiting (Rawls and O'Sullivan, 1976). Later the skin areas exposed to the contamination developed blisters and reddening. The less severely affected individuals began to develop symptoms a few days later which included skin burns and stomach pain. Children were hospitalized with skin disorders and a decrease in the lymphocytes of the peripheral blood causing concern that increased susceptibility to infections might ensue. The characteristic chloracne was seen in 3 to 4 months. In addition, over a thousand domestic animals died from exposure to the toxic chemicals.

It is too early to assess the course of disease in the exposed Italian people. Indeed, many of the acute toxic effects observed may be due to exposure to large quantities of trichlorophenol while the effects of TCDD have yet to be manifested.

Some indication of what may occur in humans following exposure to various phenols containing TCDD may be obtained from the two well documented explosions, one occurring in Ludswigshaven, Germany on October 17, 1953, at a BASF industrial site (Kimmig and Schulz, 1957) and the other in the Fine Chemical Unit of Coalite and Chemical Products Ltd. in Derbyshire, England on April 23, 1968 (May, 1973). In the German incident the chlorophenols had been exposed to reaction temperatures in excess of 200 °C for several months, thus increasing the amount of TCDD formed. During this time, many of the workers, as well as their families and domestic animals, developed chloracne. Following the explosion, many of those exposed became quite ill and in some instances death has been attributed to lesions caused by the chemicals (May, 1973). Severe liver disorders, chloracne, emphysema, myocardial degeneration, hypertension with resulting kidney damage, neurological disturbances and one intestinal carcinoma have been related to the toxic chemicals released as a result of the explosion. The chloracne has persisted in many of the patients for over 20 years (May, 1973).

The Derbyshire incident involved 79 employees who were working in the building where the explosion occurred. Those individuals exposed were devoid of the generalized effects common in the German incident with the exception of chloracne (May, 1973). The chloracne responded satisfactorily to strict hygienic practices and antibiotic therapy; however, years later many of the exposed individuals have not completely recovered from the skin disease. Following the explosion, the plant was disassembled and buried with the exception of several storage tanks that were cleaned repeatedly with high pressure steam. Three years following the explosion, two pipefitters who worked on the salvaged tanks developed chloracne. Later the son of one worker and the wife of the other experienced similar skin lesions (Jensen, 1972).

A common feature of all of the industrial accidents has been the persistence of TCDD in the contaminated areas as well as the persistence of the toxic effects experienced by the exposed persons. The long term effects of TCDD exposure are uncertain at this time.

Environmental Contamination

The levels of dioxins in the environment are increasing due to the extensive use of chemicals containing minute quantities of these compounds. The defoliant Herbicide Orange, a mixture of 2,4,5-T and 2,4-D containing levels of TCDD ranging from 0.5 to 47 parts per million, was used extensively by the United States Armed Forces in Vietnam (Department of the Air Force, 1974). There are reports of chloracne and other less specific signs such as nausea, vomiting, diarrhea, fatigue, dizziness and abortions in humans occupying the sprayed areas (Rose and Rose,

1972). There are also data showing an increased infant mortality from 30 per thousand to 47 per thousand. Animals inhabiting these sprayed areas were also affected. Large numbers of fish in sprayed ponds and rivers died. Death of domestic animals, particularly fowl and pigs, was common. Cattle experienced convulsions and skin disorders. Abortions and monstrous births were observed in many of the animal species (Rose and Rose, 1972).

Human and animal exposure to TCDD has also occurred from the improper disposal of distillate residues of chlorinated benzenes which are used as a starting material in the production of a variety of chemical products. Perhaps the most extensive occurrence of animal exposure to the dioxins (including dichloro-, trichloro-, tetrachloro-, pentachloro-, hexachloro-, heptachloro-, and octachlorodibenzodioxin) occurred in the United States in 1957 (Schmittle et al., 1958). In order to increase the caloric content of poultry rations, feed manufacturers were using a wide variety of low cost fats. These fats were collected from various sources and pooled. Some of the fats were contaminated with oily residues containing dioxins. These mixtures were sold to the feed manufacturers and incorporated in poultry feeds. As a result of the ingestion of these contaminated fats, millions of chickens died. Hundreds of thousands of the surviving contaminated chickens were processed for human consumption before the origin of the problem was discovered and steps initiated to prevent the use of this food. Unfortunately there are no data on human illnesses that may have been associated with the consumption of the adulterated poultry products.

A second reported incident in the United States occurred in 1971 and 1972 when waste oil sludge containing distillate residues of hexachlorophene production was sprayed on horse arenas to control dust (Carter et al., 1975). Heavy mortality was observed in horses, birds, dogs and cats that inhabited the sprayed areas. In one arena, 85 horses were exposed, 61 became ill and 48 died. The horses experienced emaciation, acne-like dermatitis, blood in the urine, edema, conjunctivitis, lameness and increased numbers of abortions (Case, 1972). At necropsy the horses had biliary cirrhosis with marked proliferation of the bile duct epithelium and distension of the bile ducts. The spleens were reduced in size, and the lymph nodes were small and inactive. Mice, cats and dogs that inhabited the sprayed areas also had dermatitis, loss of hair and weight loss. Human illnesses associated with the use of contaminated sludge oil included one young girl who developed kidney and urinary tract dysfunction including blood in the urine. Three other children and one adult developed chloracne. Joint pain has also been experienced by two adults who were exposed to the arena (Carter et al., 1975).

Laboratory Investigations on TCDD Exposed Nonhuman Primates

The first data on the effects of dioxins in nonhuman primates were reported by Allen and Carstens (1967). They fed rhesus monkeys a diet that contained fats adulterated with a number of dioxins including TCDD. The survival of the monkeys was inversely related to the percentage of the dioxin containing fat given the animals. Clinical and pathological changes occurring at their demise were similar regardless of concentration of material in the diet. There was also a decided decrease in the peripheral red and white blood cells. Analysis of the bone marrow showed a marked reduction in hematopoiesis. There was also decreased activity of the lymphoid tissues.

The skin changes included alopecia and subcutaneous edema which was first observed about the eyelids and later developed over the remainder of the face, trunk, and extremities. Microscopically, there was dermal edema and keratinization of the hair follicles and sebaceous glands.

In addition to a marked reduction in spermatogenesis and a dilated edematous right heart, the animals developed a hyperplastic gastritis with numerous ulcerations. The hyperplastic glandular epithelium had invaded the muscularis mucosa of the stomach and formed large clusters of epithelium in the submucosa.

Their enlarged livers contained numerous multinucleated giant hepatocytes and abundant intracytoplasmic fat vacuoles. There was also focal necrosis of hepatocytes and proliferation and stratification of the bile duct epithelium, including the common, cystic and hepatic ducts. Electron microscopically, the hepatocytes showed a decided proliferation of the smooth endoplasmic reticulum.

In subsequent experiments (Allen *et al.,* 1977) adult female rhesus monkeys were fed a diet containing 500 parts per trillion (ppt) TCDD for 9 months. During the first 3 months of exposure the animals developed periorbital edema, loss of facial hair and eyelashes, acne, accentuated hair follicles and dry scaly skin. The

TABLE 1. Hemograms of Monkeys Fed 500 Parts Per Trillion 2,3,7,8-tetrachlorodibenzo-*p*-dioxin (TCDD) for Nine Months

Animal no.	White blood cells ($\times 10^3/mm^3$)	Platelets ($\times 10^3/mm^3$)	Hemoglobin (gms%)	Hematocrit (%)
	Mean initial values			
	9.3 ± 1.2	327 ± 57	13.9 ± 0.5	42.9 ± 1.8
	Terminal values			
7	2.3	23	4.0	12.0
9	2.4	44	6.9	21.5
23	3.8	28	12.6	40.0
31	10.4	480	10.7	35.5
32	4.1	50	6.0	19.5
38	3.8	34	6.6	22.0
41	8.0	340	11.5	44.5
49	8.7	54	8.5	29.5

hemograms showed a decrease in hemoglobin and hematocrits by the sixth month and became progressively more severe with time.

After 6 months exposure to TCDD, all of the animals had developed irregular menstrual cycles. Repeated attempts to breed these animals resulted in only 3 pregnancies, two of which aborted during the initial 2 months of gestation.

One of the 8 animals developed a severe pancytopenia during the 7th month of TCDD exposure. Prior to death the peripheral blood smears were practically devoid of immature red and white blood cells. During the subsequent four months, 4 additional animals experienced similar clinical changes and died (Table 1). The total TCDD exposure of these animals at the time of death was 18.6 ± 1.8 μg per animal. Although all the surviving animals were removed from the experimental diet following 9 months of exposure, the 3 animals that survived for 12 months experienced a continuing loss of hair and periorbital edema. One of the surviving monkeys developed a severe leukopenia and thrombocytopenia by the 12th month (Table 1).

The major findings in the animals that died included a widespread loss of hair, accentuated hair follicles, and dry flaky skin. There was a loss of eyelashes, swelling of the eyelids, facial edema, and petechial hemorrhage in all animals. Subcutaneous edema was particularly prominent in the lower abdominal region and inner surface of the thighs. Irregularities in the growth of the toenails and fingernails, clubbing of the terminal digits and gangrenous necrosis of the distal phalanges were also present.

Ascitic fluid was present in the abdominal cavity of all animals. There was distension of the intrahepatic biliary system with edema and dilatation of the common bile duct. The lymph nodes throughout the abdominal cavity had undergone atrophy. In addition to the pale appearance of the abdominal organs there were foci of hemorrhage in the adrenals, pancreas, liver, endometrium, serosal and mucosal surfaces of the gastrointestinal tract and in the urinary bladder.

In the thoracic cavity the lungs exhibited focal hemorrhage in all lobes. Bilateral ventricular dilatation, cardiac enlargement and myocardial edema was present in all the animals. Focal areas of hemorrhage were present in the epicardium, myocardium, and endocardium. The skeletal musculature, particularly in the extremities, was edematous, pale and discolored by hemorrhagic foci. Extensive hemorrhage was also found in the meninges as well as isolated hemorrhagic areas in the brain parenchyma. The bone marrow had the appearance of fatty tissue in which there were focal areas of hemorrhage.

Microscopically, the bone marrow displayed a decided hypocellularity. The hematopoietic cells of the marrow were replaced primarily by fat cells with focal areas in which lymphoid appearing cells predominated. There was also a decided reduction in erythroid and myeloid stem cells as well as megakaryocytes in the marrow. The lymph nodes throughout the body were hypocellular. In addition to the absence of any distinct germinal centers, the cortical lymphocytes were sparse while the medullary cords were narrow or inapparent. The spleens also were devoid of any distinct lymphoid germinal centers and the small lymphocytes were widely dispersed.

Extensive hemorrhage was also observed microscopically. Circumscribed focal areas of hemorrhages in the heart, lung, liver, adrenals, pancreas, and skeletal muscles consisted of intact red blood cells that partially disrupted the architectural pattern of the affected tissues. Petechial areas were prominent in the dermis of the skin, submucosa of the urinary bladder and in the alimentary tract. Hemorrhages in the endometrium and meninges were more diffuse in their distribution. Hemorrhage in the brain was limited to the Verchow Robbin's space surrounding the blood vessels. The hemorrhage in the bone marrow varied considerably, being diffuse in some areas and focal in others.

Considerable modification in the morphological features of the epithelium occurred in the TCDD treated animals. Metaplastic changes characterized by numerous mucous secreting cells were present in the ductal epithelium of the salivary glands, bile ducts and pancreatic ducts. Similar changes also occurred in the bronchial epithelium and in the palpebral conjunctivae. In addition to the metaplasia, the epthelium of the bile ducts and palpebral conjunctivae developed considerable hypertrophy and hyperplasia. Squamous metaplasia of the sebaceous glands occurred more extensively in the skin of the face than in other areas.

Hypertrophy, hyperplasia, and metaplasia were observed in the gastric mucosa where the parietal and zymogenic cells were replaced by mucous secreting cells. Numerous epithelial cells invaded the submucosa through fenestrations in the muscularis mucosa. Considerable edematous fluid usually surrounded the displaced epithelium.

In addition to the squamous metaplasia of the epithelium of the sebaceous glands there was hyperkeratosis of the skin with keratinization of the hair follicles and adjacent sebaceous glands. The latter alterations were particularly prominent in the Meibomian glands of the eyelids. The pronounced thickening of the fingernails and toenails is related to excessive keratin production.

The heart muscle had hemorrhagic foci that predominated in the atria and tips of the papillary muscles. There were also foci of subendothelial and pericardial hemorrhage in the ventricles. Intra- and intercellular edema as characterized by separation of muscle fibers and myofibrils was present throughout the myocardium.

McNulty (1977) has recently reported the toxicity of TCDD to young male rhesus monkeys. The pathological changes noted by McNulty in the monkey given 2 ppb TCDD diet were similar to our findings reported above. The principal effects of TCDD in male rhesus monkeys were cachexia, mucous metaplasia, hyperplasia and ulceration of the gastric mucosa, squamous metaplasia of sebaceous glands, and atrophy of the thymus.

Laboratory Investigation on Low Level Exposure of Rats to TCDD

Van Miller and Allen (1977) and Van Miller et al. (1977) fed Sprague-Dawley male rats TCDD at concentrations varying from 5 ppb to 1 ppt for 18 months and subsequently evaluated the animals for an additional 4 months. During this time there were 28 neoplasms recorded in the 60 experimental animals. No tumors were observed in the 50 control rats that received an unmodified diet. The neoplasms in the experimental animals were found in numerous tissues throughout the body and did not seem to have any predilection for one particular tissue.

Conclusions

Extensive exposure to TCDD such as may occur in industrial accidents produces acute signs of intoxication that include nausea, vomiting, fatigue, dizziness, and skin disorders. Within a short period, humans developed hyperpigmentation and acne of exposed skin surfaces. The skin lesions usually persist for years in the more severely affected persons. Abortions, abnormal births, and liver dysfunction have been reported to occur as an aftermath of heavy TCDD exposure. Unfortunately little is known about the effects of low level exposure to TCDD that may result from contamination of the food chain. Information is available only from experimental animals that have been exposed to low levels of TCDD. These data indicate that anemia, leukopenia, gastritis and ulceration, irregularities in menstrual cycles, difficulties in conception, early abortions, abnormal births, alterations in the immune response and cancer may be an aftermath of chronic low level exposure to TCDD.

ACKNOWLEDGEMENT

This research was supported in part by U.S. Public Health Service grants ES00472 and RR00167 from the National Institutes of Health and the University of Wisconsin Sea Grant Program. Primate Center Publication No. 17-001.

References

Allen, J.R., and L.A. Carstens, 1967. Light and electron microscopic observations in *Macaca mulatta* monkeys fed toxic fat. *Amer. J. Vet. Res.,* **38:** 1513-1526.

Allen, J.R., D.A. Barsotti, J.P. van Miller, L.J. Abrahamson, and J.J. Lalich, 1977. Morphological changes in monkeys consuming a diet containing low levels of 2,3,7,8-tetrachlorodibenzo-p-dioxin. *Food Cosmet. Toxicol.,* in press.

Carter, C.D., R.D. Kimbrough, J.A. Liddle, R.E. Cline, M.M. Zack, and W.F. Barthel, 1975. Tetrachlorodibenzodioxin: an accidental poisoning episode in horse arenas. *Science,* **188:** 738-740.

Case, A.A., 1972. Toxicosis of public health interest. *Clin. Toxicol.,* **5:** 267-270.

Department of the Air Force, 1974. Disposition of Herbicide Orange by Incineration. Final Environmental Statement, November 1974, pp. 36-37.

Jensen, N.E., 1972. Chloracne: three cases. *Proc. Roy. Soc. Med.,* **65:** 687-688.

Kimmig, J., and K.H. Schulz, 1957. Berufliche akne (sog. chlorakne) durch chlorierte aromatische zylklische ather. *Dermatologica,* **114:** 540-546.

May, G., 1973. Chloracne from the accidental production of tetrachlorodibenzodioxin. *Brit. J. Ind. Med.,* **30:** 276-283.

McNulty, W.P., 1977. Toxicity of 2,3,7,8-tetrachlorodibenzo-p-dioxin for rhesus monkeys: a brief report. *Bull. Environ. Contam. Toxicol.,* **18:** 108-109.

Neubert, D., P. Zens, A. Rothenwallner, and H.-J. Merker, 1973. A survey of the embryotoxic effects of TCDD in mammalian species. *Environ. Health Persp.,* **5:** 67-79.

Rawls, R.L., and D.A. O'Sullivan, 1976. Italy seeks answers following toxic release. *Chem. Eng. News,* August 23, 1976, pp. 27-35.

Rose, H.R., and S.P.R. Rose, 1972. Chemical spraying as reported by refugees from South Vietnam. *Science,* **177:** 710-712.

Schmittle, S.C., H.M. Edwards, and D. Morris, 1958. A disorder of chickens probably due to a toxic feed -- preliminary report. *J. Amer. Vet. Med. Assoc.,* **143:** 216-219.

Sparschu, G.L., F.L. Dunn, and V.K. Rowe, 1971. Study of the teratogenicity of 2,3,7,8,-tetrachlorodibenzo-p-dioxin in the rat. *Food Cosmet. Toxicol.,* **9:** 405-412.

van Miller, J.P., and J.R. Allen, 1977. Chronic toxicity of 2,3,7,8-tetrachlorodibenzo-p-dioxin in rats. *Fed. Proc.,* **36:** 396.

van Miller, J.P., J.J Lalich, and J.R. Allen, 1977. Increased incidence of neoplasms in rats exposed to low levels of 2,3,7,8-tetrachlorodibenzo-*p*-dioxin. *Chemosphere,* **6:** 625-632.

Renal Effects of 2,3,7,8-Tetrachlorodibenzo-p-dioxin

JERRY B. HOOK, KEVIN M. McCORMACK and WILLIAM M. KLUWE

Abstract—Intoxication with 2,3,7,8-tetrachlorodibenzo-p-dioxin (TCDD) produced marked increases in cellular smooth endoplasmic reticulum content and microsomal drug metabolizing enzyme activity in kidney cortex. Therefore, it was of interest to determine the effect of TCDD on several proximal tubular functions. Adult rats were treated with 10, 25, and 50 µg/kg, TCDD. Ten µg/kg TCDD 3 or 7 days after treatment did not affect p-aminohippurate (PAH) accumulation by renal cortical slices. N-methylnicotinamide (NMN) accumulation was slightly decreased following this treatment. At 25 µg/kg, TCDD decreased the capacity of renal tissues to transport both PAH and NMN 7 days after exposure. The increase in ammoniagenesis and gluconeogenesis observed in acidosis was not significantly different in animals that had been treated with 25 µg/kg TCDD 7 days before experimentation. Glomerular filtration rate and effective renal plasma flow were decreased in rats after 25 or 50 µg/kg TCDD. Volume expansion did not alter this relationship. Fractional sodium excretion was less than 1% in both control and TCDD-treated animals. With volume expansion sodium excretion increased to approximately 5% and was not different for control and TCDD-treated animals. TCDD enhanced several hepatic and renal drug metabolizing enzyme systems but did not alter the acute toxicity of $CHCl_3$.

Introduction

2,3,7,8-Tetrachlorodibenzo-p-dioxin (TCDD) is one of the most toxic compounds known to man. Death following TCDD intoxication is delayed for several weeks and

JERRY B. HOOK, KEVIN M. McCORMACK and WILLIAM M. KLUWE • Department of Pharmacology, Michigan State University, East Lansing, Michigan 48824, U.S.A.

is preceded by progressive debilitation (Greig *et al.,* 1973; Harris *et al.,* 1973). At doses far less than lethal, TCDD is an effective inducer of hepatic and renal microsomal drug metabolizing enzymes (Fowler *et al.,* 1975; Hook *et al.,* 1975; Lucier *et al.,* 1973). In association with the increase in enzyme activity in the liver and kidney following TCDD the content of smooth endoplasmic reticulum is markedly increased (Fowler *et al.,* 1973, 1975). Interestingly, in the kidney the increase in endoplasmic reticulum appeared to be confined to the cells of the pars recta of the proximal tubule (Fowler *et al.,* 1975). Considerable information has been obtained concerning the functional significance of TCDD-induced alterations in hepatic function. However, to our knowledge little work has been done to evaluate the effect of TCDD on kidney function. Thus, our studies were designed to evaluate the effects of sublethal doses of TCDD on renal function in the rat.

Effects of Sublethal Doses of TCDD on Renal Function

The kidney is involved not only in regulation of salt and water balance, but plays an important role in control of acid-base balance and the elimination of a variety of organic compounds such as waste products and many drugs. Thus, our experiments were designed to evaluate the effect of TCDD on drug elimination, salt and water excretion and metabolic control of acid-base balance. In our first study (Pegg *et al.,* 1976) young adult rats were treated with 10 or 25 µg/kg of TCDD, intraperitoneally. The general systemic effect of the chemical was seen as a decrease in growth rate of these animals. Both doses retarded general body weight gain at 3 and 7 days. No deaths were seen following these doses (Pegg *et al.,* 1976).

One of the most sensitive indices of nephrotoxicity is the ability of renal cortical tissue to actively transport organic compounds (Berndt, 1976). Thin renal cortical slices are prepared free hand and then incubated in a dilute salt solution containing the organic compounds whose transport is to be assessed. Following 60-90 min incubation under oxygen the tissue is removed and analyzed for the compound under study. The data are reported as the final slice-to-medium (S/M) concentration ratio. In most studies of this type the slices are cut parallel to the surface of the kidney to obtain slices containing primarily proximal convoluted tubules. However, as pointed out the predominant morphological changes produced by TCDD appeared in the pars recta, located deeper in the cortex and in the outer medulla. To evaluate transport of organic compounds in both the proximal convolution and the pars recta slices were made in both parallel and perpendicular planes. TCDD appeared to have a slight depressant effect on both *p*-aminohippurate (PAH) and N-methylnicotinamide (NMN) transport whether measured in parallel or perpendicular slices. This effect was most pronounced in animals 7 days after receiving 25 µg/kg of TCDD (Pegg *et al.,* 1976).

To assess the effects of TCDD on metabolism of the kidney, the ability of renal cortical slices to form ammonia and glucose was determined using the method of Roobol and Alleyne (1974). Control and TCDD (25 µg/kg) treated animals were made acidotic by supplying 0.28 M ammonium chloride as sole drinking fluid for the duration of the experiment. Non-acidotic controls were allowed tap water. After 7 days the animals were sacrificed by cervical dislocation and thin renal cortical slices were prepared as before. The slices were incubated for 1 hr at 37 °C in a buffered salt solution with glutamine as substrate. Glucose and ammonia in the medium were assayed and production expressed as μmoles/g tissue/hr. Ammoniagenesis and gluconeogenesis in renal cortical slices were significantly increased by metabolic acidosis. TCDD had no significant effect on the ability of tissue from either non-acidotic or acidotic animals to produce ammonia or glucose (Pegg et al., 1976).

Thus, even with doses of TCDD that reduced total body growth and doses that had been reported by others to produce rather profound ultrastructural changes in the kidney, no significant changes in function could be observed in vitro. However, the bulk of renal work and energy utilization is in the reabsorption of sodium, thus it was of interest to evaluate the ability of the kidneys of TCDD intoxicated animals to reabsorb sodium and water.

Renal function was determined in vivo using rats treated one week prior to experimentation with either 25 or 50 μg/kg TCDD. Animals were anesthetized with 60 mg/kg sodium pentobarbital, intraperitoneally. Body temperature was maintained at 33-38 °C using heat lamps. A PE 50 cannula was inserted into the bladder and urine was collected under mineral oil in preweighed vials. The left femoral vein was cannulated for infusion. Both femoral arteries were cannulated to monitor blood pressure using a Statham transducer and a Beckman type RS dynograph and to obtain blood samples. The infusion solution contained 1% inulin ($[^3H]$ 0.5 μCi/ml) and 0.6% PAH ($[^{14}C]$ 0.5 μCi/ml) in normal saline. The solution was infused at 0.018 ml/min using an infusion pump. Three 30 min urine collections were then taken with blood (300 μl) being sampled at the middle of each collection. Following the 3 initial clearance periods the animals were infused with 1:4 rat plasma/saline solution (5% body weight) and 15 min allowed for equilibration, then three more 30 min clearance periods were taken (Pegg et al., 1976).

The clearance of inulin (glomerular filtration rate) and of PAH (effective renal plasma flow) were both significantly decreased 7 days after 25 and 50 μg/kg TCDD The glomerular filtration rate (GFR) and effective renal plasma flow (ERPF) remained less than control through the period of volume expansion. Filtration fraction (GFR/ERPF) was unaffected by TCDD. Urine flow rates were not different between control and TCDD treated animals during control collections. Following volume expansion however, percent increase in urine flow was greater in control animals than in TCDD treated. Fractional sodium excretion was higher during the first collection period in animals treated with 25 μg/kg but the difference decreased by the third control. Fractional sodium excretion in control and TCDD treated animals was not significantly different during volume expansion (Pegg et al., 1976).

Thus, by all measures studied it appeared that there was no specific effect of TCDD on renal function, but rather, any effects seen appeared to be secondary to general debilitation of the animals. However, this does not mean TCDD was without functionally significant effects on the kidney. Either inappropriate functions were measured or the changes in renal function were too subtle to be recognized by the relatively gross methods we used. We had previously obtained a precedent for such a conclusion. Polybrominated biphenyls (PBBs), like TCDD, are capable of inducing hepatic and renal mixed function oxidases (Dent et al., 1976; McCormack et al., 1977a). In previous work we had been unable to show any direct nephrotoxicity produced by PBBs (McCormack *et al.*, 1977b). However in animals pretreated with PBBs renal drug metabolizing enzymes were stimulated and the nephrotoxicity of chloroform was enhanced (Kluwe *et al.*, 1977). That work suggested that a stimulator of renal drug metabolizing capabilities might sensitize the kidney to toxicity caused by a nephrotoxic agent administered secondarily. Thus, experiments

TABLE 1. Effect of TCDD on Mouse Hepatic and Renal AAH, EH, BP-2-OH and BP-4-OH Activities[a]

Treatment	Hepatic			
	AHH[b]	EH[c]	BP-2-OH[d]	BP-4-OH[e]
Control	11.20 (1.53)	0.55 (0.13)	0.11 (0.01)	2.08 (0.28)
1.6 µg TCDD/kg	32.72* (2.65)	2.19* (1.11)	0.16* (0.01)	2.10 (0.20)
16 µg TCDD/kg	84.44* (3.08)	0.94* (0.11)	0.13* (0.01)	2.94* (0.42)

Treatment	Renal			
	AHH	EH	BP-2-OH	BP-4-OH
Control	0.05 (0.01)	0.10 (0.01)	0.01 (0.01)	0.02 (0.01)
1.6 µg TCDD/kg	0.09* (0.02)	0.34* (0.12)	0.01 (0.01)	0.05* (0.01)
16 µg TCDD/kg	2.78* (0.60)	0.08 (0.01)	0.04* (0.01)	0.04* (0.01)

[a]Mice were treated with 1.6 or 16 µg/kg TCDD, intraperitoneally. Seventy-two hrs later animals were sacrificed for enzyme measurements. Values are Means ± S.E.M. for 4 animals. Asterisks indicate a statistically significant difference from the respective control.
[b]Fluorescence units/mg microsomal protein/min.
[c]nmoles styrene glycol formed/mg microsomal protein/min.
[d]nmoles 2-hydroxybiphenyl produced/mg microsomal protein/min.
[e]nmoles 4-hydroxybiphenyl produced/mg microsomal protein/min.

were conducted to evaluate the effect of TCDD pretreatment on the nephrotoxicity of the halogenated hydrocarbon, chloroform.

Effect of TCDD Pretreatment on the Nephrotoxicity of Chloroform

Adult, male, ICR mice (Spartan Farms, Haslett, Mi) were treated with 1.6 or 16 μg TCDD/kg, intraperitoneally. Seventy-two hours later representative animals were sacrificed to determine the hepatic and renal microsomal enzyme activities and others were challenged with various doses (0.5 to 25 μl/kg) of chloroform. Twenty-four hrs after the dose of chloroform, BUN (urea nitrogen) as an estimate of renal function, and SGOT (serum glutamate oxaloacetate transaminase) as an estimate of liver function, were analyzed. Animals used for enzyme assays were sacrificed by cervical dislocation and kidneys were excised, weighed and chopped into ice-cold 66 mM Tris buffered to pH 7.4 with HCl (kidneys) or 1.15% KCl (livers). Kidney postmitochondrial supernatants were prepared by homogenization in 3 volumes of 66 mM Tris pH 7.4 followed by centrifugation at 10,000 x g for 20 min.

TABLE 2. Effect of TCDD Pretreatment on
Acute Toxicity of $CHCl_3$ in Mice:
Liver Weight/Body Weight[a]

Pretreatment	0	$CHCl_3$ (μl/kg) 0.50	2.50	5.0	25.0
			LW/BW × 100		
Control	6.46 (0.13)	6.41 (0.12)	6.52 (0.23)	6.71 (0.19)	6.73 (0.23)
TCDD 1.6 μg/kg	6.75 (0.23)	6.84 (0.21)	7.16 (0.29)[b]	7.24 (0.23)[b]	7.35 (0.30)[b]
TCDD 16 μg/kg	7.90 (0.25)[b]	8.00 (0.30)[b]	8.24 (0.18)[b]	7.74 (0.17)[b]	8.01 (0.30)[b]

[a]Mice were treated with 1.6 or 16 μg/kg TCDD, intraperitoneally. Seventy-two hrs later control and treated animals received $CHCl_3$ or vehicle control. Twenty-four hrs later animals were sacrificed.
[b]Significantly greater than control ($p < 0.05$).

Liver microsomes were prepared (Netter, 1960) and resuspended (Dent *et al.*, 1976). All assays were performed on the day of supernatant and microsomal preparation. Protein was measured (Lowry *et al.*, 1951) using bovine serum albumin as a standard. Enzyme activities measured were epoxide hydratase (EH) (Oesch *et al.*, 1971), aryl hydrocarbon hydroxylase (AHH) (Nebert and Gelboin, 1968; Oesch, 1976), and biphenyl-2- and -4-hydroxylase (BP-2-OH and BP-4-OH) (Creaven *et al.*, 1965). BUN was measured with Sigma reagents (Sigma Chemical Co., St. Louis, Mo.) and SGOT by the method of Reitman and Frankel (1957).

TCDD treatment significantly increased the activity of all microsomal enzymes measured in the liver and kidney tissue (Table 1). The activity of all enzymes was considerably lower in kidney than in liver. TCDD increased liver weight to body weight ratio (Table 2) without altering relative kidney weight to body weight ratio (Table 3). Neither BUN nor SGOT were increased by TCDD. Administration of chloroform had no effect on liver weight to body weight ratio in control animals but did increase liver weight to body weight ratio in animals receiving TCDD (Table 2). Chloroform did not change the kidney weight to body weight ratio even at the highest doses in TCDD treated animals (Table 3). The doses of chloroform chosen were those that were expected to have minimal effects on the liver and kidney in anticipation of potentiation of these effects. No potentiation was demonstrable.

TABLE 3. Effect of TCDD Pretreatment on Acute Toxicity of $CHCl_3$ in Mice: Kidney Weight/Body Weight[a]

Pretreatment	0	$CHCl_3$ (μl/kg) 0.50	2.50	5.0	25.0
		KW/BW × 100			
Control	1.52 (0.05)	1.51 (0.05)	1.52 (0.05)	1.58 (0.10)	1.69 (0.14)
TCDD 1.6 μg/kg	1.52 (0.03)	1.45 (0.05)	1.44 (0.04)	1.71 (0.13)	1.72 (0.10)
TCDD 16 μg/kg	1.64 (0.08)	1.51 (0.06)	1.52 (0.03)	1.68 (0.05)	1.56 (0.03)

[a]Mice were treated with 1.6 or 16 μg/kg TCDD, intraperitoneally. Seventy-two hrs later control and treated animals received $CHCl_3$ or vehicle control. Twenty-four hrs later animals were sacrificed.

Conclusions

Thus, these data demonstrate that although TCDD has profound metabolic effects on the kidney as in the liver, these metabolic changes do not appear to be accompanied by physiological alterations. Whereas PBBs appear to potentiate the action of a metabolically activated chemical such as chloroform, such potentiation was not seen with TCDD. Thus, although TCDD is an extremely toxic compound systemically and specific ultrastructural and biochemical alterations in renal cortical tissue appear to be produced by TCDD, the effects of the compound on measurable parameters of kidney function most probably can be attributed to a decline in the general health of the animal at the time of experimentation. Even though TCDD has marked effects on the cells of the proximal tubule the treatment regimens we have utilized do not produce specific changes in physiological functions of these cells.

ACKNOWLEDGMENTS

The authors' research is supported in part by USPHS Grants ES0560, AM10913 and GM01761.

References

Berndt, W.O., 1976. Renal function tests: What do they mean? A review of renal anatomy, biochemistry, and physiology. *Environ. Hlth. Persp.*, **15**: 55-75.

Creaven, P.J., D.V. Parke, and R.T. Williams, 1965. A fluorimetric study of the hydroxylation of biphenyl *in vitro* by liver preparations of various species. *Biochem. J.*, **96**: 879-885.

Dent, J.G., K.J. Netter, and J.E. Gibson, 1976. The induction of hepatic microsomal metabolism in rats following acute administration of a mixture of polybrominated biphenyls. *Toxicol. Appl. Pharmacol.*, **38**: 237-249.

Fowler, B.A., G.E.R. Hook, and G.W. Lucier, 1975. Tetrachlorodibenzo-*p*-dioxin induction of renal microsomal enzyme systems. *Soc. Toxicol. 14th Annual Meeting, Abstracts,* pp. 110-111.

Fowler, B.A., G.W. Lucier, H.W. Brown, and O.S. McDaniel, 1973. Ultrastructural changes in rat liver cells following a single oral dose of TCDD. *Environ. Hlth. Persp.*, **5**: 141-148.

Greig, J.B., G. Jones, W.H. Butler, and J.M. Barnes, 1973. Toxic effects of 2,3,7,8-tetrachlorodibenzo-*p*-dioxin. *Food Cosmet. Toxicol.*, **11**: 585-595.

Harris, M.W., J.A. Moore, J.G. Vos, and B.N. Gupta, 1973. General biological effects of TCDD in laboratory animals. *Environ. Hlth. Persp.*, **5**: 101-109.

Hook, G.E.R., T.C. Orton, J.A. Moore, and G.W. Lucier, 1975. 2,3,7,8-Tetrachlorodibenzo-*p*-dioxin induced changes in the hydroxylation of biphenyl by rat liver microsomes. *Biochem. Pharmacol.*, **24**: 335-340.

Kluwe, W.M., and J.B. Hook, 1977. Polybrominated biphenyl potentiation of acute CHCl$_3$ toxicity. *The Pharmacologist*, **19**: 199.

Lowry, O.H., N.J. Rosebrough, A.C. Fair, and R.J. Randall, 1951. Protein measurement with the folin phenol reagent. *J. Biol. Chem.*, **193**: 265-275.

Lucier, G.W., O.S. McDaniel, G.E.R. Hook, B. Fowler, B.R. Sonawane, and E. Faeder, 1973. TCDD-induced changes in rat liver mocrosomal enzymes. *Environ. Hlth. Persp.,* **5:** 199-209.

McCormack, K.M., S.Z. Cagen, D.E. Rickert, J.E. Gibson, and J.G. Dent, 1977a. Stimulation of hepatic and renal mixed function oxidase in developing rats by polybrominated biphenyls. *Drug Met. Disp.,* submitted.

McCormack, K.M., W.M. Kluwe, D.E. Rickert, V.L. Sanger, and J.B. Hook, 1977b. Renal and hepatic microsomal enzyme stimulation and renal function following three month dietary exposure to polybrominated biphenyls. *Toxicol. Appl. Pharmacol.,* submitted.

Nebert, D.W., and H.V. Gelboin, 1968. Substrate-inducible microsomal aryl hydroxylase in mammalian cell culture. I. Assay and properties of induced enzyme. *J. Biol. Chem.,* **243:** 6242-6249.

Netter, K.J., 1960. Eine Methode zur direkten Messung der O-Demethylierung in Lebermikdrosomen und ihre Anwendung auf die Mikdrosomenhemurrkung von SKF 525-A. *Naunyn-Schmiedebergs Arch. Exp. Path. Pharmak.,* **238:** 292-300.

Oesch, F., D.M. Jerina, and J. Daley, 1971. A radiometric assay for hepatic epoxide hydratase activity with [7-^{3}H] styrene oxide. *Biochim. Biophys. Acta,* **227:** 685-691.

Oesch, F., 1976. Differential control of rat microsomal "aryl hydrocarbon" monoxygenase and epoxide hydratase. *J. Biol. Chem.,* **251:** 79-87.

Pegg, D.G., W.R. Hewitt, K.M. McCormack, and J.B. Hook, 1976. Effect of 2,3,7,8-tetrachlorodibenzo-*p*-dioxin on renal function in the rat. *J. Toxicol. Environ. Hlth.,* **2:** 55-65.

Reitman, S., and S. Frankel, 1957. A colorimetric method for the determination of serum glutamic oxaloacetic and glutamic pyruvic transaminases. *Am. J. Clin. Pathol.,* **28:** 56-63.

Roobol, A., and G.A.O. Alleyne, 1974. Control of renal cortex ammoniagenesis and its relationship to renal cortex gluconeogenesis. *Biochim. Biophys. Acta,* **353:** 83-91.

Environmental Impact of Pentachlorophenol and Its Products — a Round Table Discussion

PHILIP J. CONKLIN and FERRIS R. FOX (Editors)

Introduction

A round table discussion concerning pentachlorophenol, its contaminants and its impact on the environment was moderated by Daniel Cirelli of the U.S. Environmental Protection Agency. Discussants included Robert Johnson and Eugene Kenaga of Dow Chemical Company; George Fries of the U.S. Department of Agriculture; Richard Hoos of the Canadian Environmental Protection Service and Eric Reiner of 3M Company. The general topic of discussion involved incidents in which pentachlorophenol or its possible contaminants caused health problems or were potentially hazardous. Details of the incidents were presented and the steps for control were proposed. Comments, questions and answers were directed to the discussants as well as the audience.

Pentachlorophenol-Related Incidents in the United States: Michigan Cattle

The discussion was initiated by Donald Isleib of the Michigan Department of Agriculture who presented a summary of the pentachlorophenol (PCP) and dioxin contamination problem in the State of Michigan. During the last few years, there has been a widely publicized polybrominated biphenyl (PBB) contamination in Michigan. The State Department of Agriculture initiated a survey of PBB in cattle herds and their quarters after persistent reports of health problems in humans and animals from low levels of these chemicals. Out of 1,100 farmsteads surveyed, only

---

PHILIP J. CONKLIN and FERRIS R. FOX • Faculty of Biology, University of West Florida, Pensacola, Florida 32504, U.S.A.

100 farmers reported problems. Only 13 of these cases were diagnosed to be of a toxic nature by a team of veterinary diagnosticians. These 13 cases were subjected to a complete analysis by a joint team from Michigan State University and the Michigan Department of Agriculture. Cattle from nine of these herds were found to contain PCP in concentrations of 2 ppb to 12 ppm. Since these were all dairy herds, milk sales were stopped from the affected herds. The detection of dioxins in the liver and adipose tissue of animals from one herd caused the Michigan Department of Agriculture to quarantine all nine herds based on the suspicion that these additional herds would show dioxin contamination. Milk from all the herds was analyzed for dioxin content and found to be free of any detectable amount. Therefore, milk from these herds was restored to the market but the herds were kept under beef quarantine.

In cooperation with the U.S. Department of Agriculture, Food and Drug Administration, Dow Chemical Company and Wright State University samples from the nine cattle herds were collected and analyzed. Dioxins were reported in only two of these herds. One of these herds, the George Lemunion herd, had 50 ppb octachlorodibenzo-p-dioxin (OCDD) and 12 ppm PCP while the other one, the Heiss herd, had OCDD and 570 ppb PCP. Both herds were destroyed by the owners.

Because dioxins have been found associated with PCP in animals, the Michigan Department of Agriculture has initiated a new regulation under the state's pesticide law. This new regulation will require chemical manufacturers to reduce the dioxin composition of their products in order to be eligible for sale in Michigan.

Since the occurrence of this incident, the Animal, Plant and Health Inspection Survey is doing a nationwide survey of beef from slaughterhouses for PCP and dioxin content. The Food and Drug Administration is also doing some short-term chronic tests on these compounds. The Environmental Protection Agency is noting any occurrences of pesticides in animals and action will be taken if a nationwide problem exists.

The source of PCP and dioxin contamination in the Michigan herds needs to be thoroughly investigated. The original Lemunion herd had been destroyed because of PBB contamination, and the barn had burned. A new barn was constructed using treated lumber, and new cattle were confined to this barn. Suggested means for uptake of PCP were from licking treated wood, inhalation of the chemicals and skin contact. The Heiss herd may have obtained PCP from the treated wood of silage and feed bins. Since the Wood Preservation Institute has estimated the content of PCP and dioxin in the entire Lemunion barn to be too low to cause the health problems and death found there, some aspect of this incident has been overlooked.

Would cows re-exposed to the Lemunion barn take up PCP? No answer can be given at this time. However, cattle removed from the barn did show declining levels of PCP. Also, in relation to PCP toxicity studies on cattle, there have been no tests made on lactating animals where toxic effects might be found at lower PCP concentrations. In the cattle herds exposed to PCP and dioxins, typically there were no overt health problems until calving when the adults died. This is probably due to a stress situation.

Assuming that OCDD, hepta-CDD and hexa-CDD are essentially no more toxic than PCP and that TCDD is not found in PCP what is causing the health problems? The assumption that some dioxins are no more toxic than PCP should be clarified as to acute or chronic toxicity. Dow Chemical Company has examined the toxicity of dioxins found in PCP and reported the results at the Dioxin Conferences in 1970 and 1971 which were published in *Advances in Chemistry*. Some dioxins were found to be more toxic than PCP alone, whereas, the relationship of these dioxins to PCP with dioxins present shows a different level of toxicity. Dioxins should be more suspect because they can be accumulated over a period of time and present greater health problems. Animals exposed to dioxin and then removed from exposure have continued to exhibit toxic effects. The half-life for dioxin in cattle is unknown.

How was the PCP analysis done? If gas chromatography were used and the samples were methylated then any pentachloroanisole (PCA) would not be detected separately from PCP. Since PCA is known to be very persistent, it may also contribute to animal health problems.

Pentachlorophenol and Related Incidents in Europe

In Sweden the manufacture of pentachlorophenol (PCP) does not appear to be carried out under strict control and the commercial product labelled PCP may contain 80% tetrachlorophenol. European manufacture of PCP is generally by direct chlorination of phenol or alkaline hydrolysis of the appropriate chlorobenzene. Approximately equal use is made of both processes. In the alkaline hydrolysis method high temperature favors the side reaction towards dioxin formation. Two separate incidents in Sweden have focused attention on the presence of dioxin in the commercial product. The mass death of chickens exposed to materials treated with commercial PCP and several industrial accidents involving numerous employees have led to an investigation of contaminants in PCP that may form dioxins.

The careless handling of PCP-treated lumber in Swedish sawmills has led to the withdrawal of licensing of chlorophenols for wood preservation as of January 1978. The continual exposure of some employees in the sawmill to contaminated sawdust is one health problem. Most of the lumber in Sweden is dipped and therefore handling of the timber and inhalation of sawdust have presented hazards.

Another health problem contributing to license withdrawal is the use of treated sawdust and wood waste for heating in sawmills. Waste wood is put into combustion furnaces which vary in temperature and are normally below the temperature for combustion of the dimeric compounds such as dioxins.

Have there been any known health problems related to combustion of waste woods? There have been no known health problems from the combustion facilities, but several people have been poisoned from handling of sawn lumber. However, several dimeric impurities have been found in the gases from the combustion furnaces.

What is being advocated for replacement of PCP in wood preserving? Bifluorides have been considered as a PCP substitute as well as benzimidazoles. However, these substitutes are not as effective as PCP in preserving the wood. Whether these substitutes for PCP are likely to cause similar problems is not yet known.

Have there been specific problems in the workers handling the PCP-treated wood? Yes, these workers have had symptoms of dioxin and mostly dibenzofuran poisoning.

What are the contaminants found in commercial PCP in Sweden? Commercial PCP contains 1-5% chlorinated phenoxy-phenols, 50-200 ppm of dibenzofurans, 100-1,000 ppm of chlorinated phenyl ethers and 1-2,000 ppm of chlorinated dioxins. The hepta- and octachlorodibenzo-*p*-dioxins are in the highest while the lowest concentrations are of tri- and tetrachlorodibenzo-*p*-dioxins.

Other incidents of industrial exposure to PCP and its contaminants have occurred in Germany and Italy. The most recent occurrence was late last year (1976) when a chemical plant exploded in Italy. Humans and animals from a nearby village were exposed to dioxin. The effects of the incident are still being recorded. Cases of chloracne have been identified with dioxin poisoning.

Pentachlorophenol and Related Incidents in Canada

Pentachlorophenol (PCP) usage in Canada is mainly involved with wood treatment. Most of the cases of PCP hazard have been associated with the wood preservation industry. In Western Canada all wood is treated by dipping in a tank of PCP for short-term protection against mold and bacteria. Until recently most of these dip tanks were open to the weather allowing overflow following heavy rains. One incident of PCP pollution occurred after a four inch rainfall at a lumber treatment facility. The treatment tank, which usually contains 10,000 ppm penta- and tetrachlorophenol, overflowed into a stream killing 5,000 adult coho salmon. As a result of this and similar incidents, the Environmental Protection Service has encouraged the use of enclosed tanks or putting roofs over open tanks. Also encouraged is the construction of drainage areas around the tanks in order that overflow can be collected into secondary tanks for holding purposes.

Large quantities of pentachlorophenol solution have been dumped by wood-treating companies into landfill sites in Canada. These companies are being encouraged to recycle the PCP for use either alone or mixed with other compatible preservatives instead of dumping. The solids that are filtered from the tanks are combusted above 1,500 °C to destroy PCP residues.

Another incident of PCP being implicated in a fish kill occurred when telephone poles were resprayed. In 1975 a company routinely sprayed the base of existing telephone poles to prevent destruction. During one month 15 fish kills were reported and all were traced to wash-off from the resprayed poles. All of the poles were near streams and heavy rains washed the solution a few yards to the stream. The company complied with the order to stop spraying.

An unusual incident involving PCP with no known resultant hazards was the case of a privately owned United States oil well drilling ship. The ship was brought into Vancouver to be refitted for drilling in the Arctic. The ship had a permanent type ballast consisting of barite ($BaSO_4$) with a water solution of Na-PCP and paraformaldehyde. The concentration of PCP in the ballast was 55 ppm. The ship's hull had to be restructured to withstand the severe operating conditions. Along with the structural changes, 2,000 tons of the ballast had to be removed and antifreeze solution added to the remaining ballast. After all possible alternatives for dumping the material were analyzed, the decision was made to dispose of the ballast at sea. The solution was diluted and dumped over 2,000 square miles of ocean 250 miles off the coast.

Why was antifreeze put in the ballast? The drill ship was to be kept in the North and frozen into a harbor over the winter. Without antifreeze there was a chance of the hull rupturing when water in the ballast froze and expanded.

What is the usage of PCP in oil well drilling in Canada? As of now no PCP is being used in oil well drilling over water in Canada. However, the use of PCP products as antimicrobial agents in oil-drilling operations in the marine and coastal areas of the United States of America has been documented.

Fish-Kills in Japan

It is of interest to recall the extensive fish mortalities in Japan resulting from the use of PCP and its salts as herbicides and molluscicides. The Japanese Government has restricted the use of PCP which led to a marked decline in the overall production and usage of PCP in Japan.

Hexachlorobenzene as a Source of Pentachlorophenol

Hexachlorobenzene (HCB) is used by European manufacturers as a precursor of PCP, while pentachlorobenzene is a waste product of the industrial synthesis of chlorinated compounds. Pentachlorobenzene is used as a fire retardant. Chlorobenzenes have been found in human tissues and in human milk. Although PCP in the United States is solely produced by chlorination of phenol, small amounts of HCB may be found in the commercial products. HCB is an almost ubiquitous impurity in chlorinated solvents during manufacture. It is also a registered pesticide and is used as a fungicide on wheat seed (not for consumption). HCB is the most commonly found chlorinated hydrocarbon in meat, and it has been found at below tolerance levels in 30% of the samples analyzed. Because of the ease of biotransformation of HCB to PCP, residues of PCP may in part be due to HCB pollution.

Pentachlorophenol Usage: Summary and Conclusions

Technical grade PCP does not consist solely of PCP. The driving force of the production reaction is not solely towards PCP. The producers intentionally under-chlorinate to give a certain required amount of tetrachlorophenol which also results in traces of trichlorophenol. PCP is an extremely versatile toxicant and a product of choice in many applications. It is a bactericide, an algicide, a fungicide and in general an effective herbicide. It is also a desiccant and a soil sterilant where a total kill of vegetation may be required. As a fungicide and bactericide it has been used in adhesives, oils, paints and rubber; it has been used for mildew control in textiles and carpet shampoos and as a mold preventative in food processing plants. It has been added to fabrics for mothproofing, and is used in the construction industry for control of wood-boring insects. The sodium salt of PCP (sodium pentachlorophenate) is used as a preservative in ammonium alginate, in the manufacture of gaskets for food containers and as a wood preservative in crates for packaging raw agricultural products. It has clearance from the U.S. Department of Agriculture on a "no residue" basis for use on agricultural crops and agricultural premises such as poultry houses, beehives and greenhouse benches.

The persistence of PCP and its contaminants in the environment is of concern. Since all the users of PCP are not identified the fate of all the residues is not known. The long-term effect of PCP, its contaminants and its residues have not been completely examined. Although PCP is lost from animal tissues readily, many of its precursors and metabolites are persistent. Pentachloroanisole (PCA), a major product of PCP under aerobic conditions, appears to be a more persistent compound. Pentachlorophenoxyphenol, a predioxin, has been found to have an LC_{50} of 70-800 ppb in 96 hours for three species of fish. This is the only known data on predioxin toxicity. Very little is known of the toxicity and pharmacology of contaminants such as tetrachlorophenols, trichlorophenols, hexa-, hepta-, and octachlorodibenzo-p-dioxins, and chlorinated dibenzofurans. Further work on these compounds would contribute greatly towards evaluating the environmental impact of chlorophenols and their contaminants.

Future decisions concerning the use of PCP must be made with great discretion. One important consideration is the economic impact of banning PCP. Since Michigan initiated its new regulation (reducing allowable dioxin content) the cost of PCP in that state has risen 10% and the ultimate cost to the consumer has risen by 1%. In the United States, treatment of wood with PCP results in a 7.5 billion dollar savings by lengthening the useful life of wood. This also helps to conserve our timber resources by sparing an estimated 43 million dollars worth of additional timber. Another consideration is the availability of replacement compounds, and their threat to the environment. Undoubtedly, PCP has already become rather widespread in the environment. Preliminary results of a national survey in the United States indicate the presence of PCP in 85% of human urine samples analyzed at a mean concentration of 6 ppb. This ubiquitous nature of PCP and the fact that PCP (including its salts) is the second most heavily used pesticide in the United States suggest a need for a careful and thorough evaluation of its impact on environmental health.

Index